MOUNTAIN DIVERSITY

MOUNTAIN DIVERSITY

Dynamics of High-Altitude Environments

ROBERT MARCHANT
Professor, Department of Environment and Geography,
University of York, North Yorkshire, United Kingdom

ELSEVIER

ACADEMIC PRESS
An imprint of Elsevier
elsevier.com/books-and-journals

Academic Press is an imprint of Elsevier

125 London Wall, London EC2Y 5AS, United Kingdom
50 Hampshire Street, 5th Floor, Cambridge, MA 02139, United States

ISBN: 978-0-443-22251-1

For Information on all Academic Press publications visit our website at https://www.elsevier.com/books-and-journals

Publisher: Andre Wolff
Acquisitions Editor: Jessica Mack
Editorial Project Manager: Apoorva Shrivastava
Production Project Manager: Omer Mukhtar Moosa
Cover Designer: Arumugam Kothandan

Typeset by Aptara, New Delhi, India

Contents

Acknowledgements

My own approach to working across mountains and being able to connect to mountain peoples, places, and history has been profoundly influenced by the voices I have encountered over the past years. Since working in the Rukiga Highlands of south-west Uganda for my PhD I have been inspired by magnificent mountains around the world, particularly by their uniqueness but by their shared character. Numerous people have offered friendship, support, and comments on the work that has underpinned this book; without these contributions, this work would not have been possible. I am deeply indebted to the people I have been fortunate to meet along the way! The open generosity, particularly from those with the 'least', has been quite humbling, and I thank the countless people who have helped with roadside breakdowns, provided plenty of tea and food, and provided great conversation, insight, and depth over the years. Thanks must go to the people who have helped me during some fantastic experiences, who are far too many to list here. With the fear of missing people out you know who you are, and I am indebted to working with you! Particularly I thank friends, colleagues, and family who have shared these adventures; in particular Antje Ahrends, Oli Boles, Mike Bunguard, Claudia Capitani, Colin Courtney Mustaphi, Aida Cuni-Sanchez, Nicholas Dere, Jemma Finch, Esther Githumbi, Robert Halcrow, Nwabueze, Igu, Tabatha Kabora, Rebecca Kariuki, Rahab Kinyanjui, Paul Lane, Carol Lang, Edem Mahu, Esther Makinde, Amos Majule, Andrew Marshall, Colin McClean, Cassian Mumbi, Veronica Muiruri, Rebecca Newmann, Tobias Nyumba, Dickens Odeny, Anthony Onyekuru, Dan Olago, Peter Omeny, Olivia Norfolk, Paramita Punwong, Pattrawut Pushingha, Marion Pfeifer, Phil Platts, Suzi Richer, Stephen Rucina Mathai, Hamidu Seki, Evans Sitati, Daryl Stump, Jessica Thorn, Lucy Waruingi, Jonah Western, Jeff Worden, and Pius Yanda. My family deserves special mention; Laura helped to produce some of the figures in the book whereas I was helped greatly in edits and research for this book by Chloë and Caroline – indeed none of what I am would have been possible without you!

Institutions such as The National Museums of Kenya, the British Institute in Eastern Africa, the Africa Conservation Centre, World Conservation Society, Institute for Tropical Forest Conservation, WWF-Tanzania and WWF-Kenya, Universities of Amsterdam, Hull, Nairobi, Nelson Mandela

Institute of Science and Technology, Leicester, Mahidol, Makerere, Dar es Salaam, Sokinie University of Agriculture, Trinity College, Dublin, and York where some of these ideas started to be developed and have been nurtured over the years. Particularly instrumental has been the Mountain Research Initiative and the amazing work the MRI has done to galvanise the mountain research community across the world. Over the past 25 years, MRI staff and the wider community have worked tirelessly to ensure that the impact on mountain ecosystems, environments, and societies is far greater than the sum of the parts.

Numerous funders have supported the work over the years particularly the European Commission, the European Research Council, Past Global Changes (PAGES), Formas, SIDA, The Royal Society, the Natural and Environmental Research Council, Economic and Social Research Council, the Arts and Humanities Research Council, the Swiss National Science Foundation, The Mountain Research Initiative, WorldWild Fund, World Conservation Society, Liz Claiborne Foundation, and the Percy Sladen Memorial Trust.

CHAPTER 1

Overview: What is a mountain, and where are they located?

1.1 Overview to the book

Mountains are vital; they are the source of most rivers, the food basket of the world, and they influence climates from global to local scales while forming cradles, barriers, and bridges that shape ecosystem composition and their distribution. Mountains sustain both societies living in highland regions, and those downstream in river catchments. They are amongst the most ecologically diverse and captivating landscapes on Earth. Their towering slopes and elevations create unique environments that support biodiversity ranging from microscopic bacteria to charismatic megafauna. In recent decades, the world's mountains have been increasingly recognised as essential components of global biophysical, ecological, and socio-economic systems. As mountains come under increasing social, economic, biodiversity, and, not least, environmental pressures their critical role in underpinning sustainable development is now more widely acknowledged than ever.

The world's mountain systems cover a relatively small portion of Earth's surface but are highly important (Fig. 1.1). They harbour distinct ecosystems, contain most of the snow and ice outside the polar regions and provide water (life) for billions of people (Pepin et al., 2022) (Fig. 1.2). Land use in mountain areas is highly varied, ranging from recreational playgrounds for mountaineers, skiers, walkers and climbers to providing key resources that sustain life - including timber, energy, water, food, minerals, and medicinal plants – as well as underpinning aesthetic and spiritual values that create society (Adler et al., 2022; Alfthan et al., 2018; Cuni-Sanchez et al., 2021). Like many regions worldwide, mountains are experiencing increasing pressure from agricultural expansion, driven largely by population growth and the associated development of infrastructure such as roads, pipelines, and power lines. These pressures are often played out in terms of changes in land use. Land is an important lens that will be used throughout the book to understand changes to mountain systems. In the same way land use choices can cause problems, land use is the one place where society has

Mountain Diversity: Dynamics of High-Altitude Environments.
DOI: https://doi.org/10.1016/B978-0-443-22251-1.00001-0

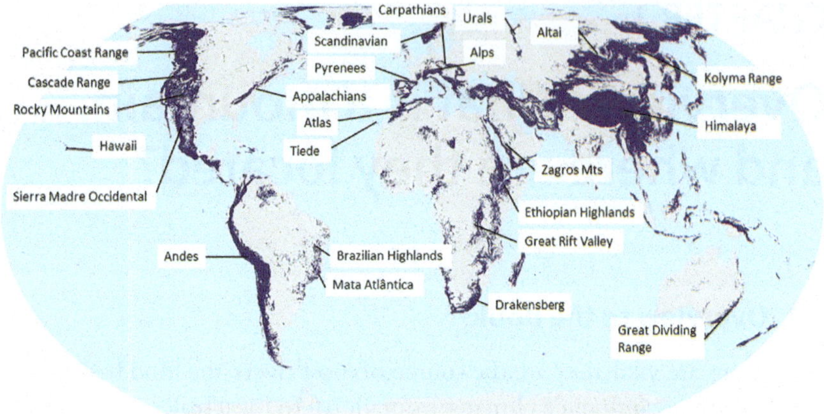

Figure 1.1 The world's main mountain systems based on Kapos et al. (2000) are shown on an 'Equal Earth (EPSG 8857) projection. *Based on Thornton et al. (2022).*

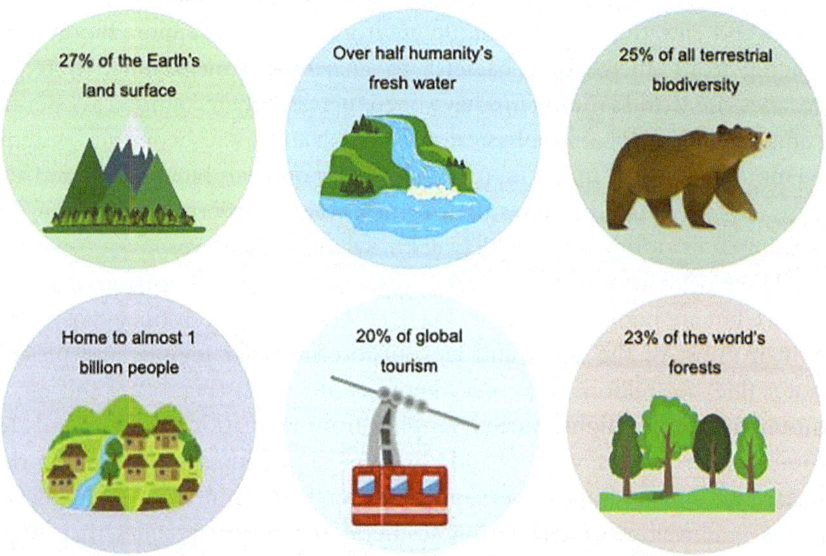

Figure 1.2 Mountain systems, although covering a relatively small portion of Earth's surface, are highly important as storehouses of biodiversity, water, and drivers of tourism.

agency; choices can make a difference and help navigate the challenges that sometimes feel insurmountable.

Mountains are immensely important for people, although they only account for approximately 25% of Earth's land area, and only about 12%

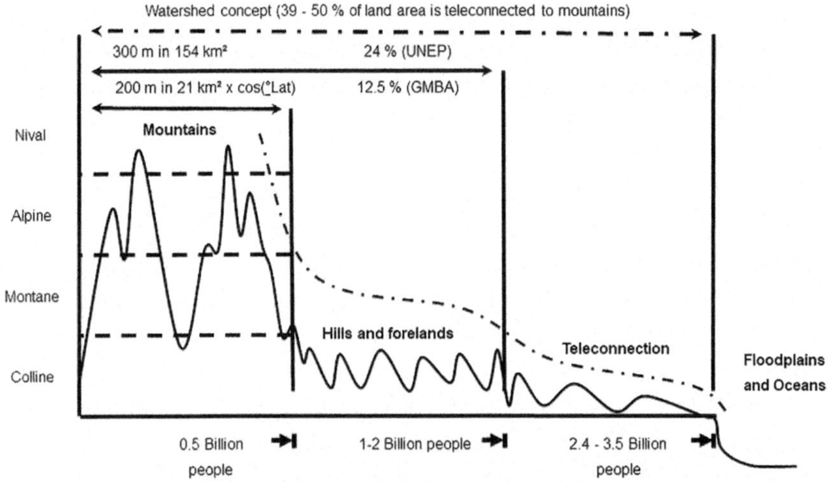

Figure 1.3 A comparison of two broad definitions of mountainous land and the four major mountain zones. The UNEP definition identifies mountains as covering 24% of all land outside Antarctica (Kapos et al., 2000), whereas the GMBA definition classifies only 12.5% as mountainous (Körner et al., 2011). The greater extent under the UNEP definition is primarily due to the exclusion of land below 300 m and the inclusion of all tablelands above 2500 m (regardless of ruggedness). Also showing the estimated number of people living in, close to, or under the influence of mountain areas.

of the global population lives directly in mountainous regions; more than 50% of the world's population is directly or indirectly dependent on mountain resources (Ives & Messerli, 1997). Over 671 million people live in high mountain regions (Jones & O'Neill, 2016), a number that rises to nearly 1.2 billion when including seasonal visitors (Körner et al., 2017). This number can be tripled when encompassing communities that rely on mountain resources. For example, the Himalayas alone support the livelihoods of over 1 billion people living in the surrounding lowlands (Xu et al., 2009). Populations are often concentrated in the highly productive hills and forelands adjacent to high mountain areas (Fig. 1.3). Mountains are characterised by both high biological and cultural diversity. While mountain communities are often adaptive, resourceful, and independent, they inhabit highly fragile environments. Mountain communities, and those closely tied to mountain systems, often face significant barriers in coping with climate impacts due to low incomes and limited opportunities to diversify their livelihoods (McDowell et al., 2013). In many regions, mountain communities also face difficult socio-economic conditions, which further increase their vulnerability to climate change (Alfthan et al., 2018). It has become increasingly apparent that mountain ecosystems are fragile and highly susceptible to global environmental and socio-economic changes, but also that mountain communities are adapting, as they have had to do over thousands of years (Cuni-Sanchez et al., 2025).

This book provides an insight into the essential aspects of mountain origins, climates, environments, and their ecosystems. The major drivers of mountain building and erosion, and the fundamental processes of mountain systems, are presented. How mountains have changed, are changing, and may continue to change throughout the Anthropocene will be explored. By addressing both the physical and human dimensions of mountain systems, as well as the responses of mountain environments to recent climate change, globalisation, and land use transformations, this book aims to bridge the gap between science, policy, and communities living within mountains. The book's content is organised into three main sections: understanding mountain systems (Chapters 2, 3, 4, and 5), exploring current threats (Chapter 6), and then identifying opportunities and future solutions for sustainable mountain socio-ecological systems (Chapters 7 and 8).

1.2 What is a mountain?

Many definitions of mountains exist, each based on different criteria and all valid within their contexts. These criteria include topographical parameters such as elevation, relative relief, and slope angle (Kapos et al., 2000; Platts et al., 2011), hypsometric curves (Elsen & Tingley, 2015), geomorphological processes (Price et al., 2013), ecological zonation (Körner et al., 2017), political or management frameworks (Price & Byers, 2017), and the presence of snow and ice (Beniston et al., 2018). Depending on the definition and method of delineation used, between 13% and 30% of Earth's land surface (excluding Antarctica) can be considered mountainous (Pepin et al., 2022).

Whether an area is classified as a mountain has substantial conservation and biogeographic implications (Körner et al., 2021; Snethlage et al., 2022). A critical feature of mountain landscapes is land inclination. The resulting topography, in turn, influences climate factors such as solar radiation, wind patterns, snow distribution, and the allocation of water, soil, and isolation – together combining to create a rich diversity of habitats and ecosystems (Chapter 2). In recent decades, significant efforts have been made to define and map mountainous areas globally (Kapos et al., 2000; Platts et al., 2011;

Plate 1.1 Mountains take many shapes and forms, from magnificent peaks to rounded hills, from free-standing volcanoes to connected mountain chains. There are many different types of mountain characterised by steep slopes, a confined summit, and significant elevation relative to the surrounding terrain. Clockwise from top left. Shira plateau, Kilimanjaro, Tanzania, Ethiopian Highlands, Yeha, Ethiopia, Mount Tiede, Tenerife, the Mathews Range, northern Kenya, Ol Doinyo Lengai, northern Tanzania, the Chamonix Valley, France, Cathedral Peak, Drakensburg, South Africa, Obergurgl, Austria. *Central picture Parincota, Chile (All Photographs Rob Marchant).*

Price et al., 2013; Snethlage et al., 2022). Despite this, mountains have always been, and continue to be, difficult to define precisely, particularly when trying to determine where mountains transition into hills (Plate 1.1). What one person regards as a mountain, another might perceive as a hill (Linnaeus, 1781; Smith & Mark, 2003), and some mountains, for example, those in the English Lake District, are clearly quite different in stature from an extensive mountain range such as the Alps, the Andes, or a free-standing volcano (Plate 1.1). Elevation alone is not a reliable criterion, as high-elevation regions such as tablelands and plateaus do not necessarily exhibit mountainous topography, although they can be well over 2000 m and exhibit mountain character. Similarly, climate cannot serve as a defining factor, since mountains span nearly all climatic zones. The most common approaches involve combinations of parameters such as elevation, slope, ruggedness, or relief – usually derived from remotely sensed digital elevation models – along

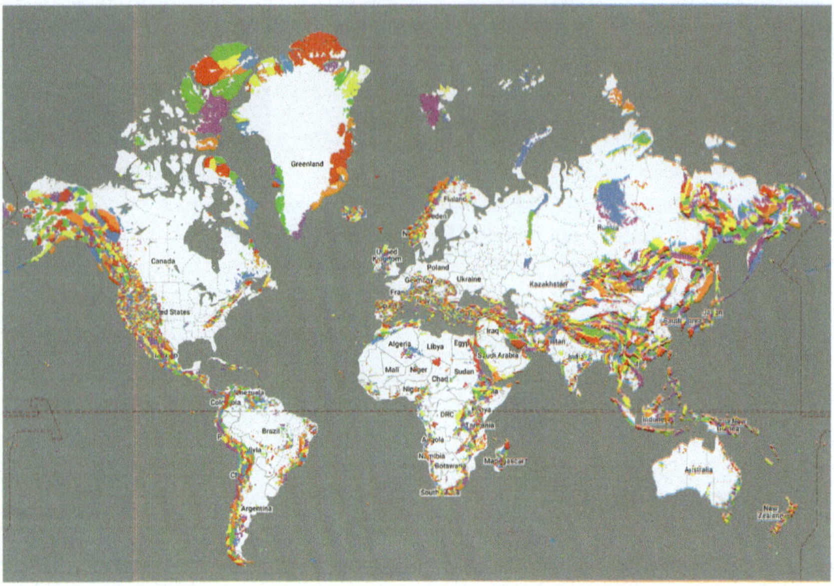

Figure 1.4 The Global Mountain Biodiversity Assessment Mountain Inventory v2.0 is a hierarchically organised spatial vector layer in an ESRI GIS shapefile format. The polygons represent thousands of mountain ranges across various scales. Depending on the region, these polygons range from very small (e.g., High Tatras) to very large (e.g., Carpathians) (Snethlage et al., 2022).

with a suite of specific threshold values to delineate mountainous terrain (Kapos et al., 2000; Karagulle et al., 2017; Körner et al., 2011; Platts et al., 2011; Snethlage et al., 2022). However, the global approach to delineate mountains loses the details that many of these mountains are discreet entities with deep-rooted environment and social constructions (Fig. 1.1). While 'Mountains' do provide a unifying connection, and there is something unique about mountains that makes such a book worthwhile, we should not lose sight that mountains are a series of discrete and bounded places that connect the world (Fig. 1.4).

A mountain is typically defined as a landform characterised by steep slopes, a confined summit, and a significant rise in elevation (e.g., > 300 m in elevation) relative to the surrounding terrain (Kapos et al., 2000). Two widely used classification systems are one that was developed by the United National World Conservation Monitoring group (UNEP-WCMC) over 25 years ago, and a more recent one from the Global Mountain Biodiversity Assessment (GMBA) (Snethlage et al., 2022) (Fig. 1.4). Both these systems

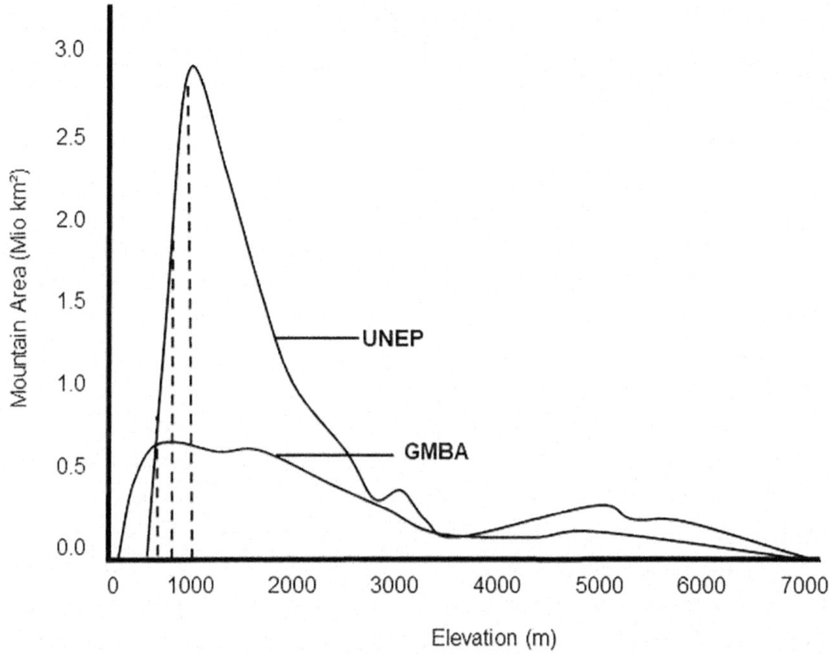

Figure 1.5 The definition of mountains drives the outcome of global mountain area via ruggedness: UNEP definition applies a 300 m across 3 × 3 cells of 30" applied to 154 km^2 GMBA, applied 200 m across 3 × 3 cells of 30" applied to 21 km^2. The dotted lines indicate elevation thresholds of 600, 800, and 1000 m. The main differences between the UNEP and GMBA definitions arise in lower elevation zones. The area <300 m elevation is included in the GMBA definition but excluded by UNEP, though this difference has minimal impact on the total area classified as mountainous. *Adapted from Körner et al. (2021).*

typically rely on a combination of elevation (commonly >300 m above sea level) and ruggedness criteria (Fig. 1.5). However, earlier definitions sometimes included areas such as plateaus or hilly lowlands that some would argue lack true mountain characteristics. To address this, more refined methodologies have been developed, incorporating finer spatial resolution and precise metrics – such as local elevation range. For example, Tito et al. (2020) proposed using the maximum elevation difference within a 3 × 3 grid of 30 arc seconds, identifying mountainous terrain where the elevation difference exceeds 200 m. Using this method, only 12% of Earth's terrestrial land surface (excluding Antarctica) is classified as mountainous (Fig. 1.5). A clear and quantitative definition of mountain boundaries, such as that proposed by the GMBA, offers the added benefit of distinguishing mountain forelands as a separate category (Körner et al., 2021; Snethlage et al., 2022).

1.3 Why are mountains important?

In addition to topographic, cultural, and environmental diversity, mountains are geologically diverse and comprise a wide variety of rock types. This geological diversity influences local topography, soil composition, and hydrology, resulting in heterogeneous landscapes that support diverse plant communities and associated wildlife. It is this diversity that lies at the heart of the mountain system's beauty – and also why mountains are such complex and challenging places to live in, work in, and understand. Mountains are intrinsically unstable systems, undergoing substantial changes in response to tectonic, erosional, and climatic processes over a range of geological timescales. Following mountain construction, there is a whole series of interacting processes with the ultimate aim of mountain destruction (Chapters 2 and 4). The interaction between mountain substrates, life forms, and climate systems, across a range of spatial scales, gives rise to highly diverse and distinct mountain environments (Rahbek et al., 2019). Despite their significance and global importance, mountain ecosystems continue to be among the most endangered in the world. While mountain regions support resilient and adaptive communities, they are increasingly vulnerable to climate change, land use change, and socio-economic pressures.

Mountains may cover only ~25% of the Earth's continental land surface, yet they are home to more than 85% of the world's amphibians, birds, and mammals, many of which are endemic to mountain regions (Chapter 3). In addition to biological diversity, high-altitude regions are also culturally diverse and economically important. They are highly valued for their recreational, aesthetic, and spiritual significance and provide a wide range of ecosystem services, including agriculture, tourism, water regulation, and energy (Chapter 5). Mountains play a fundamental role in the global hydrological cycle (Chapter 2). They receive more precipitation than lowland areas, experience less evaporation, and store water in the form of snow and ice, which melts seasonally to feed major river systems (Barnett et al., 2005; Viviroli et al., 2011). However, globally changing climates and the diminishing cryosphere are threatening this role, with serious implications for water supply in downstream regions (Viviroli et al., 2020) (Chapter 4). In the main mountain areas, slight temperature increases are leading to major hydrological changes (Haeberli & Weingartner, 2020; Musselman et al., 2017) with associated risks (Chapter 6). Mountain regions are particularly vulnerable to human interventions and climate change, which threaten both biodiversity and the critical services they provide (Chapter 6).

Well-functioning mountain ecosystems can buffer climatic extremes, thus absorbing shocks such as intense rainstorms and prolonged droughts (Chapter 7). Besides being vital to the hydrological system, they also act as significant carbon stores and regulators of global biogeochemical cycles. Yet, despite their global importance, mountain systems remain underrepresented in academic research and in the policy agenda and severely underfunded at national and international levels. Alarmingly, many mountain areas remain unprotected, with key biodiversity areas and some 250 mountain regions worldwide currently having no formal protection (Li et al., 2022). Such gaps also provide great opportunities for mountains to be increasingly valued and to be placed at the heart of an agenda for a sustainable future (Chapter 8). This book will present a comprehensive overview of current understanding of biodiversity across the world's mountains (Chapter 3), and propose priorities for rethinking how we understand, value, protect, and conserve high-altitude regions (Chapter 8). By adopting an interdisciplinary approach that integrates temporally and spatially diverse datasets, this book aims to inform development planning and conservation efforts for more sustainable mountain social-ecological systems (Chapter 8). Understanding how mountain environments, and the benefits they provide, have changed and are changing will help guide adaptation strategies and enhance preparedness for future challenges the world and humanity are facing – creating a future where mountains can be front and centre rather than on the edge and peripheral, in support of a more sustainable future.

References

Adler, C., Wester, P., Bhatt, I., Huggel, C., Insarov, G., Morecroft, M., Muccione, V., Prakash, A., Alcántara-Ayala, I., Allen, S. K. and Bader, M., 2022. Cross-chapter paper 5: Mountains. In: Climate change 2022: Impacts, adaptation and vulnerability. contribution of working group II to the sixth assessment report of the intergovernmental panel on climate change. H.-O. Pörtner, D.C. Roberts, M. Tignor, E.S. Poloczanska, K. Mintenbeck, A. Alegría, M. Craig, S. Langsdorf, S. Löschke, V. Möller, A. Okem, B. Rama (eds.). Cambridge University Press, Cambridge, UK and New York, NY, USA, pp. 2273–2318, doi:10.1017/9781009325844.022.

Alfthan, B., Gjerdi, H., Schoolmeester, T., Agrawal, N. K., Andresen, M., Gjerdi, H. L., Jurek, M., & Gupta, N. (2018). *Mountain adaptation outlook series: Outlook on climate change adaptation in the Hindu Kush Himalaya.* United Nations Environment Programme, GRID-Arendal, and the International Centre for Integrated Mountain Development.

Barnett, T. P., Adam, J. C., & Lettenmaier, D. P. (2005). Potential impacts of a warming climate on water availability in snow-dominated regions. *Nature, 438*(7066), 303–309. https://doi.org/10.1038/nature04141.

Beniston, M., Farinotti, D., Stoffel, M., Andreassen, L. M., Coppola, E., Eckert, N., Fantini, A., Giacona, F., Hauck, C., Huss, M., Huwald, H., Lehning, M., López-Moreno, J. I.,

Magnusson, J., Marty, C., Morán-Tejéda, E., Morin, S., Naaim, M., Provenzale, A., & ..., Vincent, C. (2018). The European mountain cryosphere: A review of its current state, trends, and future challenges. *Cryosphere, 12*(2), 759–794. https://doi.org/10.5194/tc-12-759-2018.

Cuni-Sanchez, A., Aneseyee, A. B., Baderha, G. K. R., Batumike, R., Bitariho, R., Imani, G., Jha, N., Kaganzi, K. R., Kaplin, B. A., Klein, J. A., Leite, A., Marchant, R. A., Martin, E. H., Mcharazo, F., Mwangi, B., Ngute, A. S. K., Nkengurutse, J., Nkurunziza, A., Olaka, L., & ..., Zafra-Calvo, N. (2025). Perceived climate change impacts and adaptation responses in ten African mountain regions. *Nature Climate Change, 15*(2), 153–161. https://doi.org/10.1038/s41558-024-02221-w.

Cuni-Sanchez, A., Sullivan, M. J. P., Platts, P. J., Lewis, S. L., Marchant, R., Imani, G., Hubau, W., Abiem, I., Adhikari, H., Albrecht, T., Altman, J., Amani, C., Aneseyee, A. B., Avitabile, V., Banin, L., Batumike, R., Bauters, M., Beeckman, H., Begne, S. K., & ..., Zibera, E. (2021). High aboveground carbon stock of African tropical montane forests. *Nature, 596*(7873), 536–542. https://doi.org/10.1038/s41586-021-03728-4.

Elsen, P. R., & Tingley, M. W. (2015). Global mountain topography and the fate of montane species under climate change. *Nature Climate Change, 5*(8), 772–776. https://doi.org/10.1038/nclimate2656.

Haeberli, W., & Weingartner, R. (2020). In full transition: Key impacts of vanishing mountain ice on water-security at local to global scales. *Water Security, 11*, 100074. https://doi.org/10.1016/j.wasec.2020.100074.

(1997). In Ives, J., & Messerli, B. (Eds.), *Mountains of the world: A global priority* (p. 495). Parthenon Publishing Group.

Jones, B., & O'Neill, B. C. (2016). Spatially explicit global population scenarios consistent with the shared socioeconomic pathways. *Environmental Research Letters, 11*(8), 084003. https://doi.org/10.1088/1748-9326/11/8/084003.

Kapos, V., Rhind, J., Edwards, M., Price, M. F., & Ravilious, C. (2000). Developing a map of the world's mountain forests. In *Forests in sustainable mountain development: A state of knowledge report for 2000* (pp. 4–19). Task force on forests in sustainable mountain development. https://doi.org/10.1079/9780851994468.0004.

Karagulle, D., Frye, C., Sayre, R., Breyer, S., Aniello, P., Vaughan, R., & Wright, D. (2017). Modeling global Hammond landform regions from 250-m elevation data. *Transactions in GIS, 21*(5), 1040–1060. https://doi.org/10.1111/tgis.12265.

Körner, C., Jetz, W., Paulsen, J., Payne, D., Rudmann-Maurer, K., & Spehn, E. M. (2017). A global inventory of mountains for bio-geographical applications. *Alpine Botany, 127*(1), 1–15. https://doi.org/10.1007/s00035-016-0182-6.

Körner, C., Paulsen, J., & Spehn, E. M. (2011). A definition of mountains and their bioclimatic belts for global comparisons of biodiversity data. *Alpine Botany, 121*(2), 73–78. https://doi.org/10.1007/s00035-011-0094-4.

Körner, C., Urbach, D., & Paulsen, J. (2021). Mountain definitions and their consequences. *Alpine Botany, 131*(2), 213–217. https://doi.org/10.1007/s00035-021-00265-8.

Li, B. V., Jenkins, C. N., & Xu, W. (2022). Strategic protection of landslide vulnerable mountains for biodiversity conservation under land-cover and climate change impacts. *Proceedings of the National Academy of Sciences, 119*(2). https://doi.org/10.1073/pnas.2113416118.

Linnaeus, C. 1781. On the increase of the habitable earth. Amoenitates Academicae 2: 17–27.

McDowell, G., Ford, J. D., Lehner, B., Berrang-Ford, L., & Sherpa, A. (2013). Climate-related hydrological change and human vulnerability in remote mountain regions: A case study from Khumbu, Nepal. *Regional Environmental Change., 13*(2), 299–310. https://doi.org/10.1007/s10113-012-0333-2.

Musselman, K. N., Clark, M. P., Liu, C., Ikeda, K., & Rasmussen, R. (2017). Slower snowmelt in a warmer world. *Nature Climate Change,* 7(3), 214–219. https://doi.org/10.1038/nclimate3225.

Pepin, N. C., Arnone, E., Gobiet, A., Haslinger, K., Kotlarski, S., Notarnicola, C., Palazzi, E., Seibert, P., Serafin, S., Schöner, W., Terzago, S., Thornton, J. M., Vuille, M., & Adler, C. (2022). Climate changes and their elevational patterns in the mountains of the world. *Reviews of Geophysics, 60*(1). https://doi.org/10.1029/2020rg000730.

Platts, P. J., Burgess, N. D., Gereau, R. E., Lovett, J. C., Marshall, A. R., McClean, C. J., Pellikka, P. K. E., Swetnam, R. D., & Marchant, R. (2011). Delimiting tropical mountain ecoregions for conservation. *Environmental Conservation, 38*(3), 312–324. https://doi.org/10.1017/S0376892911000191.

Price, M. F., & Byers, A. C. (Eds.). (2017). *Mountain geography: Physical and human dimensions*. University of California Press.

Price, M. F., Byers, A. C., Friend, D. A., Kohler, T., & Price, L. W. (2013). *Mountain geography: Physical and human dimensions*. (pp. 1–378). University of California Press. https://www.ucpress.edu/books/mountain-geography/hardcover.

Rahbek, C., Borregaard, M. K., Colwell, R. K., Dalsgaard, B., Holt, B. G., Morueta-Holme, N., Nogues-Bravo, D., Whittaker, R. J., & Fjeldså, J. (2019). Humboldt's enigma: What causes global patterns of mountain biodiversity? *Science, 365*(6458), 1108–1113. https://doi.org/10.1126/science.aax0149.

Smith, B., & Mark, D. M. (2003). Do mountains exist? Towards an ontology of landforms. *Environ Plann B Plann Des, 30*, 411–427. https://doi.org/10.1068/b12821.

Snethlage, M. A., Geschke, J., Ranipeta, A., Jetz, W., Yoccoz, N. G., Körner, C., Spehn, E. M., Fischer, M., & Urbach, D. (2022). A hierarchical inventory of the world's mountains for global comparative mountain science. *Scientific Data, 9*(1). https://doi.org/10.1038/s41597-022-01256-y.

Thornton, J. M., Snethlage, M. A., Sayre, R., Urbach, D. R., Viviroli, D., Ehrlich, D., Muccione, V., Wester, P., Insarov, G., Adler, C., & Freeman, J. (2022). Human populations in the world's mountains: Spatio-temporal patterns and potential controls. *PLoS One, 17*(7), e0271466. https://doi.org/10.1371/journal.pone.0271466.

Tito, R., Vasconcelos, H. L., & Feeley, K. J. (2020). Mountain ecosystems as natural laboratories for climate change experiments. *Frontiers in Forests and Global Change, 3*. https://doi.org/10.3389/ffgc.2020.00038.

Viviroli, D., Archer, D. R., Buytaert, W., Fowler, H. J., Greenwood, G. B., Hamlet, A. F., Huang, Y., Koboltschnig, G., Litaor, M. I., López-Moreno, J. I., Lorentz, S., Schädler, B., Schreier, H., Schwaiger, K., Vuille, M., & Woods, R. (2011). Climate change and mountain water resources: Overview and recommendations for research, management and policy. *Hydrology and Earth System Sciences, 15*(2), 471–504. https://doi.org/10.5194/hess-15-471-2011.

Viviroli, D., Kummu, M., Meybeck, M., Kallio, M., & Wada, Y. (2020). Increasing dependence of lowland populations on mountain water resources. *Nature Sustainability, 3*(11), 917–928. https://doi.org/10.1038/s41893-020-0559-9.

Xu, J., Grumbine, R. E., Shrestha, A., Eriksson, M., Yang, X., Wang, Y., & Wilkes, A. (2009). The melting Himalayas: Cascading effects of climate change on water, biodiversity, and livelihoods. *Conservation Biology, 23*(3), 520–530. https://doi.org/10.1111/j.1523-1739.2009.01237.x.

CHAPTER 2

Mountain foundations: Geological, climate, soil, and fire regimes

2.1 Introduction

This chapter provides a foundation for understanding mountain systems, exploring their geological origins and the ensuing topography, climatic regimes, soil characteristics, and fire environments. These foundational elements are key to controlling the distribution of mountain plants and animals (Chapter 3) and thus the more applied aspects of mountain systems, discussed in subsequent chapters. A solid grasp of these physical underpinnings is essential for identifying both the challenges facing mountain environments, the links through the catchment (Fig. 2.1), and the potential solutions. Despite their amazing variety, mountain regions often share core features: high geological diversity, complex landscape structures, and diverse climatic and fire regimes, both in the past and at the present, all contributing to their ecological and geological uniqueness.

2.2 Geological foundations and the formation of mountain systems

One of the most fundamental geological processes on Earth is mountain building. Movement of the planet's crust, through mechanisms of collision and divergence, creates new land (Fig. 2.2). Without plate tectonics, Earth would be a largely static, flat surface, lacking the diverse age, origin, and composition that shape biodiversity and ecosystems. Mountain geology encompasses the formation, structure, and evolution of mountain ranges. The process of orogeny – or mountain building – is driven by plate tectonics, creating climatic, landscape, and ecological gradients at a planetary scale. Orogeny includes the uplift of oceanic crust, emplacement of intrusive magmatic bodies, and the formation of surface volcanic structures. Many of the world's mountains originate from plates of the Earth's crust colliding into

Mountain Diversity: Dynamics of High-Altitude Environments.
DOI: https://doi.org/10.1016/B978-0-443-22251-1.00002-2

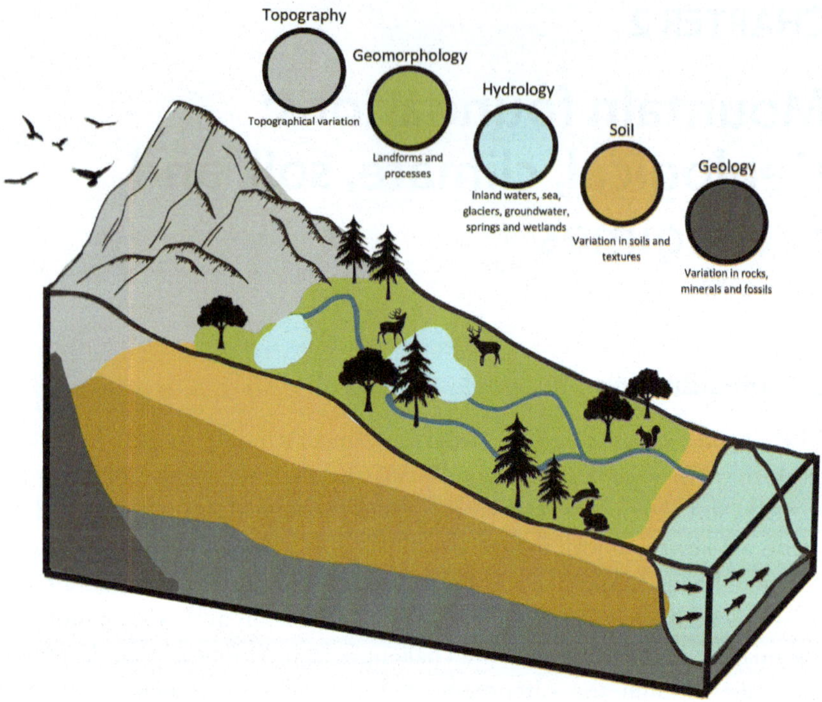

Figure 2.1 A grasp of physical underpinnings is essential for identifying the functioning of mountain environments and the links through the catchment.

each other, sending 'ripples' of Earth's crust to be uplifted, or new crust to be formed through volcanic activity. For example, the African plate is currently moving towards the European plate, forming the Alps, whilst the Indian plate continues to push into the Eurasian plate, uplifting the Himalayas over the last 55 to 50 ± 10 million years (Ding et al., 2022). This continuous movement of tectonic plates has resulted in the present-day arrangement of continents, and continues to shape Earth's surface, creating both the highest mountain ranges and the deepest oceanic trenches. Mountain formation typically results from crustal uplift, folding, faulting, and volcanic activity.

In addition to tectonic forces, isostatic uplift, which is the vertical movement of Earth's crust in response to changes in surface load, also contributes to mountain building. This process is particularly evident when large ice sheets melt, such as the Laurentide ice sheet that expanded across northern latitudes at the last glacial maximum (Chapter 4). The removal of the heavy ice mass (Godbout et al., 2023) allows the crust to rebound and rise, a phenomenon most common in high-latitude regions. More

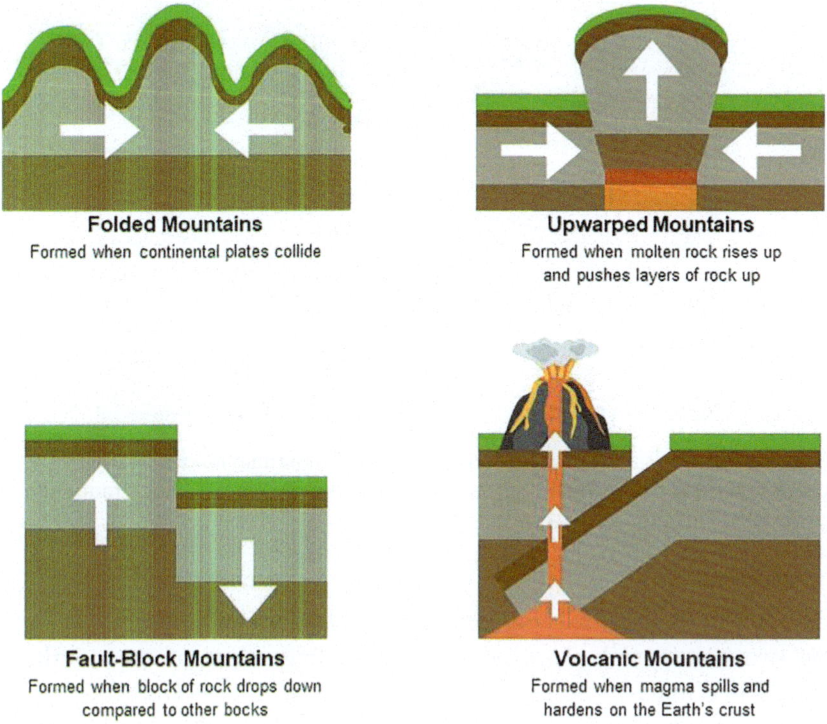

Figure 2.2 Schematic showing the main different types of mountain formation.

locally, landslides and the rapid removal of surface load can have a similar impact (Gargani, 2022). Cosmogenic radionuclide analysis, in combination with thermochronology and palaeoaltimetry, has vastly improved models on global mountain building and our understanding of their formation, evolution, and age (Perrigo et al., 2020). There are a series of main mountain types depending on the character of the plate tectonics.

2.2.1 Folded mountain ranges

Folded mountain ranges are formed through the collision and compression of tectonic plates, leading to intricate folding of rock layers, reflecting the immense forces exerted during tectonic convergence. At a convergent plate boundary, two lithospheric plates move towards each other and collide; neither plate is subducted due to their similar densities. Instead, the colliding plates crumple and buckle, leading to the formation of extensive mountain ranges. Convergent plate boundaries may involve various combinations of

Plate 2.1 The Alps are a classic example of folded mountain ranges where high compression has led to the folding of sedimentary rocks, Chamonix Valley, France. *(Photograph Rob Marchant).*

continental-continental, continental-oceanic, or oceanic-oceanic collisions. As two continental plates collide, the leading edges of the continents experience intense compressional forces. These forces cause the crust to deform, resulting in horizontal shortening and the folding and faulting of rock layers. The Alps are a classic example of a thrust belt formed by the collision of the African and Eurasian plates (Plate 2.1). The Himalayas are an example of a mountain range formed by the collision of the Indian and Eurasian plates. The ongoing collision continues to uplift the region, creating some of the highest peaks, including Mount Everest. The Himalayas have uplifted quickly and relatively recently (in the past 15 million years or so (Perrigo et al., 2020). Folded mountain ranges are characterised by anticlines (upward-arching folds) and synclines (downward-curving folds) arranged in parallel patterns. Anticlines typically form ridges or peaks, while synclines often coincide with valleys or basins. In some folded mountain ranges, thrust faults occur when rocks are pushed horizontally along a fault plane, resulting in the stacking of rock layers and the formation of thrust sheets. These geological features are characterised by dramatic topographic relief and can include overthrust belts, regions where older rocks are displaced over younger rocks along low-angle thrust faults.

The massive compressive forces generated by plate motion cause horizontal shortening and deformation of the Earth's crust, leading to the folding of rock layers. Ductile deformation, where the rocks bend and flow

Plate 2.2 Rocks deform, bend, and flow without fracturing under intense pressure and temperature. Alps aerial view. *(Photograph Rob Marchant)*.

without fracturing, occurs under intense pressure and temperatures. Ductile deformation results in the formation of large-scale folds, characterised by smooth, curving shapes. Brittle deformation can happen particularly near the Earth's surface and produces fractures, faults, and smaller-scale folds, contributing to the overall complexity of folded mountain structures. The Appalachians in North America and the Alps in Europe result from the compression and folding of sedimentary rocks during tectonic collision, leading to the layers' surface geology folded onto each other (Plate 2.2).

2.2.2 Fault-block mountains

Fault-block mountains are formed through a combination of tectonic forces acting along geological faults. These mountains typically exhibit steep, rugged slopes on one side and relatively gentler slopes on the other side as tensional forces act on deforming the Earth's crust along the fault (Lensen, 1958). Blocks of crust may undergo vertical displacement along fault planes, as one block of crust, known as a horst, is uplifted relative to the surrounding blocks, while adjacent blocks subside, forming grabens. The uplift of the horst block results in the formation of the fault-block mountain, characterised by steep fault scarps or escarpments on the uplifted side. Meanwhile, the subsidence of the graben blocks may lead to the formation of valleys or basins between the fault-block mountains. An iconic example

Plate 2.3 The Andes are a good example of a mountain range formed at a convergent plate boundary and comprise a mix of volcanoes and uplifted rock. Parque Nacional Lauca, Chile. *(Photograph Rob Marchant).*

of fault-block mountains is the Basin and Range Province in the western United States. This region is characterised by alternating mountain ranges (horsts) and elongated valleys (grabens), resulting from extensional tectonic forces associated with the stretching of the Earth's crust.

2.2.3 Subduction-generated mountain chains

As an oceanic plate collides with a continental plate, the denser oceanic plate is subducted beneath the continental plate. This subduction process generates intense heat and pressure, leading to the formation of molten rock (magma) that accumulates, creating pressure, causing the overlying rocks to uplift and form mountain ranges. The Andes in South America (Plate 2.3) are a good example of a mountain range formed at a convergent plate boundary, where one plate is forced below the other with a combination of uplifted rocks and volcanoes.

2.2.4 Formation at divergent plate boundary

Mountains also form at divergent plate boundaries, where two plates move away from each other, creating a gap. As the plates separate, magma rises from the mantle to fill the gap, solidifying and forming a new crust. Repeated volcanic activity and the accumulation of solidified magma led to the creation of volcanic mountain ranges. The East African Rift is an example of a divergent plate boundary, where new mountains are forming along

Plate 2.4 Kilimanjaro is an iconic, relatively young, free-standing mountain. Kilimanjaro view from Eduata Emma, Amboseli, Kenya. *(Photograph Rob Marchant).*

the divergent plates. This includes iconic free-standing mountains such as Mount Kilimanjaro (Plate 2.4), which is relatively young at only around 2 million years old.

2.2.5 Transform plate boundaries

As two plates slide past each other horizontally, they do not typically create large mountain ranges but can lead to the formation of fault-block mountains such as the Sierra Nevada in the United States. Thermochronologic studies have shown that an ancestral Sierra was uplifted (i.e., surface uplift) during the late Cretaceous to earliest Cenozoic period, followed by a much lower rate of uplift in the middle to late Cenozoic period. The higher tectonic activity in the southwestern Sierras and southern San Joaquin Valley is a consequence of San Andreas and Sierra Nevada fault interactions that began in the late Pliocene and are continuing to the present (Figueroa & Knott, 2010).

2.3 Shaping and denuding mountain landscapes

Once a mountain has been created, a whole series of processes operate to shape and denude the mountain, eroding it through ice, water, and mass movement. Understanding these dynamic processes and responses to climate change is essential for assessing the impacts of environmental change on mountain ecosystems and human societies. These processes, and the speed

Plate 2.5 Snow cover plays a critical role in regulating the Earth's surface energy balance, particularly during the transitional melt and accumulation periods of the snow season. View down to the Mer de Glace from Aiguille du Midi. *(Photograph Rob Marchant).*

at which they operate, have changed as global climate systems have changed (Chapter 4), leading to the unique character of every mountain range (Tovar et al., 2022).

2.3.1 Mountain cyrosphere

Glaciers are one of the most powerful agents of landscape change; their massive erosive power is able to shape mountain landscapes. Mountain glaciers (Plate 2.5), controlled by the interaction of temperature, precipitation, and topography, not only carve landscapes but also regulate hydrology, influence ecosystems, and affect human livelihoods. The cryosphere, comprising snow, glaciers, permafrost, and lake and river ice, is a fundamental component of high mountain regions. In high-elevation zones, extensive ice fields and caps can dominate the upper reaches of mountain ranges. These frozen lands profoundly affect physical, biological, and human systems, influencing not only mountain ecosystems and surrounding lowlands but also distant marine environments. The changes in cryosphere, particularly in snow and glacier dynamics, have altered the quantity and timing of runoff in snow-dominated and glacier-fed basins, impacting local downstream water availability and agricultural functioning (Hock et al., 2019). Through snow-albedo feedback, snow cover plays a critical role in regulating the Earth's surface energy balance (Huang et al., 2022). When snow cover is reduced, often exacerbated by aerosol deposition and interactions with vegetation, more solar radiation is absorbed by the land, leading to localised atmospheric warming (Armstrong & Brun, 2008). This positive feedback loop is most pronounced during the transitional periods of the snow season (Scherrer et al., 2012). Although we know the importance of snow feedback, the global-scale climate feedback from high mountain cryosphere changes remains poorly understood (Hock et al., 2019), a concern in this era of rapid glacier declines around the world.

Mountain glaciers are dynamic bodies of ice that flow downhill, driven by gravity and the internal deformation of ice (Plate 2.5). Glaciers form through the accumulation and compression of snow, which over time transforms into dense firn and then to glacial ice, which then flows downslope under gravity. Glaciers define the nival zone in many mountain systems and are categorised into three primary types of increasing size. The smallest are the Cirque glaciers – these small, bowl or armchair-shaped glaciers with a steep headwall, a gently sloping or over-deepened floor, and often convex slopes are separated from the lower valley. Cirques occupy uppermost valleys, often forming sharp ridges called arêtes between glacial valleys that can feed ice into these (Plate 2.6). The presence of cirques is recognised as an indicator of past glaciation, and thus of former climate regimes when ice was more dominant (Chapter 4). Cirque metrics have also been used to infer past variations in solar radiation, cloud cover, and prevailing wind directions, as well as the extent and magnitude of past glaciations (Li et al., 2024). The morphological characteristics of cirques have been studied in many places around the world, including Britain and Ireland, the Alps, Spain, Scandinavia, Siberia, the Rocky Mountains, the Andes, Papua New Guinea, Australia, and New Zealand (listed in detail in Barr and Spagnolo, 2015). The variations of cirque metrics are affected by temperature, precipitation, and the types of glaciation (Li et al., 2024). Cirque morphology can be used in palaeo-climate reconstructions and provide important clues relevant to solar radiation, palaeo wind direction, cloud cover, and the extent of glaciations associated with the development of the cirques. For example, Cirques usually have a strong NNE-facing bias in the Northern Hemisphere because under cloud-free conditions, north-facing slopes are shadier, and in the morning when temperatures are relatively low, solar radiation is less effective on slopes with an eastward component (Li et al., 2024).

The second main type of glaciers is valley glaciers that form in mountain valleys, flow downslope and carve distinctive U-shaped valleys, and over-steepened walls and rock pinnacles (Plate 2.7). Their movement rate depends on several factors, including glacier size, slope steepness, and basal conditions.

Plate 2.6 Different types of glaciers: Bottom left is the glacier Gígjökull in southern Iceland, Bottom right the glacier Gaisbergferner above the village of Obergurgl, Austria. These valley glaciers are constrained by the surrounding topography, where snow often accumulates in depressions and builds up thickness and starts to flow over the cirque glaciers (*bottom right*). These constrained glaciers are in contrast to ice sheets that cover landscapes and mountains before terminating in crevasses *(top left)* and the outlet of the Greenland Ice Sheet, east of the town of Kangerlussuaq. *(Photographs David Rippin [www.PhotoRippin.com]).*

Plate 2.7 Glacial Landscapes are characterised by U-shaped valleys, hanging valleys, cirques, and cliffs. View across the Mer de Glace, Chamonix, France. *(Photograph Rob Marchant).*

A glacier's mass balance – the difference between accumulation (mainly snowfall) and ablation (melting and sublimation) – determines whether it is advancing or retreating (Barandun et al., 2021). A positive mass balance leads to glacier growth, while a negative mass balance results in retreat with most glaciers around the world in a state of retreat (Chapter 6). Critically, valley glaciers are sensitive indicators of climate change: many are experiencing rapid retreat and mass loss in recent decades due to rising temperatures and shifting precipitation patterns. Several notable mountain glaciers can be used to illustrate their global diversity. The Siachen Glacier in the eastern Karakoram of the Himalayas stretches 76 km, making it one of the longest outside the polar regions (Liang & Wang, 2023). In the French Alps, the Mer de Glace, located on the northern slopes of Mont Blanc, extends about 7 km and is France's largest glacier (Plate 2.6). In the southern Andes of Argentina, the Perito Moreno Glacier in Los Glaciares National Park is renowned for its dramatic calving events into the Pacific Ocean.

As glaciers move, they erode and deposit material, leaving behind glacial landforms such as moraines: extensive accumulations of rocky debris eroded by the glacier as it scours the mountain. Terminal moraines mark the furthest extent of glacier advance, while lateral and medial moraines trace the glacier's sides and centre. These glacial processes shape dramatic features like U-shaped valleys, hanging valleys, cirques, and cliffs (Plate 2.7). Mountain glaciers are crucial to regional hydrology, supplying a continuous flow of freshwater to rivers and streams – especially during dry seasons. They also influence local climates and drive erosion and sediment transport, shaping mountain landscapes. In addition to glaciers, 'snowfields', large, persistent accumulations of snow, also characterise high mountain environments. However, as glaciers around the world are generally in a state of retreat and ice accumulation is often reduced, the force of these major mountain shapers is also being reduced. The long-term impact of this will be crucial, as the power of possibly the most important of the elements that shape mountain topography is curtailed. Indeed, one has only to look across most mountain systems to see the impact glaciers have had, in order to understand the impact of removing these mountain shapers is going to have.

The largest accumulations of ice are found in ice caps and icefields that cover a region. These are largely restricted to the Earth's poles and are currently not a dominant influence in shaping mountain regions, summits, and plateaus. However, in the past, large ice caps have covered entire mountain ranges such as the Alps, with only the highest summits like Mount Blanc remaining ice-free. Ice caps generate ice streams or outlet glaciers, shaping vast areas of the landscape (Sierra-Hernández et al., 2025). The main ice cap on the world's mountains is on the Himalayas, sometimes referred to as the third pole, but again, this is much restricted in the present day relative to the past glacial periods (Chapter 4).

2.3.2 Mountain hydrosphere

High mountains are complex hydrological systems that serve as critical 'water towers' for both local communities and downstream populations (Viviroli et al., 2020). All major rivers – including the Amazon (Andes), Nile (East African Highlands), and Ganges (Himalayas) – originate in mountain ranges. Mountain rivers, characterised by steep gradients and variable discharge patterns, are also powerful agents of erosion, shaping mountainous landscapes through the removal and transport of rock and sediment and feeding into the soil-forming process in the surrounding lowlands. These processes create classic mountain landforms, including V-shaped valleys, river terraces, and interlocking spurs (Plate 2.8).

Mountain rivers erode terrain through hydraulic action, whereas turbulent water dislodges rock during snowmelt or heavy rainfall. One of the main actions is that of abrasion, where sediment particles grind against each other, the riverbed, and banks, dislodging and breaking particles into smaller fragments. Sediment transport, especially during high discharge events, plays a crucial role in landscape evolution; redistributing material downstream and reshaping channel morphology and valley structure (Plate 2.8). Sediments in the upper reaches tend to be large, angular, and rocky, with the particles quickly becoming smaller and more rounded as they shape the channel, and in themselves are shaped by the hydraulic action. In addition to the physical abrasion and carving out of river valleys, the process of solution erosion can occur, particularly in carbonate-rich limestone areas. Water acts as a weak acid (carbonic acid) to dissolve soluble minerals and leads to distinctive landforms such as caves, stalagmites, stalactites, sinkholes, and disappearing streams (Plate 2.9).

Mountains are so important for global freshwater supply as they receive more precipitation, experience lower evapotranspiration, and store vast amounts of water in snow and ice. However, climate change is reducing snowpack and glacial storage, leading to increasingly unpredictable discharge patterns and significant downstream impacts on water supply, hydropower, agriculture, and cultural practices (Carey et al., 2017). Water revived on permeable limestone mountains often surfaces along spring lines, which

Plate 2.8 Mountain rivers systems create classic V-shaped valleys, river terraces, and interlocking spurs. View down a valley close to Les Deux Alps, France. *(Photograph Rob Marchant).*

Plate 2.9 In limestone areas, water acts as a weak acid (carbonic acid) to dissolve soluble minerals and leads to distinctive landforms such as caves, stalagmites, stalactites, sinkholes, and disappearing streams. White Scar Cave, North Pennines. *(Photographs Rob Marchant).*

Plate 2.10 Water revived on permeable limestone mountains often surfaces along spring lines that are vital to maintain steady and consistent flows into lowlands such as this Blue Lake spring in Krabi Province, Thailand. *(Photograph Rob Marchant).*

are vital to maintain steady and consistent flows into lowlands (Plate 2.10). Groundwater, although invisible, is a massively under-researched component of mountain hydrology. It provides resilience to fluctuating surface water supply, but is also increasingly threatened by long-term climate change (Somers et al., 2019). Human land use also affects groundwater recharge; for instance, overgrazing in the Ecuadorian Páramo compacts soil and increases surface run-off, leading to less infiltration and groundwater recharge (Buytaert et al., 2006). Despite recent progress, high mountain groundwater systems remain poorly understood, constrained by limited field data and the extreme heterogeneity of mountain landscapes (Somers & McKenzie, 2020). Observation wells are scarce in many mountain regions, leading to high uncertainty in groundwater modelling and management. Addressing these knowledge gaps requires sustained investment and collaboration amongst scientists, local communities, and policymakers (Chapter 7).

2.3.3 Mass wasting processes

In high-elevation mountain areas, exposed rock faces, cliffs, and ridges dominate the landscape. These areas are subject to intense physical weathering, especially freeze-thaw cycles, where water infiltrates rock fractures, freezes, expands, and progressively breaks the rock apart. This process leads to the formation of unstable scree slopes (Plate 2.11) that can be a major

Plate 2.11 Unstable scree slopes are formed from large amounts of small rock fragments accumulated by free-thaw weather on rock, such as these scree deposits on Noordkette Klettersteig, Austria, and Blà Bheinn, Skye. *(Photographs Rob Marchant).*

trigger for rock falls, which are common in alpine environments. Warming temperatures further exacerbate rockfall risk by reducing the cohesion: as permafrost melts, the eroded rocks that were 'cemented' together by frozen water become mobile and thus increase slope instability (Gruber & Haeberli, 2007).

A variety of mass movements, such as landslides, rockfalls, debris flows, and avalanches, all contribute to slope instability and landscape evolution in mountainous areas. Mountain slopes often lie near a critical threshold of stability. An alteration in conditions, such as increased slope angle, water saturation, vegetation loss, or seismic activity, can trigger the downslope movement of soil, rock, and debris under gravity (Crozier, 2010). These movements range from slow, imperceptible soil creep to rapid and catastrophic events like landslides and debris flows. Landslides are one of the most prevalent forms of mass movement in mountains and can be triggered by any combination of intense rainfall, rapid snowmelt, glacial retreat, seismic activity, or human-induced disturbances such as deforestation, road construction, and mining (Crozier, 2010). These events can result in severe consequences, including loss of life, infrastructure damage, downstream flooding, and, in extreme cases, where this coincides with human settlement, destruction (Esfahan et al., 2025). Debris flows and mudflows, which occur when saturated loose material flows rapidly downslope, often follow extreme storms or glacial lake outburst floods. These fast-moving flows can impact large areas at the mountain base, posing serious hazards to downstream communities

(Jakob & Hungr, 2005; Schwanghart, Worni, Huggel, Stoffel, & Korup, 2016. The Himalayas, Andes, Alps, and East African volcanic mountains are particularly prone to landslides due to a combination of high relief, tectonic activity, and seasonal hydrological extremes.

The impacts of high magnitude and low frequency mass movements extend beyond direct physical destruction and mountain shaping. They influence sediment transport, river dynamics, and habitat structure, thereby altering entire ecosystems. In populated mountain areas, these risks are amplified by land use changes, expanding infrastructure, and climate change. Rising global temperatures, increasingly variable precipitation, and thawing permafrost are expected to intensify mass movement hazards in the coming decades (IPCC, 2022). Effective risk management in mountainous regions requires a combination of hazard mapping, early warning systems, community engagement, and climate-resilient infrastructure planning. As environmental and societal pressures grow, proactive and integrated strategies will be essential to reduce vulnerability and build resilience (Chapter 6).

2.4 Mountain climate systems

Mountain climate systems are characterised by their complexity and variability; elevation, latitude, topography, and geographical location result in a wide range of climatic conditions. This combination of elevation and topographic complexity produces a mosaic of microclimates, which can vary dramatically within just a few meters. Mountain climate systems create unique macro and microclimates that support diverse ecosystems (Chapter 3) and human activities (Chapter 5). Mountain climate regimes are primarily characterised by variations in temperature, precipitation, and wind patterns, all of which are influenced by factors such as altitude, slope aspect, and orientation. Large mountain ranges also play a crucial role in modifying large-scale atmospheric circulation patterns. For example, the Andes, Rocky Mountains, and the Tibetan Plateau influence upper-tropospheric flow, altering mid-latitude storm tracks and modifying regions of atmospheric subsidence and drying (Insel, Poulen, & Ehlers, 2010). Mountain climates are also highly sensitive to changes, particularly those induced by global climate change, both past and present (Chapter 4), which poses significant challenges for the future (Chapter 6). Understanding and adapting to these dynamic climate systems is essential for managing the environmental and societal impacts associated with mountains.

Plate 2.12 Cap clouds, or pileus clouds, form as thin, veil-like layers directly over mountain summits like this one over Aiguille du Diable. *(Photograph Rob Marchant).*

Elevation is the most critical factor in determining the mountain climatic regime. Sharp temperature contrasts across elevations generate pressure gradients, increasing wind speeds and change of direction, which often align with the orientation of valleys (Zhang et al., 2020). Mountains also significantly affect cloud formation. When moist air is forced to rise over mountainous terrain, orographic clouds form – typically stratiform in nature. In more stable atmospheric conditions, lenticular clouds (altocumulus or cirrocumulus) appear as smooth, lens-shaped formations produced by standing waves in airflow over peaks, indicators of strong upper-level winds and atmospheric stability (Houze, 2014). Cap clouds, or pileus clouds, form as thin, veil-like layers directly over mountain summits, created by the rapid uplift and cooling of moist air (Plate 2.12). In contrast, cumulus and cumulonimbus clouds often develop from convective heating when the sun warms mountain slopes, commonly leading to afternoon thunderstorms. Mountains also shape non-precipitating orographic clouds and modify preexisting cloud systems, such as frontal systems and tropical cyclones. The interaction between mountains and incoming air masses varies depending on whether the low-level flow is blocked or lifted over the terrain. For instance, atmospheric rivers, which are narrow corridors of concentrated moisture, are significantly altered as they pass over mountain ranges, determining the impact of orographic precipitation (Wang et al., 2022).

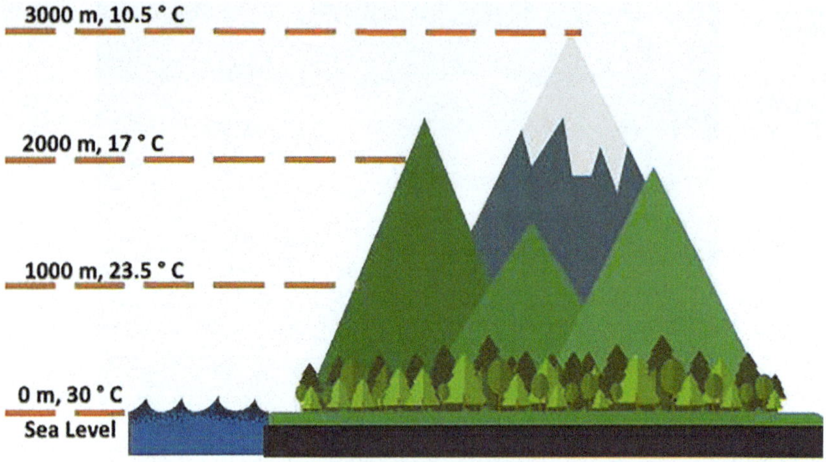

Figure 2.3 Mountain view showing a combination of factors, including elevation, latitude, and topography. Elevation is the most critical factor in shaping mountain climates.

2.4.1 Temperature

The altitudinal temperature gradient, the rate at which temperature decreases with increasing elevation, varies across seasons, mountain ranges, and specific elevation zones (Wagner, 1930). As altitude increases, air pressure decreases, resulting in cooler temperatures due to the thinning atmosphere. Elevation is the primary driver of temperature variation in mountains, a relationship commonly described by the lapse rate. The average environmental lapse rate is approximately 6.5°C per 1000 m of elevation gain (Fig. 2.3), though this rate is influenced by local conditions, especially atmospheric moisture. In tropical and subtropical regions, the mean tropospheric lapse rate closely follows the saturated adiabatic lapse rate, averaging around 6°C per 1000 m of elevation (Stone & Carlson, 1979). In contrast, due to variations in air pressure and density, mid to high latitudes weaken the lapse rate, resulting in more gradual temperature changes with altitude (Pepin et al., 2025).

Mountain ranges exhibit dramatic temperature gradients across their vertical profiles. Latitude plays a key role in determining the general climatic regime of a mountain range – tropical, temperate, or polar. In low latitude mountains, the Himalayas or Mount Kilimanjaro, for example, tropical conditions prevail at the base, while constantly freezing temperatures occur near the summits. As dramatic as the changes in temperature with altitude are, the changes in temperature through the day can also be remarkable:

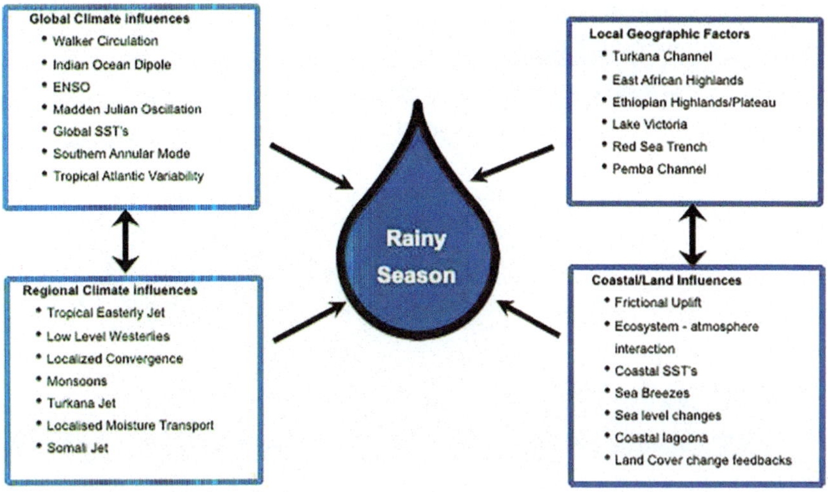

Figure 2.4 Precipitation patterns in mountainous regions are shaped by more than just elevation but also reflect complex interactions with local, regional, and global factors. All of these interactions clearly lead to a highly dynamic and variable rainfall pattern.

a temperature range of 30°C can be common. These steep temperature gradients shape distinct ecosystems across elevations (see Chapter 3). These altitudinal gradients play a fundamental role in shaping mountain climate systems, biodiversity (Chapter 3), and human adaptation (Chapter 5), underscoring the need for elevation-specific climate data.

2.4.2 Precipitation as rainfall

Orographic precipitation is a dominant driver of Earth's water cycle, particularly in controlling seasonal water supply in high-elevation areas. Mountains frequently receive more precipitation than surrounding lowlands due to the orographic effect: as moist air masses ascend mountain slopes, they cool, condense, and precipitate, typically on the windward side. Individual mountain ranges are often characterised by steep precipitation gradients (Fig. 2.3). These gradients are influenced by elevation, slope orientation, topographic complexity, and regional air circulation patterns. In addition to the common temperature lapse rate, some studies identify a precipitation lapse rate, reflecting the systematic change in precipitation with elevation (Avanzi et al., 2021). Precipitation patterns in mountainous regions are shaped by more than just elevation but also reflect complex interactions with local, regional, and global factors (Figs. 2.4 and 2.5), including slope

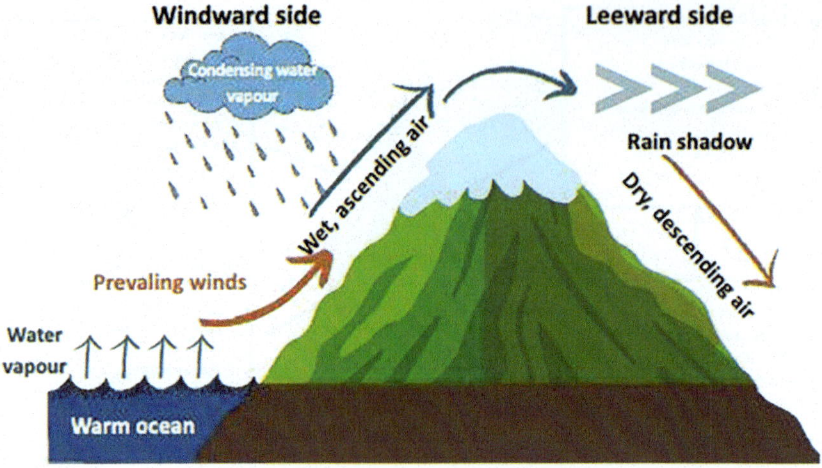

Figure 2.5 A schematic model mountain showing how topography shapes the movement of moist air and its associated precipitation, resulting in a wet side of the mountain opposite a rain shadow. Many mountains block moist air, leading to increased precipitation on the windward slope and a dry climate on the leeward side. Consequently, the two slopes of the mountain differ in terms of vegetation type, as well as species diversity, along their elevational gradients (Chapter 3).

direction, air pressure changes, atmospheric moisture, mountain orientation (strike), and proximity to moisture sources (Basist et al., 1994).

Topographic relief plays a critical role in the formation of rain shadows, especially in mountain systems near large bodies of water (Plate 2.13). The distribution, intensity, and duration of orographic precipitation are governed by three main factors (Rotunno & Houze, 2007): synoptic circulation patterns, intensity of mesoscale atmospheric lifting along slopes, and the timing and location of condensation, driven by convection and turbulence. A good example is the Himalayas, where south-facing slopes are often hot and dry, while north-facing slopes are cooler and moister (Wambulwa et al., 2021). Similarly, the Andes show distinct differences in rainfall, with heavy precipitation on their western slopes and dry conditions in the eastern rain shadow. On Mount Kilimanjaro, rainfall on the southeastern slopes is significantly greater than on the drier northwestern slopes (Plate 2.13).

Changes in orographic precipitation patterns and the processes that control them are expected to significantly affect future hydrology. However, these phenomena remain poorly understood at a global scale (Pepin et al., 2025). Climate change is already disrupting Earth's hydrological cycles, altering not only the amount of precipitation but also its frequency, intensity,

Plate 2.13 The rain shadow on Mount Kilimanjaro results in significantly greater rainfall on the Southeastern slopes in comparison to the northwestern areas, as demonstrated by this boundary of this moisture-laden cloud rolling in from the Indian Ocean located some 300 km to the west. *(Photograph Rob Marchant).*

and seasonality (Sun et al., 2024). The Tibetan Plateau is a major source of Asia's largest rivers, including the Yangtze, Yellow, Yarlung Zangbo, Indus, and Ganges, which support over 1.4 billion people downstream (Immerzeel et al., 2010). The precipitation regime on the Tibetan Plateau is influenced by the westerlies, the South Asian monsoon, and the East Asian monsoon (Li et al., 2021). These systems transport moisture, which interacts with the plateau's complex topography to shape stark, spatial, and temporal precipitation gradients. The Himalayas can be climatically divided into Western, Central, and Eastern regions; the Eastern and Central Himalayas are usually characterised by a warm and wet climate owing to the moisture-laden air supplied by the Indian summer monsoon, while the western sector has a warm and dry climate. The Southern Himalayas receive up to 10 times more precipitation than the northern slopes, due to orographic uplift and the blocking of moist monsoonal air by the Himalayan Mountains (Ahsan et al., 2023). The Himalayas are characterised by decreasing precipitation from east to west, where the windward southern sides experience rainy and warm conditions, and the leeward northern sides are characterised by dry and cold climates (Fig. 2.6). Precipitation across the plateau can vary from over 600 mm $year^{-1}$ in the southeastern foothills to less than 50 mm $year^{-1}$ in the arid interior (Li et al., 2021).

Traditional ground-based precipitation measurements are limited in spatial coverage, especially in the Global South, where underinvestment

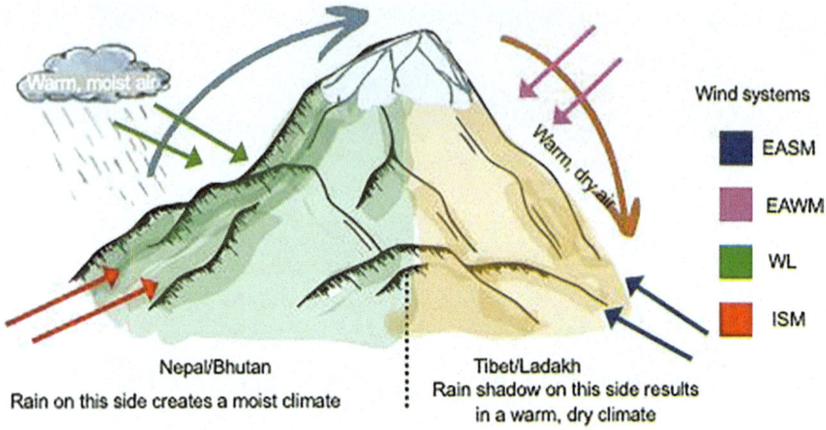

Figure 2.6 Geographical location of Himalayas and the atmospheric circulation systems in East Asia after (Wambulwa et al., 2021); the arrows indicate the wind systems: East Asian summer monsoon (EASM), East Asian winter monsoon (EAWM), Westerlies (WL), Indian summer monsoon (ISM).

in meteorological infrastructure has resulted in significant data gaps, particularly on mountains due to issues of access and maintenance. This lack of observational data impedes the accurate precipitation assessment, the understanding of change, and hydrological modelling. Satellite-based precipitation products, such as NASA's Tropical Rainfall Measuring Mission (TRMM), offer high-resolution data that improve monitoring over remote and rugged mountain areas (Qin et al., 2014). These products are invaluable for applications in hydrological forecasting, climate modelling, and water resource management, especially in inaccessible high-altitude regions with harsh environmental conditions (Li et al., 2021). However, we ultimately still need ground-truthing to calibrate and improve these Earth observation products – direct measurements remain invaluable.

2.4.3 Precipitation as snowfall

Falling snow plays a vital role in mountain hydrology by transferring moisture and latent heat between the atmosphere and the land surface (Plate 2.14). It modifies surface properties by altering albedo and emissivity, where accumulated snow acts as both a thermal insulator, regulating sensible heat fluxes, and a surface water reservoir (Daloz et al., 2020). Globally, snow accumulation in high mountain regions is estimated at 1773 km^3, representing approximately 5% of total global snowfall (Daloz et al., 2020).

Plate 2.14 Snow plays a vital role in mountain hydrology. *(Photograph Rob Marchant).*

This stored solid water is critical to downstream hydrology, indirectly supporting the water security of ~1.9 billion people who rely on mountain 'water towers' (Immerzeel et al., 2020). Snowfall and snow accumulation processes continue to undergo rapid transformation under global warming. Snow courses also serve as natural precipitation gauges and can be invaluable for calibrating hydrological models and assessing the precipitation lapse rate – the variation of snowfall with elevation (Avanzi et al., 2021). In the Pyrenees and Alps, snowfall is projected to decrease below 2000 m and increase above 2000 m, demonstrating a vertical shift in snow regimes (Le Roux et al., 2021). While heavy snowfall is driven primarily by large-scale atmospheric circulation, its local intensity and duration are often amplified by terrain-specific thermodynamic feedback (Daloz et al., 2020).

Whilst global warming is reducing average snowfall, it also increases the risk of extreme snowfall events by raising atmospheric moisture content (Sun et al., 2024). Warmer air holds more moisture, creating favourable conditions for high-intensity snowfall (Quante et al., 2021). This increased volatility poses new challenges for water resource management, affecting not only water availability but also the predictability of water supply. In the Italian Alps, lower elevations have seen significant declines in snowfall due to rising mean temperatures, while higher elevations during mid-winter show slight increases, associated with greater overall precipitation (Bertoldi et al., 2023). Across the Himalayas, snowfall amount, frequency, and fractional contribution have declined significantly, with a shorter snowfall season also observed (Sun et al., 2024).

Extreme snowfall can lead to natural disasters such as flooding, snow avalanches, and infrastructure damage (Chapter 6). Heavy snowfall impacts transportation networks, electricity grids, communication systems, and tourism-dependent economies. For example, in January 2017, a cold spell in Europe combined with earthquakes in central Italy triggered a massive avalanche in Rigopiano, Abruzzo, causing significant loss of life (Frigo et al., 2018). One particular phenomenon of rain-on-snow is a major driver of flood events in mountain catchments, combining both rainfall dynamics and snowpack energy balance (López-Moreno et al., 2021). The hydrological response to rain-on-snow depends on preexisting snow conditions, snow cover extent, and concurrent weather variables. Warming temperatures are expected to shift the rain-on-snow season, decreasing its frequency in late spring but increasing it in late winter and early spring. The geographic variability of rain-on-snow trends explains why studies from North America and the Alps have reported inconsistent findings regarding future changes in frequency and associated flood risks (Hock et al., 2019). This underscores the need for region-specific snow forecasting and water management strategies, vital under a changing climate.

2.4.4 Precipitation as nonprecipitating clouds

Mountains located near oceans are often characterised by the rapid uplift of moisture-laden air, which cools and condenses as it rises, leading to the formation of mists and low-level clouds. This process results in 'occult precipitation', whereby moisture condenses directly onto vegetation, contributing to the formation of distinctive cloud forests. These forests are especially prevalent in tropical and temperate mountainous regions influenced by warm, humid oceanic winds. The influence of adiabatic lifting in ocean-facing mountain systems, which significantly shapes cloud and vegetation dynamics, can help to sustain diverse and often endemic plant communities (Chapter 3). Globally, cloud forests cover approximately 215,000 km^2, or about 0.14%, of Earth's land surface. However, more than 46% of their

Plate 2.15 Mount Marsabit in northern Kenya. *(Photograph Rob Marchant).*

original extent has been lost due to land-use changes (Gutiérrez-García et al., 2020; Scatena et al., 2011), which is leading to loss of biodiversity and, perhaps more importantly, loss of water engines, vitally important, particularly to mountains surrounded by drylands such as Mount Marsabit in northern Kenya (Plate 2.15).

Occult precipitation includes both deposited forms – such as dew, white dew, hoar frost, and soft rime (a white ice deposition that forms when the water droplets in light freezing fog or mist freeze to the outer surfaces of vegetation) and collected forms – including fog, cloud water, and hard rime (Tesař et al., 2000). Occult precipitation in high mountain environments contributes directly to the catchment water balance; intercepted water drips from vegetation to the ground after exceeding leaf storage capacity, and also reduces plant transpiration by maintaining surface moisture on the foliage (Cárdenas et al., 2017) (Plate 2.16). This moisture input can be especially important during dry seasons, helping to limit water loss under high evaporative demand (Aparecido et al., 2018; Nair et al., 2011).

The quantitative contribution of occult precipitation can be locally significant. For example, in the Krkonoše Mountains (Czech Republic), cloud and fog water accounts for 9%–15% of total annual vertical precipitation (Hůnová et al., 2020). In some mountain climates, small high-altitude valleys can trap air under temperature inversion layers, where cool air and low wind activity prevent circulation. These isolated air pockets are often fed by humid air masses originating from river valleys located hundreds of kilometres away. Such long-distance moisture transportation and topographically isolated

Plate 2.16 Occult precipitation is often characterised by misty elfin-like forest shrouded in lichens, mosses, and bromeliads. Cloud forests capture moisture from the air, slow down surface runoff, and release water gradually, ensuring a steady supply to lower elevations. *(Photograph Rob Marchant)*.

conditions may promote stable microclimates, contributing to the long-term persistence of certain species during periods of past climatic change (Chapter 4). However, due to its elusive nature and the difficulty of direct measurement, occult precipitation is often under-represented in regional rainfall records.

While the role of fog in tropical montane ecosystems is documented on some mountains (Bruijnzeel et al., 2011), there remains a knowledge gap concerning fog inputs in páramo ecosystems – high tropical alpine environments typical of the Andes. Studies in Colombia estimate water inputs via occult precipitation in páramos at 120–212 mm year^{-1}, though these values vary substantially across and within sites, and can be very site specific (Cárdenas et al., 2017). Moreover, fog presence is influenced by connections to downslope ecosystems that generate humid air masses, as observed in the cloud forests of Costa Rica (Lawton et al., 2001). Beyond its hydrological importance, fog and cloud water are also important pathways for atmospheric pollutant deposition. These waters typically have higher ionic concentrations and are more acidic than rainwater. The chemical composition of fog water varies with local emissions, topography, and meteorological conditions, making it difficult to generalise without considering site-specific factors (Hůnová et al., 2020).

These different sources of moisture combine to form a whole range of mountain water systems, such as lakes, rivers, and waterfalls (Plate 2.17), that

Plate 2.17 Lakes, rivers, and waterfalls are a vital part of the functioning of the mountain hydrological system. *(Photographs Rob Marchant)*.

are a vital part of the functioning of the mountain hydrological system that connects the highlands to the lowlands.

2.4.5 Wind patterns

Winds influence local weather conditions, including temperature and cloud cover, and can have regional-scale impacts. A notable example is the Foehn wind, a warm, dry downslope wind on the leeward side of the Alps, which can rapidly increase temperatures and lower humidity, affecting weather patterns across Central Europe. Whether atmospheric flow is lifted over or diverted around mountains depends on several factors: wind speed, static stability, and the height and form of the mountain (Jackson et al., 2012). In the mid latitudes, such interactions can influence or modulate the jet stream. Mountain shape and orientation determine how air masses move, leading to variations in wind behaviour, precipitation patterns, and temperature. Valleys, ridges, and plateaus all further contribute to this complexity. For instance, the Himalayas block the monsoon winds, causing heavy rainfall on their southern slopes, while the Tibetan Plateau to the north remains arid.

The Eastern Carpathians act as a barrier to maritime (moist) air flows over Central and Eastern Europe (Kłapyta et al., 2020).

Mountaintops can be extreme places and are often characterised by very high wind speeds. In the nival zone, winds are especially strong and persistent, contributing to significant wind erosion by transporting loose particles of rock and ice across the landscape. Mountain topography plays a crucial role in shaping wind patterns, giving rise to katabatic (downslope) and anabatic (upslope) winds. Downslope winds are mesoscale meteorological phenomena observed globally (Abatzoglou et al., 2021). These winds, characterised by adiabatic warming and turbulent mixing, can create favourable conditions for wildfire spread. Their strong, dry gusts are hazardous, promoting fast-moving wildfires that threaten communities and contribute to unique regional fire regimes. Additionally, downslope winds impact air quality and human health. They can entrain and transport dust and smoke, especially from wildfires, and exacerbate heat stress in valleys and coastal areas, where the winds terminate (Abatzoglou et al., 2021; Schwarz et al., 2020).

2.4.6 Mountain microclimates

The complex terrain of mountains creates diverse microclimates, where one area can have significantly different climatic conditions from an adjacent area. Factors such as aspect, vegetation cover, and proximity to water bodies contribute to these microclimates. For example, in the Rocky Mountains, south-facing slopes receive more sunlight and are warmer and drier, supporting different vegetation types than the cooler, moister north-facing slopes. The Himalayan southern slopes are influenced by the Indian monsoon, leading to heavy rainfall, while the northern slopes are in the rain shadow, resulting in arid conditions. The diversity of the topography provides numerous microhabitats for wildlife, with sheltered areas (such as depressions or lee sides of ridges) being warmer and less exposed to wind than open or exposed sites. The structure of forests creates a variety of microclimates. For example, shaded areas under dense canopies are cooler and moister, while open spaces receive more sunlight and can be warmer and drier; deadwood on the forest floor offers shelter for small mammals and insects, while the canopy supports numerous epiphytes. The diverse structure of subalpine forests provides a range of microhabitats, from sun-exposed rocky outcrops to shaded forest floors. Despite the overall harsh environmental regime, microhabitats exist in sheltered cracks in the rock (Plate 2.18) or beneath the snowpack. These microhabitats provide more favourable conditions for life, such as

Plate 2.18 Microclimates within rock cracks are colonised by plants in exposed areas of rock, such as in Stubai Valley (*left*) and the Moloti-Drakensburg (*right*). *(Photographs Rob Marchant).*

slightly warmer temperatures, a concentration of plant-available moisture, or protection from wind. These microhabitats support a wide variety of species adapted to specific conditions (Plate 2.18), including specialised insects, birds, and mammals (Chapter 3).

2.4.7 Diurnal and seasonal climate regimes

Mountains are characterised by extreme diurnal fluctuations in climate, particularly temperature (Gallou et al., 2023) (Fig. 2.7). Short-term climatic oscillations – yearly and diurnal variation of temperature, seasonality, and growing season length – combine in ways that are unique to mountains. For example, temperature can vary on a daily basis from freezing to plus 30°C, as the night frost transitions to extreme heat. In complex montane landscapes, east-facing slopes (with morning sun) will provide different conditions from adjacent, cold and foggy west-facing slopes. Shadows from neighbouring mountains can drive differences among foothills with otherwise similar aspects. These aspects of spatial and temporal variation, as well as geophysical

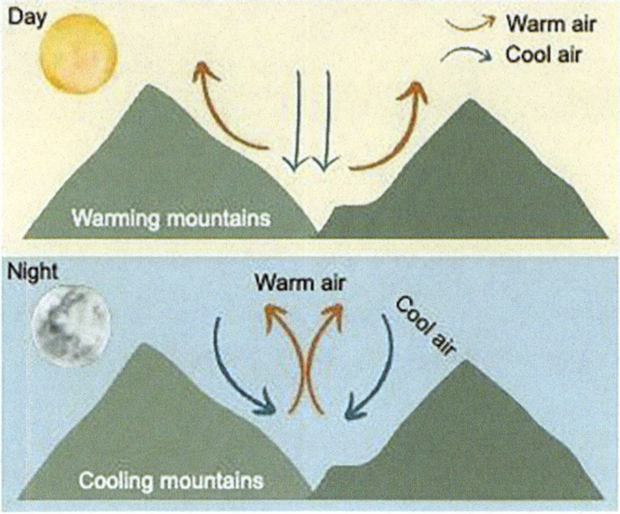

Figure 2.7 Each day, there are significant variations in temperature depending on season, height of sun, and cloud cover.

variation of mountains in terms of differing slopes and aspects, interact to create many rare and unique climate regimes. The effect of longer-term climatic oscillations is unique in mountains and results in a periodic reconfiguration of connectivity among mountain habitat patches that will be explored in Chapter 4.

2.4.8 Climate volume of mountains – Capturing extreme diversity

As seen from the previous sections, mountains are characterised by distinct three-dimensional complex topography and a pronounced variation of climatic and edaphic conditions, operating over multiple temporal and spatial scales, resulting in numerous ecological niches that often lead to biodiversity hotspots (Bobrowski et al., 2021). This diverse heterogeneous climate is magnified, as mountains are not the idealised triangles so often depicted in textbooks, including this one! Mountains are full of nooks and crannies and diverse topography, producing a highly intricate web of environments that ultimately leads to high species diversity, novel climates, and hence novel species (Chapter 4). Clearly, there are multiple climate parameters all interacting independently but also in combination, leading to a highly complex and multi-dimensional climate regime that gives mountains their

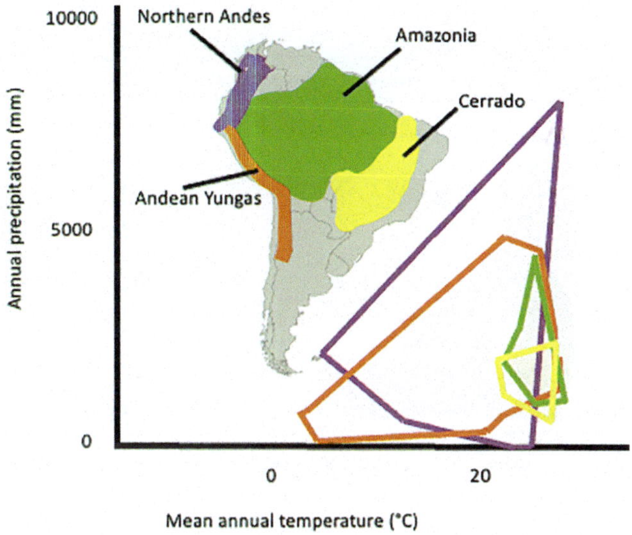

Figure 2.8 The climate spaces of particular mountain regions (Northern Andes, purple, and the Andean Yungas, orange) encompass a large proportion of all climate space and are disproportionately larger than the climate space of neighbouring lowland regions (Amazonia, green, and the Cerrado, yellow). *Modified from Rahbek et al. (2019a).*

special character. Topographically complex tropical mountains may encompass a remarkable volume of different climate types, equivalent to the climate volume of vast geographical expanses within lowland regions (Rahbek et al., 2019b) (Fig. 2.8).

Some mountain climates are rare only in the sense that they are restricted to mountains, but in having many combinations of aspect, location, and local weather pattern, even small mountain ranges may create globally exceptional combinations of climatic factors, provided that the topography is sufficiently rugged. The geographical position of the mountain chain will also have a significant impact on the overall climatic character of the mountains, with those located close to lakes having very different climates from those close to deserts, and the mountains in continental settings being very different from oceanic mountains. This overall diversity of climate can be captured within the concept of climate volume (Fig. 2.5).

2.5 Mountain soils

Mountain soils have their own character. Some 100 years ago, Jenny (1929) first noted that soils of alpine areas are darker and richer

Plate 2.19 Fertile andosols form on volcanic rocks in the Andes, Chile. *(Photograph Rob Marchant).*

in humus than those of lowlands. Mountain soils cover 22%–27% of the Earth's total soil surface (Bevington & Brideau, 2020). Soil properties vary strongly with elevation, but not necessarily in the same way across different regions. The wide range of altitudes and slope aspects allows for an extraordinary diversity of physical and chemical soil properties compared to the lowlands (Aguado-Norese et al., 2023). Mountain soils are formed through a complex factor that combines climate, geology, the orographic situation, vegetation cover, the nature of human activities, and time (Bäumler & Zech, 1994). Factors such as pH, temperature, water availability, oxygen levels, and nutrient availability vary with location; thereby impacting the composition of microbial populations and overall soil character (Aguado-Norese et al., 2023). In mountainous regions, the parent material varies widely, comprising igneous, sedimentary, and metamorphic rocks, which influence soil mineral content and texture. For example, in the Andes and on East African mountains, volcanic rocks have contributed to the formation of fertile andosols, which are rich in minerals and support diverse and productive agriculture at high altitudes (Plate 2.19). The steep slopes of mountains hinder the accumulation of deep soil profiles, leading to shallow, stony, thinner, and more nitrogen-limited soils, particularly at high elevations (Tito et al., 2020). For example, in the Andes, soils at higher elevations are generally thin and rocky, limiting agricultural potential to specific crops adapted to these conditions. Conversely, soils are older, deeper, and more phosphorus-limited at low elevations (Vitousek & Sanford, 1986; Vitousek & Farrington, 1997).

Physical and chemical soil properties vary considerably within a mountain range; indeed, there is often great diversity across a single mountain (Tito et al., 2020). Small changes in landscape position produce changes in the frequency and duration of wet periods, affecting the conditions under which the soils develop. The texture of mountain soils can vary widely, depending on the parent material, weathering processes, and organic matter content. Soils often contain a mix of sand, silt, and clay, with a significant presence of coarse fragments like gravel and rocks. For example, soils in the Rocky Mountains can range from sandy loams in lower elevations to more clay-rich soils in sheltered valleys, reflecting the diverse parent materials and microclimates (Liptzin & Seastedt, 2010). Flooding and drainage characteristics are crucial determinants for the presence of anaerobic microorganisms in soils with lower drainage capacity (Tang et al., 2020). The steep slopes of mountains affect soil development by controlling the movement of water, the deposition of sediments, and the rate of erosion. Flat areas such as plateaus and valley floors tend to have deeper soils due to slower erosion rates and greater accumulation of organic matter. In the Alps, soils on south-facing slopes tend to be thinner and drier, while those on north-facing slopes are deeper and more developed due to differences in solar radiation and moisture retention.

Another important factor determining the type and character of soil is time: soils take a long time to form. For example, in the Alps, the formation of a stable and homogeneous substrata is estimated to require more than 200 years at an elevation of 2000 m (Freppaz et al., 2010). The rates of soil-forming processes, which characterise soil development, generally decrease with increasing time (Bäumler & Zech, 1994). This temporal element is interlinked with the climate that plays a critical role in soil formation by influencing weathering processes, organic matter decomposition, and erosion. Similarly to how climate varies with elevation in mountain regions, there are different soil types along the altitudinal gradient. For example, the Sierra Nevada in California ranges from fertile soils in the lower foothills to rocky, less developed soils at higher elevations. Soils in the nival zone, or on newly deposited volcanic ash (Plate 2.20) or lava, are poorly developed or absent, as the constant freezing and thawing, along with glacial activity, prevent the accumulation of organic material and the formation of soil layers. What little soil exists is often thin and composed of coarse weathered rock fragments.

Mountain soils support unique ecosystems and high levels of biodiversity, providing habitats for a wide range of plant and animal species

Plate 2.20 Thin young soils forming on recent ash on the slopes of Mount Teide, Tenerife. *(Photograph Rob Marchant)*.

Plate 2.21 Mountain soils are crucial for a variety of ecological, hydrological, and socioeconomic functions, as well as storing large amounts of carbon. Mountain soils support unique ecosystems and high levels of biodiversity, providing habitats for a wide range of plant and animal species. *(Photograph Rob Marchant)*.

(Plate 2.21). Between soil profiles and horizons, there are patterns of diversity that form a clear structure in the soil microbial community (Aguado-Norese et al., 2023). Vegetation, microorganisms, and soil fauna contribute to the formation of mountain soils by adding organic matter from the surface, promoting nutrient cycling, and enhancing soil structure. The vegetation type in particular influences the organic content and acidity of the soil. In the Himalayan region, the dense Oak and *Rhododendron* forests contribute to the formation of humus-rich soils, which support diverse plant communities (Wang et al., 2022). Conversely, under coniferous forests such as on the slope of Mount Teide in Tenerife, soils are often acidic due to the accumulation of acidic leaf litter and the leaching of basic cations.

Soil organic matter represents an important carbon reservoir in the terrestrial biosphere. It is estimated that soil organic matter contains roughly double the amount of carbon compared to the atmosphere or aboveground vegetation (Schimel, 1995), although many studies continue to focus solely on the aboveground carbon (Cuni-Sanchez et al., 2021). Soil carbon stocks in alpine areas, as in other cold ecosystems, are particularly important, since low temperatures limit decomposer activity and keep carbon immobilised in soils for long periods of time (Körner, 2003) (Plate 2.21).

Mountain slopes, particularly those of Central Europe, have been intensively used for centuries. Mountain soils on slopes are usually well-drained, relatively easy to work, and productive. Due to socio-economic pressures, the changes in mountains in many regions around the world are dramatic. Marginal land is abandoned while land on productive sites, such as in valley bottoms, is intensified (Hagedorn et al., 2010). For instance, in the Swiss Alps, the forested area above 1400 m is increasing by 7% per decade, twice as much as at lower altitudes (Bolliger et al., 2008). In the Eastern Central Alps, between 35 and 70% of the steep and high-altitude areas, compared to only 7% of the favourable valleys, have been abandoned during the last 150 years and reforested (Tasser et al., 2007).

Mountain soils play a key role in regulating water flow by absorbing and storing precipitation, releasing it gradually into rivers and streams. This function is essential for maintaining the hydrological balance and providing water and soil resources to downstream areas. Accelerated soil erosion is a common and environmentally destructive consequence of development, especially in mountain regions (Harden, 2001). Road building, trail use, excavation, extractive activities, and construction can also cause severe soil erosion (Plate 2.22). Due to the steep terrain, mountain soils are particularly prone to erosion by wind, water, and gravity. This erosion can lead to soil loss, affecting soil depth, fertility, and stability, which in turn influences vegetation patterns and land use. Indeed, mountain soils are crucial for a variety of ecological, hydrological, and socio-economic functions. Soil conservation technologies are relatively simple and well known, but often they are not applied where they could be most effective, because the connections between different elements of the landscape (e.g., roads and cultivated fields)

Plate 2.22 Severe soil erosion can be common in mountain systems, such as these deep gullies formed adjacent to the Pare Mountains of northern Tanzania due to rapid erosion from steep slopes. Soil conservation technologies are relatively simple and well-known, but often they are not applied or maintained. *(Photograph Rob Marchant).*

are little understood (Harden, 2001) and frequently applied out of context. Soil loss lowers crop productivity, changes on and off-site hydrology, and creates off-site deposits that may be damaging (Harden, 2001; Pimentel et al., 1995). Soil is vulnerable to erosion where it is exposed to moving water or wind and where conditions of topography or human use, such as steep slopes, compacted surfaces, removal of vegetation, or years of ploughing, increase the force of the moving fluid or decrease the cohesion of the soil (Harden, 2001).

2.6 Mountain fire dynamics

Mountains, particularly volcanic mountains and those supporting human population, can be dominated by fire regimes that are highly influential in shaping the ecosystem, soils, land use, and stability of the mountain social ecological systems. Fire regime refers to fire frequency, fire intensity, average fire size, and time of burning. Fire regime is controlled primarily by the amount, type, arrangement, and condition (i.e., dryness) of fuels (Rocca et al., 2014) with variables such as relative humidity, temperature, wind speed, timing of snowmelt, and precipitation events, further affecting fuel moisture and flammability. Changing climates that yield less precipitation and lower humidity can dry vegetation and soil and increase fire frequency in regions having fire-prone vegetation (Eskandari et al., 2020).

Climate is undoubtedly a major factor in past and present fire regimes (Finch et al., 2014, 2017; Rodman et al., 2020), but humans are also important drivers of fire regime change (Knight et al., 2022). There has to be material to burn, and the amounts of fuel vary on time scales, from years to centuries, and also with rates of biomass decomposition. Fuels have to be dry enough to burn, and changes in daily to seasonal weather conditions are the primary controls on whether ignition can occur and how extensive or intensive a fire will be. The type of fire is not just determined by the amount of fuel, but horizontal and vertical fuel structure greatly affect fuel connectivity across landscapes, and the ability of fire to move from surface to canopy layer (Rocca et al., 2014). Forests in cool and wet sites, such as the subalpine forests, tend to be relatively productive with higher fuel loads and connectivity. However, the fuels in these forests are rarely dry enough for extensive burning to occur, so fire frequency and expanse are relatively low. When fires occur, they tend to burn at high severity, and the resulting active crown fires cause extensive areas of tree mortality (Schoennagel et al., 2004).

Fire is one of the most important natural disturbances in ecosystems around the world and can have an important positive role, but also causes major environmental, social, and economic losses (Eskandari et al., 2020). Fire has shaped the diversity of life on Earth for millions of years. Variation in fire regimes continues to be a source of biodiversity across the globe, and many plants, animals, and ecosystems depend on particular temporal and spatial patterns of fire. On mountains where there is long-term exposure to fires, such as volcanic mountains, the ecosystem often becomes 'fire adapted'. For example, on the Canary Islands or the Cape Mountain systems, some species have multiple physiological adaptations to life on a fiery mountain, such as thick protective bark, the ability to resprout from the trunk following fire, cones and seeds that require fire to release the seeds, and copious amounts of resin that make the wood difficult to burn (Plate 2.23). Indeed, some ecosystems are not only adapted to withstand fires but depend on the fire (Plate 2.23) for regeneration and functioning.

Although people have been using fire to modify environments for millennia, the combined effects of human activities are now changing patterns of fire at a global scale – to the detriment of human society, biodiversity, and ecosystems (Kelly et al., 2020). People intentionally and accidentally influence the likelihood of fires in forests, with land use and land cover having a strong relationship with wildfire (Kariuki et al., 2018, 2021). Past fire management practices can greatly complicate efforts to project ecosystem response to future climate and fire management alternatives. For

Plate 2.23 Fire-adapted species such as the Canary pine *Pinus canariensis* (*left*) can resprout from the tree following a fire. Some ecosystems depend on fire for functioning, such as the *Protea*-dominated heath on many African mountains. *(Photographs Rob Marchant).*

some forest types, fire exclusion has caused profound changes in forest composition and structure, the type and amount of fuel, and the resulting fire behaviour (Rocca et al., 2014). Dendroecological studies have provided abundant information on the nature of historical fire regimes, and these have helped define the historic ranges of variability in disturbance processes and forest conditions before the widespread impacts of Euro-American settlement (Friederici, 2003).

In many parts of the world, there are projected to be quite dramatic changes to the fire seasons. These seasons will start earlier and end later, but mountains will likely also be drier due to the earlier snowmelt and the direct impacts of the temperature increases on fuel dryness (Rocca et al., 2014). Wildland fires in the United States have the potential to release large amounts of carbon to the atmosphere through the combustion of wildland fuels and subsequent decomposition of dead woody biomass (Rocca et al., 2014). Smoke generated from wildland fires is an emerging concern because of its impacts on atmospheric visibility, human health, and air quality. The interaction between climate, fuels, and the frequency and severity of wildfires across Rocky Mountain forests, as with all mountain areas, is complex (Schoennagel et al., 2004). The annual area burned by wildfires has increased during the last few decades in the southern Rocky Mountain region across North America, possibly in response to recent climate change and the gradual accumulation of fuels following decades of effective fire suppression (Schoennagel et al., 2004). Historically, short-interval, low-severity surface fires maintained sparse open stands of vegetation that would not burn at high intensity. Denser stands provide 'ladder fuels' at intermediate heights that carry fire up into the canopy, promoting unprecedentedly large, catastrophic fires that may be increasing in size and frequency (Schoennagel et al., 2004). High-severity fires, in combination with a changing climate, interrupt system inertia and 'break the legacy lock' (Johnstone et al., 2010) of the prior system state, driving rapid shifts in local vegetation. Fuel-reduction treatments (mechanical thinning and prescribed burning) may effectively reduce fire severity (Schoennagel et al., 2004).

Increasing wildfire activity and declining forest resilience to wildfire have the potential to drive widespread conversions, from forested to non-forested vegetation types (e.g., grasslands to xeric shrublands to mesic shrublands) across many montane ecosystems of western North America, and much of this increase has been attributed to decreases in summer precipitation (Holden et al., 2018). A warming climate also hinders post-fire forest recovery, and severely burned areas are less likely to recover to a forested state when burning is followed by warm, dry periods (Davis et al., 2019). With the increasing frequency of wildfires in the Himalayas and their surroundings (You et al., 2018), modelling the effect of wildfires on Himalayan taxa could be an important research direction in the near future.

Fires ignited by lightning and Indigenous people have influenced the structure and composition of forest ecosystems in the American West for millennia, although the extent of the impact of indigenous burning on Californian ecosystems continues to be contested (Knight et al., 2022). In common with water and soil management, we can look to history for insights into effective fire management. In the Jemez Mountains of northern New Mexico in the United States, large communities of Pueblo farmers lived in *Ponderosa* pine forests, which have adapted to frequent (every 5–10 years), low-severity surface fires. Indeed, major features of human impacts in the agricultural landscape include applied fire management for agricultural production, and fuelwood collecting to support contained uses of fire (Roos et al., 2021). Over the past century, California's fire-prone conifer forests have been altered by the curtailment of Indigenous stewardship, the exclusion of fire, and the harvest of merchantable trees (Knight et al., 2022). The forest structure and composition encountered by Euro-American colonists in the mid-1800s were thus shaped by cultural burning practices developed over the last millennia (Knight et al., 2022). Specifically, watersheds encompassed the routine deployment of small situational burns (i.e., patch burns of <10 ha) along with larger broadcast burns (i.e., >10 ha) during the late winter/early spring and late summer/early autumn, with general boundaries constrained by topography (i.e., slope and aspect) and fuels (i.e., loads and connectivity) (Knight et al., 2022). These traditional management practices have been removed or eroded, with consequences for those living today within mountain areas.

2.7 Conclusion

At the heart of Mountain systems is diversity at all levels; this diversity starts with their geological origins and the ensuing diverse topography. In turn, the erosional process of ice, water, and wind then works on eroding uplifted mountain systems to become grains of sand. The climatic regimes, soil characteristics, and fire environments are all influenced by the underlying geology that form cornerstones for determining mountain character. Despite their variety, mountain regions often share core features: high geological diversity, complex landscape structures, diverse climatic and fire regimes – both past and present – further contributing to their ecological and geological uniqueness. Understanding these mountain environmental foundations is key to interpreting both the distribution of mountain ecosystems and the applied aspects of mountain systems explored in subsequent chapters. Indeed, understanding the underlying physical foundations is essential for identifying both the challenges facing mountain environments and potential solutions – we have to know the mountain system to work with it, rather than against it.

References

Abatzoglou, J. T., Hatchett, B. J., Fox-Hughes, P., Gershunov, A., & Nauslar, N. J. (2021). Global climatology of synoptically-forced downslope winds. *International Journal of Climatology, 41*(1), 31–50. https://doi.org/10.1002/joc.6607.

Aguado-Norese, C., Cárdenas, V., Gaete, A., Mandakovic, D., Vasquez-Dean, J., Hodar, C., Pfeiffer, M., & Gonzalez, M. (2023). Topsoil and subsoil bacterial community assemblies

across different drainage conditions in a mountain environment. *Biological Research, 56*(1). https://doi.org/10.1186/s40659-023-00445-2.
Ahsan, S., Bhat, M. S., Alam, A., Sheikh, H. A., & Farooq, H. (2023). Hydrological extremes and climatic controls on streamflow in Jhelum basin, NW Himalaya. *Theoretical and Applied Climatology, 151*(3-4), 1729–1752. https://doi.org/10.1007/s00704-022-04346-4.
Aparecido, L. M. T., Teodoro, G. S., Mosquera, G., Brum, M., Barros, F. d. V., Pompeu, P. V., Rodas, M., Lazo, P., Müller, C. S., Mulligan, M., Asbjornsen, H., Moore, G. W., & Oliveira, R. S. (2018). Ecohydrological drivers of Neotropical vegetation in montane ecosystems. *Ecohydrology, 11*(3). https://doi.org/10.1002/eco.1932.
Armstrong, R. L., & Brun, E. (2008). *Snow and climate: Physical processes, surface energy exchange and modelling* (p. 256). Cambridge University Press.
Avanzi, F., Ercolani, G., Gabellani, S., Cremonese, E., Pogliotti, P., Filippa, G., Morra DI Cella, U., Ratto, S., Stevenin, H., Cauduro, M., & Juglair, S. (2021). Learning about precipitation lapse rates from snow course data improves water balance modeling. *Hydrology and Earth System Sciences, 25*(4), 2109–2131. https://doi.org/10.5194/hess-25-2109-2021.
Barandun, M., Pohl, E., Naegeli, K., McNabb, R., Huss, M., Berthier, E., Saks, T., & Hoelzle, M. (2021). Hot spots of glacier mass balance variability in Central Asia. *Geophysical Research Letters, 48*(11). https://doi.org/10.1029/2020GL092084.
Barr, I. D., & Spagnolo, M. (2015). Glacial cirques as palaeoenvironmental indicators: Their potential and limitations. *Earth-Science Reviews, 151*, 48–78. https://doi.org/10.1016/j.earscirev.2015.10.004.
Basist, A., Bell, G. D., & Meentemeyer, V. (1994). Statistical relationships between topography and precipitation patterns. *Journal of Climate,* 7(9), 1305–1315. https://doi.org/10.1175/1520-0442(1994)007⟨1305:srbtap⟩2.0.co;2.
Bäumler, R., & Zech, W. (1994). Characterization of andisols developed from nonvolcanic material in Eastern Nepal. *Soil Science, 158*(3), 211–217. https://doi.org/10.1097/00010694-199409000-00007.
Bertoldi, G., Bozzoli, M., Crespi, A., Matiu, M., Giovannini, L., Zardi, D., & Majone, B. (2023). Diverging snowfall trends across months and elevation in the northeastern Italian Alps. *International Journal of Climatology, 43*(6), 2794–2819. https://doi.org/10.1002/joc.8002.
Bevington, A., Brideau, M. A., & Geertsema, M. (2020). *Mountain environments. Encyclopedia of Earth Science Series*. Springer https://doi.org/10.1007/978-3-319-12127-7_209-1/FIGURES/1.
Bobrowski, M., Weidinger, J., Schwab, N., & Schickhoff, U. (2021). Searching for ecology in species distribution models in the Himalayas. *Ecological Modelling, 458*, 109693. https://doi.org/10.1016/j.ecolmodel.2021.109693.
Bolliger, J., Hagedorn, F., Leifeld, J., Böhl, J., Zimmermann, S., Soliva, R., & Kienast, F. (2008). Effects of land-use change on carbon stocks in Switzerland. *Ecosystems, 11*(6), 895–907. https://doi.org/10.1007/s10021-008-9168-6.
Bruijnzeel, L. A., Scatena, F. N., & Hamilton, L. S. (2011). *Tropical montane cloud forests: Science for conservation and management*. Cambridge University Press https://doi.org/10.1017/CBO9780511778384.
Buytaert, W., Célleri, R., De Bièvre, B., Cisneros, F., Wyseure, G., Deckers, J., & Hofstede, R. (2006). Human impact on the hydrology of the Andean páramos. *Earth-Science Reviews, 79*(1-2), 53–72. https://doi.org/10.1016/j.earscirev.2006.06.002.
Cárdenas, M. F., Tobón, C., & Buytaert, W. (2017). Contribution of occult precipitation to the water balance of páramo ecosystems in the Colombian Andes. *Hydrological Processes, 31*(24), 4440–4449. https://doi.org/10.1002/hyp.11374.
Carey, M., Molden, O. C., Rasmussen, M. B., Jackson, M., Nolin, A. W., & Mark, B. G. (2017). Impacts of glacier recession and declining meltwater on mountain societies. *Annals of the American Association of Geographers, 107*(2), 350–359. https://doi.org/10.1080/24694452.2016.1243039.

Crozier, M. J. (2010). Landslides: Causes, consequences and environment. Routledge.
Cuni-Sanchez, A., Sullivan, M. J. P., Platts, P. J., Lewis, S. L., Marchant, R., Imani, G., Hubau, W., Abiem, I., Adhikari, H., Albrecht, T., Altman, J., Amani, C., Aneseyee, A. B., Avitabile, V., Banin, L., Batumike, R., Bauters, M., Beeckman, H., Begne, S. K., & …, Zibera, E. (2021). High aboveground carbon stock of African tropical montane forests. *Nature, 596*(7873), 536–542. https://doi.org/10.1038/s41586-021-03728-4.
Daloz, A. S., Mateling, M., L'ecuyer, T., Kulie, M., Wood, N. B., Durand, M., Wrzesien, M., Stjern, C. W., & Dimri, A. P. (2020). How much snow falls in the world's mountains? A first look at mountain snowfall estimates in A-train observations and reanalyses. *Cryosphere, 14*(9), 3195–3207. https://doi.org/10.5194/tc-14-3195-2020.
Davis, K. T., Dobrowski, S. Z., Higuera, P. E., Holden, Z. A., Veblen, T. T., Rother, M. T., Parks, S. A., Sala, A., & Maneta, M. P. (2019). Wildfires and climate change push low-elevation forests across a critical climate threshold for tree regeneration. *Proceedings of the National Academy of Sciences of the United States of America, 116*(13), 6193–6198. https://doi.org/10.1073/pnas.1815107116.
Ding, L., Kapp, P., Cai, F., Garzione, C. N., Xiong, Z., Wang, H., & Wang, C. (2022). Timing and mechanisms of Tibetan plateau uplift. *Nature Reviews Earth and Environment, 3*(10), 652–667. https://doi.org/10.1038/s43017-022-00318-4.
Esfahan, R., Campillo, M., Seydoux, L., Nishida, K. and Favre-Bulle, G. (2025). Seismic precursors to the Blatten, Switzerland landslide revealed by unsupervised machine learning. eartharxiv.org.
Eskandari, S., Pourghasemi, H. R., & Tiefenbacher, J. P. (2020). Relations of land cover, topography, and climate to fire occurrence in natural regions of Iran: Applying new data mining techniques for modeling and mapping fire danger. *Forest Ecology and Management, 473*. https://doi.org/10.1016/j.foreco.2020.118338.
Figueroa, A. M., & Knott, J. R. (2010). Tectonic geomorphology of the southern Sierra Nevada Mountains (California): Evidence for uplift and basin formation. *Geomorphology, 123*(1-2), 34–45. https://doi.org/10.1016/j.geomorph.2010.06.009.
Finch, J., Marchant, R., & Courtney Mustaphi, C. J. (2017). Ecosystem change in the south pare mountain bloc, eastern arc mountains of Tanzania. *Holocene, 27*(6), 796–810. https://doi.org/10.1177/0959683616675937.
Finch, J., Wooller, M., & Marchant, R. (2014). Tracing long-term tropical montane ecosystem change in the eastern arc mountains of Tanzania. *Journal of Quaternary Science, 29*(3), 269–278. https://doi.org/10.1002/jqs.2699.
Freppaz, M., Godone, D., Filippa, G., Maggioni, M., Lunardi, S., Williams, M. W., & Zanini, E. (2010). Soil erosion caused by snow avalanches: A case study in the Aosta Valley (NW Italy). *Arctic, Antarctic, and Alpine Research, 42*(4), 412–421. https://doi.org/10.1657/1938-4246-42.4.412.
(2003). Ecological restoration of southwestern ponderosa pine forests. *Ecological Restoration, 21*, 39–41. https://doi. org/10. 3368/er. 21. 1. 39.
Frigo, B., Chiaia, B., Chiambretti, I., Bartelt, P., Maggioni, M., & Freppaz, M. (2018). The January 18th 2017 Rigopiano disaster in Italy—Analysis of the avalanche dynamics. In *Proceedings, international snow science workshop* (pp. 6–10). Austria: Innsbruck.
Gallou, A., Jump, A. S., Lynn, J. S., Field, R., Irl, S. D. H., Steinbauer, M. J., Beierkuhnlein, C., Chen, J. C., Chou, C. H., Hemp, A., Kidane, Y., König, C., Kreft, H., Naqinezhad, A., Nowak, A., Nuppenau, J. N., Trigas, P., Price, J. P., Roland, C. A., & …, Grytnes, J. A. (2023). Diurnal temperature range as a key predictor of plants' elevation ranges globally. *Nature Communications, 14*(1). https://doi.org/10.1038/s41467-023-43477-8.
Gargani, J., & Adjustment, I. (2022). Vertical motion rate variation and potential detection of past abrupt mass unloading. *Geosciences, 12*(8), 302. https://doi.org/10.3390/geosciences12080302.

Godbout, P. M., Brouard, E., & Roy, M. (2023). 1-km resolution rebound surfaces and paleotopography of glaciated North America since the last glacial maximum. *Scientific Data, 10*(1). https://doi.org/10.1038/s41597-023-02566-5.

Gruber, S., & Haeberli, W. (2007). Permafrost in steep bedrock slopes and its temperature-related destabilization following climate change. *Journal of Geophysical Research: Earth Surface, 112*(F2). https://doi.org/10.1029/2006jf000547.

Gutiérrez-García, G., Beramendi-Orosco, L. E., & Johnson, K. R. (2020). Climate-growth relationships of Pinus pseudostrobus from a tropical mountain cloud forest in northeast Mexico. *Dendrochronologia, 64*. https://doi.org/10.1016/j.dendro.2020.125749.

Hagedorn, F., Mulder, J., & Jandl, R. (2010). Mountain soils under a changing climate and land-use. *Biogeochemistry, 97*(1), 1–5. https://doi.org/10.1007/s10533-009-9386-9.

Harden, C. P. (2001). Soil erosion and sustainable mountain development. *Mountain Research and Development, 21*(1), 77–83. https://doi.org/10.1659/0276-4741(2001)021[0077:seasmd]2.0.co;2.

Hock, R., Rasul, G., Adler, C., Cáceres, B., Gruber, S., Hirabayashi, Y., Jackson, M., Kääb, A., Kang, S., Kutuzov, S., & Milner, A. (2019). High mountain areas. In H.-O. Pörtner, D. C. Roberts, V. Masson-Delmotte, P. Zhai, M. Tignor, E. Poloczanska, K. Mintenbeck, A. Alegría, M. Nicolai, A. Okem, J. Petzold, B. Rama, & N. M. Weyer (Eds.), *IPCC special report on the ocean and cryosphere in a changing climate* (pp. 131–202). Cambridge University Press.

Holden, Z. A., Swanson, A., Luce, C. H., Jolly, W. M., Maneta, M., Oyler, J. W., Warren, D. A., Parsons, R., & Affleck, D. (2018). Decreasing fire season precipitation increased recent western US forest wildfire activity. *Proceedings of the National Academy of Sciences of the United States of America, 115*(36), E8349. https://doi.org/10.1073/pnas.1802316115.

Houze, R. A. (2014). Is climate change responsible for changing landslide activity in high mountains? Earth Surface Processes and Landforms. In C. Huggel, J. J. Clague, & O. Korup (Eds.). In *Cloud Dynamics: 37* (pp. 77–91). Academic Press.

Huang, H., Qian, Y., He, C., Bair, E. H., & Rittger, K. (2022). Snow albedo feedbacks enhance snow impurity-induced radiative forcing in the Sierra Nevada. *Geophysical Research Letters, 49*(11). https://doi.org/10.1029/2022GL098102.

Hůnová, I., Hanusková, D., Jandová, K., Tesař, M., Květoň, J., & Kukla, J. (2020). Estimates of the contribution of fog water to wet atmospheric deposition in Czech mountain forests based on its stable hydrogen and oxygen isotope composition: Preliminary results. *European Journal of Environmental Sciences, 10*(2), 89–97. https://doi.org/10.14712/23361964.2020.10.

Immerzeel, W. W., Lutz, A. F., Andrade, M., Bahl, A., Biemans, H., Bolch, T., Hyde, S., Brumby, S., Davies, B. J., Elmore, A. C., Emmer, A., Feng, M., Fernández, A., Haritashya, U., Kargel, J. S., Koppes, M., Kraaijenbrink, P. D. A., Kulkarni, A. V., Mayewski, P. A., Viviroli, D., Wada, Y., Xiao, C., Yao, T., & …, Baillie, J. E. M. (2020). Importance and vulnerability of the world's water towers. *Nature, 577*(7790), 364–369. https://doi.org/10.1038/s41586-019-1822-y.

Immerzeel, W. W., Van Beek, L. P. H., & Bierkens, M. F. P. (2010). Climate change will affect the asian water towers. *Science, 328*(5984), 1382–1385. https://doi.org/10.1126/science.1183188.

Insel, N., Poulen, C. J., & Ehlers, T. A. (2010). Influence of the Andes on South American precipitation. *Climate Dynamics, 35*(7), 1477–1492.

IPCC. (2022). Climate Change 2022: Impacts, adaptation, and vulnerability. *Contribution of Working Group II to the sixth assessment report of the intergovernmental panel on climate change.* Cambridge University Press.

Jackson, P. L., Mayr, G., Vosper, S., Jackson, P. L., Mayr, G., & Vosper, S. (2012). Dynamically-driven winds. In F. K. Chow, S. F. J. De Wekker, & B. Snyder (Eds.), *Mountain weather research and forecasting* (pp. 121–218). Springer. https://doi.org/10.1007/978-94-007-4098-3_3.

Jakob, M., & Hungr, O. (2005). *Debris-flow Hazards and related phenomena*. Springer.
Jenny, H. (1929). *Vegetations-Entwicklung und Bodenbildung. Zweiter Teil: Die alpinen Bo den* (pp. 297–340). Denkschrift der Schweizerischen Naturforschenden Gesellschaft LXIII.
Johnstone, J. F., Hollingsworth, T. N., Chapin, F. S., & Mack, M. C. (2010). Changes in fire regime break the legacy lock on successional trajectories in Alaskan boreal forest. *Global Change Biology, 16*(4), 1281–1295. https://doi.org/10.1111/j.1365-2486.2009.02051.x.
Kariuki, R. W., Western, D., Willcock, S., & Marchant, R. (2021). Assessing interactions between agriculture, livestock grazing and wildlife conservation land uses: A historical example from East Africa. *Land, 10*(1), 1–19. https://doi.org/10.3390/land10010046.
Kariuki, R., Willcock, S., & Marchant, R. (2018). Rangeland livelihood strategies under varying climate regimes: Model insights from southern Kenya. *Land,* 7(2), 47. https://doi.org/10.3390/land7020047.
Kelly, L. T., Giljohann, K. M., Duane, A., Aquilué, N., Archibald, S., Batllori, E., Bennett, A. F., Buckland, S. T., Canelles, Q., Clarke, M. F., Fortin, M. J., Hermoso, V., Herrando, S., Keane, R. E., Lake, F. K., McCarthy, M. A., Morán-Ordóñez, A., Parr, C. L., Pausas, J. G., & …, Brotons, L. (2020). Fire and biodiversity in the Anthropocene. *Science, 370*(6519). https://doi.org/10.1126/science.abb0355.
Kłapyta, P., Krzemień, K., Gorczyca, E., Krąż, P., Dubis, L., Kłapyta, P., Krzemień, K., Gorczyca, E., Krąż, P., & Dubis, L. (2020). Geomorphologic effects of human impact across the Svydovets Massif in the Eastern Carpathians in Ukraine. *Studia Geomorphologicacar Patho-balcanica, LIII–LIV,* 85–111.
Knight, C. A., Anderson, L., Bunting, M. J., Champagne, M., Clayburn, R. M., Crawford, J. N., Klimaszewski-Patterson, A., Knapp, E. E., Lake, F. K., Mensing, S. A., Wahl, D., Wanket, J., Watts-Tobin, A., Potts, M. D., & Battles, J. J. (2022). Land management explains major trends in forest structure and composition over the last millennium in California's Klamath Mountains. *Proceedings of the National Academy of Sciences of the United States of America, 119*(12). https://doi.org/10.1073/pnas.2116264119.
Körner, C. (2003). *Alpine plant life: Functional plant ecology of high mountain ecosystems*. Springer.
Lawton, R. O., Nair, U. S., Pielke, R. A., & Welch, R. M. (2001). Climatic impact of tropical lowland deforestation on nearby montane cloud forests. *Science, 294*(5542), 584–587. https://doi.org/10.1126/science.1062459.
Le Roux, E., Evin, G., Eckert, N., Blanchet, J., & Morin, S. (2021). Elevation-dependent trends in extreme snowfall in the French Alps from 1959 to 2019. *Cryosphere, 15*(9), 4335–4356. https://doi.org/10.5194/tc-15-4335-2021.
Lensen, G. J. (1958). A method of Graben and Horst formation. *The Journal of Geology, 66*(5), 579–587. https://doi.org/10.1086/626537.
Li, J., Li, Q., Wu, Y., Ye, L., Liu, H., Wei, J., & Huang, X. (2021). Mountains act as museums and cradles for hemipteran insects in China: Evidence from patterns of richness and phylogenetic structure. *Global Ecology and Biogeography, 30*(5), 1070–1085. https://doi.org/10.1111/geb.13276.
Li, Y., Evans, I. S., Spagnolo, M., Pellitero, R., Barr, I. D., & Ely, J. C. (2024). ACME2: An extended toolbox for automated cirque metrics extraction. *Geomorphology, 445*. https://doi.org/10.1016/j.geomorph.2023.108982.
Liang, Q., & Wang, N. (2023). Mountain glacier flow velocity retrieval from ascending and descending sentinel-1 data using the offset tracking and MSBAS technique: A case study of the Siachen Glacier in Karakoram from 2017 to 2021. *Remote Sensing, 15*(10), 2594. https://doi.org/10.3390/rs15102594.
Liptzin, D., & Seastedt, T. R. (2010). Regional and local patterns of soil nutrients at Rocky Mountain treelines. *Geoderma, 160*(2), 208–217. https://doi.org/10.1016/j.geoderma.2010.09.022.
López-Moreno, J. I., Pomeroy, J. W., Morán-Tejeda, E., Revuelto, J., Navarro-Serrano, F. M., Vidaller, I., & Alonso-González, E. (2021). Changes in the frequency of global high

mountain rain-on-snow events due to climate warming. *Environmental Research Letters, 16*(9), 094021. https://doi.org/10.1088/1748-9326/ac0dde.

Nair, U. S., Ray, D. K., Lawton, R. O., Welch, R. M., Pielke, R. A., & Calvo-Alvarado, J. (2011). *The impact of deforestation on orographic cloud formation in a complex tropical environment. Tropical Montane Cloud Forests: Science for Conservation and Management* (pp. 538–548). Cambridge University Press. https://doi.org/10.1017/CBO9780511778384.057.

Pepin, N., Apple, M., Knowles, J., Terzergo, S., Arnone, E., Hanchon, L., & Zardi, D. (2025). Elevation- dependent climate change in mountain areas. *Nature Reviews Earth and Environment*, 1–17.

Perrigo, A., Hoorn, C., & Antonelli, A. (2020). Why mountains matter for biodiversity. *Journal of Biogeography, 47*(2), 315–325. https://doi.org/10.1111/jbi.13731.

Pimentel, D., Harvey, C., Resosudarmo, P., Sinclair, K., Kurz, D., McNair, M., Crist, S., Shpritz, L., Fitton, L., Saffouri, R., & Blair, R. (1995). Environmental and economic costs of soil erosion and conservation benefits. *Science, 267*(5201), 1117–1123. https://doi.org/10.1126/science.267.5201.1117.

Qin, J., et al. (2014). Evaluating TRMM rainfall data in mountainous terrain. *Journal of Hydrometeorology, 15*(3), 1232–1244.

Quante, L., Willner, S. N., Middelanis, R., & Levermann, A. (2021). Regions of intensification of extreme snowfall under future warming. *Scientific Reports, 11*(1). https://doi.org/10.1038/s41598-021-95979-4.

Rahbek, C., Borregaard, M. K., Antonelli, A., Colwell, R. K., Holt, B. G., Nogues-Bravo, D., Rasmussen, C. M. Ø., Richardson, K., Rosing, M. T., Whittaker, R. J., & Fjeldså, J. (2019a). Building mountain biodiversity: Geological and evolutionary processes. *Science, 365*(6458), 1114–1119. https://doi.org/10.1126/science.aax0151.

Rahbek, C., Borregaard, M. K., Colwell, R. K., Dalsgaard, B., Holt, B. G., Morueta-Holme, N., Nogues-Bravo, D., Whittaker, R. J., & Fjeldså, J. (2019b). Humboldt's enigma: What causes global patterns of mountain biodiversity? *Science, 365*(6458), 1108–1113. https://doi.org/10.1126/science.aax0149.

Rocca, M. E., Brown, P. M., MacDonald, L. H., & Carrico, C. M. (2014). Climate change impacts on fire regimes and key ecosystem services in Rocky Mountain forests. *Forest Ecology and Management, 327*, 290–305. https://doi.org/10.1016/j.foreco.2014.04.005.

Rodman, K. C., Veblen, T. T., Battaglia, M. A., Chambers, M. E., Fornwalt, P. J., Holden, Z. A., Kolb, T. E., Ouzts, J. R., & Rother, M. T. (2020). A changing climate is snuffing out post-fire recovery in montane forests. *Global Ecology and Biogeography, 29*(11), 2039–2051. https://doi.org/10.1111/geb.13174.

Roos, C. I., Swetnam, T. W., Ferguson, T. J., Liebmann, M. J., Loehman, R. A., Welch, J. R., Margolis, E. Q., Guiterman, C. H., Hockaday, W. C., Aiuvalasit, M. J., Battillo, J., Farella, J., & Kiahtipes, C. A. (2021). Native American fire management at an ancient wildland–urban interface in the Southwest United States. *Proceedings of the National Academy of Sciences of the United States of America, 118*(4). https://doi.org/10.1073/pnas.2018733118.

Rotunno, R., & Houze, R. A. (2007). Lessons on orographic precipitation from the MAP project. *Bulletin of the American Meteorological Society, 88*(4), 527–540.

Scatena, F. N., Bruijnzeel, L. A., Bubb, P., Das, S., Scatena, F. N., Bruijnzeel, L. A., Bubb, P., & Das, S. (2011). Setting the stage. In L. A. Bruijnzeel, F. N. Scatena, & L. S. Hamilton (Eds.), *Tropical montane cloud forests: Science for conservation and management* (pp. 3–13). Cambridge University Press. https://doi.org/10.1017/CBO9780511778384.003.

Scherrer, S. C., Ceppi, P., Croci-Maspoli, M., & Appenzeller, C. (2012). Snow-albedo feedback and Swiss spring temperature trends. *Theoretical and Applied Climatology, 110*(4), 509–516. https://doi.org/10.1007/s00704-012-0712-0.

Schimel, D. S. (1995). Terrestrial ecosystems and the carbon cycle. *Global Change Biology, 1*(1), 77–91. https://doi.org/10.1111/j.1365-2486.1995.tb00008.x.

Schoennagel, T., Veblen, T. T., & Romme, W. H. (2004). The interaction of fire, fuels, and climate across Rocky Mountain forests. *BioScience, 54*(7), 661–676. https://doi.org/10.1641/0006-3568(2004)054[0661:TIOFFA]2.0.CO;2.

Schwanghart, W., Worni, R., Huggel, C., Stoffel, M., & Korup, O., et al. (2016). Risks associated with landslide-dammed lakes in high mountains of Asia. *Natural Hazards and Earth System Sciences, 16*(3), 643–659.

Schwarz, L., Malig, B., Guzman-Morales, J., Guirguis, K., Ilango, S. D., Sheridan, P., Gershunov, A., Basu, R., Benmarhnia, T., Schwartz, L., Malig, B. J., Guzman Morales, J., Guirguis, K., Gershunov, A., Basu, R., & Benmarhnia, T. (2020). The health burden of fall, winter and spring heat waves in South- ern California and contribution of Santa Ana winds. *Environmental Research Letters, 15*(5). https://doi.org/10.1088/1748-9326/ab7f0e.

Sierra-Hernández, M. R., Marcantonio, F., & Thompson, L. G. (2025). Lead (Pb) isotope ratios from ice core samples from the Puruogangri ice cap and the Dasuopu glacier. *Scientific Data, 12*(1). https://doi.org/10.1038/s41597-025-05145-y.

Somers, L. D., & McKenzie, J. M. (2020). A review of groundwater in high mountain environments. *Wiley Interdisciplinary Reviews: Water,* 7(6). https://doi.org/10.1002/wat2.1475.

Somers, L. D., McKenzie, J. M., Mark, B. G., Lagos, P., Ng, G. H. C., Wickert, A. D., Yarleque, C., Baraër, M., & Silva, Y. (2019). Groundwater buffers decreasing glacier melt in an andean watershed—But not forever. *Geophysical Research Letters, 46*(22), 13016–13026. https://doi.org/10.1029/2019GL084730.

Stone, R. S., & Carlson, T. N. (1979). Atmospheric lapse rates in the tropics. *Journal of Atmospheric Sciences, 36*(3), 415–423.

Sun, F., Chen, Y., Li, Y., Duan, W., Li, B., Fang, G., Li, Z., Zhu, Z., & Feng, M. (2024). Decreasing trends of mean and extreme snowfall in High Mountain Asia. *Science of The Total Environment, 921*, 171211. https://doi.org/10.1016/j.scitotenv.2024.171211.

Tang, M., Li, L., Wang, X., You, J., Li, J., & Chen, X. (2020). Elevational is the main factor controlling the soil microbial community structure in alpine tundra of the Changbai Mountain. *Scientific Reports, 10*(1). https://doi.org/10.1038/s41598-020-69441-w.

Tasser, E., Walde, J., Tappeiner, U., Teutsch, A., & Noggler, W. (2007). Land-use changes and natural reforestation in the Eastern Central Alps. *Agriculture, Ecosystems & Environment, 118*(1-4), 115–129. https://doi.org/10.1016/j.agee.2006.05.004.

Tesař, M., Fottová, D., Eliáš, V., & Šír, M. (2000). Occult precipitation as an important contribution to the wet deposition in Bohemian Forest. *Silva Gabreta, 4*, 87–96.

Tito, R., Vasconcelos, H. L., & Feeley, K. J. (2020). Mountain Ecosystems as Natural Laboratories for Climate Change Experiments. *Frontiers in Forests and Global Change, 3*. https://doi.org/10.3389/ffgc.2020.00038.

Tovar, C., Carril, A. F., Gutiérrez, A. G., Ahrends, A., Fita, L., Zaninelli, P., Flombaum, P., Abarzúa, A. M., Alarcón, D., Aschero, V., Báez, S., Barros, A., Carilla, J., Ferrero, M. E., Flantua, S. G. A., Gonzáles, P., Menéndez, C. G., Pérez-Escobar, O. A., Pauchard, A., & ..., Hollingsworth, P. M. (2022). Understanding climate change impacts on biome and plant distributions in the Andes: Challenges and opportunities. *Journal of Biogeography, 49*(8), 1420–1442. https://doi.org/10.1111/jbi.14389.

Vitousek, P. M., & Farrington, H. (1997). Nutrient limitation and soil development: Experimental test of a biogeochemical theory. *Biogeochemistry, 37*(1), 63–75. https://doi.org/10.1023/A:1005757218475.

Vitousek, P. M., & Sanford, R. L. (1986). Nutrient cycling in moist tropical forest. *Annual Review of Ecology and Systematics, 17*(1), 137–167. https://doi.org/10.1146/annurev.es.17.110186.001033.

Viviroli, D., Kummu, M., Meybeck, M., Kallio, M., & Wada, Y. (2020). Increasing dependence of lowland populations on mountain water resources. *Nature Sustainability, 3*(11), 917–928. https://doi.org/10.1038/s41893-020-0559-9.

Wagner, F. H. (1930). Altitudinal temperature gradients in mountainous terrain.

Wambulwa, M. C., Milne, R., Wu, Z. Y., Spicer, R. A., Provan, J., Luo, Y. H., Zhu, G. F., Wang, W. T., Wang, H., Gao, L. M., Li, D. Z., & Liu, J. (2021). Spatiotemporal maintenance of flora in the Himalaya biodiversity hotspot: Current knowledge and future perspectives. *Ecology and Evolution, 11*(16), 10794–10812. https://doi.org/10.1002/ece3.7906.

Wang, K., Zhou, X. H., Liu, D., Li, Y., Yao, Z., He, W. M., & Liu, Y. (2022). The uplift of the Hengduan Mountains contributed to the speciation of three Rhododendron species. *Global Ecology and Conservation, 35*. https://doi.org/10.1016/j.gecco.2022.e02085.

You, C., Yao, T., & Xu, C. (2018). Recent increases in wildfires in the Himalayas and surrounding regions detected in central Tibetan ice core records. *Journal of Geophysical Research: Atmospheres, 123*(6), 3285–3291. https://doi.org/10.1002/2017JD027929.

Zhang, J., Zhang, M., Li, Y., Qin, J., Wei, K., & Song, L. (2020). Analysis of wind characteristics and wind energy potential in a complex mountainous region in southwest China. *Journal of Cleaner Production, 274*, 123036. https://doi.org/10.1016/j.jclepro.2020.123036.

CHAPTER 3

Mountain ecosystems: Plant and animal distributions

3.1 Introduction

Mountain geological, geophysical, and environmental diversity begets a wide range of ecosystems. Mountain biodiversity is multidimensional; that is, it includes species richness, endemism, and phylogenetic and structural diversity (Wang et al., 2022; Yu et al., 2021). Mountains contribute disproportionately to the terrestrial biodiversity of Earth, especially in the tropics, where they form hotspots of extraordinary and sometimes puzzling richness (Testolin et al., 2021) (Fig. 3.1). This diversity is a product of four key processes (speciation, dispersal, persistence, and extinction) that underpin the diversity of life on Earth, and its distribution. Covering only about 25% of all land area, mountain regions support more than 85% of the world's species of amphibians, birds, and mammals, including rare plants and iconic animals such as the snow leopard, giant panda, and mountain gorilla. The impact of mountain regions on biodiversity extends far beyond their topographical limits (Chapter 1), often affecting entire continental biotas. For example, although the Andes region is in itself the most biodiverse region on Earth, Andean orogeny is also considered a key driver for the buildup of biodiversity across the whole of South America (Esquerré et al., 2019).

As explored in Chapter 2, mountain systems are incredibly diverse geophysical entities. Ascending a mountain, temperature, oxygen levels, ultraviolet (UV) radiation, diurnal temperature ranges and extremes, and atmospheric pressures change, combining to create multiple distinct environments (Fig. 3.2). Mountain building and associated formation of heterogeneous environments promote rapid species proliferation through reduced dispersal and gene flow, while long-term ecological and climatic stability helps maintain global diversity by buffering against extinction in formerly widespread groups (Suissa et al., 2021). Many mountain-dwelling animals and plants have evolved specific adaptations to cope with the challenges of mountain living, leading to high levels of endemism

Mountain Diversity: Dynamics of High-Altitude Environments.
DOI: https://doi.org/10.1016/B978-0-443-22251-1.00003-4

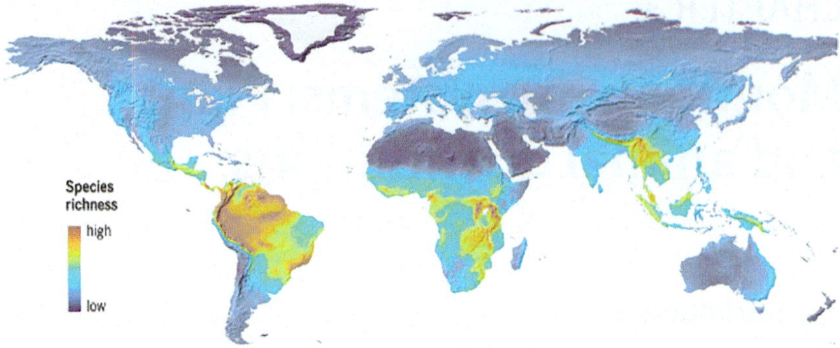

Figure 3.1 Global diversity has a strong congruence with mountain locations. Mountain regions (including adjacent lowland foothills) host roughly 87% of terrestrial global biodiversity. *Modified from Rahbek et al. (2019a).*

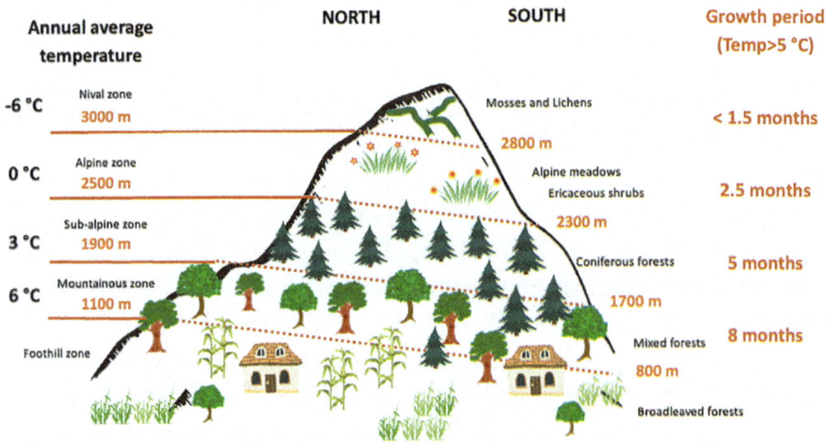

Figure 3.2 Ascending a mountain, temperature, oxygen levels, ultraviolet (UV) radiation, diurnal temperature ranges and extremes, and atmospheric pressure change: combining to create multiple distinct habitats as seen here for a typical mountain of temperate latitudes.

(Rahbek et al., 2019a). Ranging from snow leopards with thick pads on their paws to power through snow fields, to agile mountain goats able to navigate cliffs, from plants adapted to withstand fire, to others having antifreeze in their sap to prevent freezing to death. While aquatic species can be uniquely adapted to fast-flowing cold waters, many plants and animals have developed specialised adaptive features to survive and thrive in mountainous environments and contribute to overall mountain biodiversity.

Keystone species can play a critical role in maintaining the structure and function of the wider mountain ecosystems, with some species having a disproportionately large effect across and beyond the mountain system. For example, apex predators like wolves (Ausilio et al., 2021) and ecological engineers like beavers (Fedyń et al., 2024) have a cascading influence on plants and animals around them, often with impacts far away from mountains.

In addition to supporting fascinating and diverse ecosystems, mountains are excellent systems for studying species' responses to changing conditions. For over 250 years, mountains have served as natural laboratories to conduct experiments. Generations of ecologists have tested biodiversity gradient theories, sought underlying ecological mechanisms behind ecosystem functioning, and predicted biodiversity loss under global change across taxonomic groups (Rahbek et al., 2019a; Wang et al., 2022). On Mountains, climatic gradients are steep, and environmental conditions and related productivity measures change rapidly over short distances (Körner, 2007). This characteristic of rapidly changing environmental conditions over short distances and extreme diurnal and seasonal variation can be used as a model of substituting space for time. Mountains are also *relatively* less disturbed by people compared to the lowlands (Vetaas, 2021); however, as we will see, mountains and mountain ecosystem composition and distribution have been shaped by a long history of human interaction (Chapter 5).

3.2 Why are mountains so biodiverse?

The foundation for supporting extremely high diversity is that mountains are fundamentally characterised by high environmental heterogeneity. Mountains, with their high topographic and environmental complexity, support one-third of all terrestrial biodiversity (Spehn et al., 2011) (Fig. 3.3), but our understanding of the spatiotemporal maintenance of this high biodiversity remains poor, or at best controversial (Wambulwa et al., 2021). The variation in plants and animals across mountain climate zones was first characterised by Carolus Linnaeus in 1743 (Linnaeus, 1781). Another founding father of natural science, Alexander von Humboldt, followed with his work describing the extraordinarily high biodiversity and vegetation zonation on Mount Chimborazo, Ecuador, in 1802 (Birks, 2021; Rahbek et al., 2019b). Amazingly, the 1802 expedition by Alexander von Humboldt to Mount Chimborazo in the Cordillera Occidental range of the Andes led him to develop the first map of biodiversity that is still being used and tested today (Muellner-Riehl et al., 2019; Von Humboldt & Bonpland, 1826).

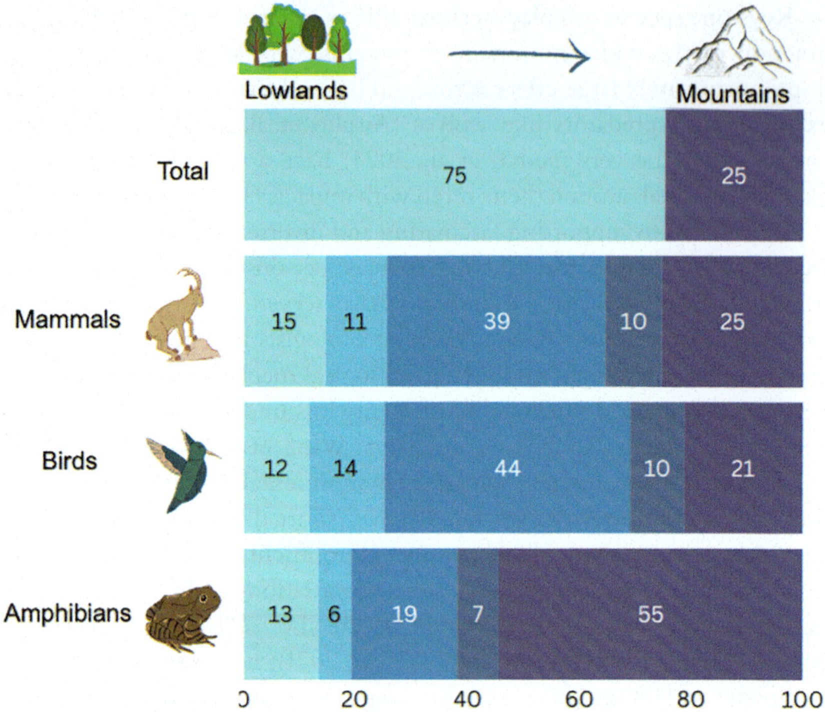

Figure 3.3 Proportion of species' range in mountains and lowlands, mountains are highly diverse per unit area, particularly for Amphibians; after Rahbek et al. (2019a).

Over the past centuries, global comparisons on mountain biodiversity have continued to draw natural scientists to use mountain ecosystems as great natural laboratories, particularly due to the distinct vertical zonation. This zonation is driven by factors such as temperature, precipitation, soil conditions, and exposure to sunlight (UV) (Antonelli et al., 2018; Lu et al., 2018). Based on a comprehensive dataset of 137 plant families in the Himalayas, Manish (2019) showed that temperature was the best explanatory variable for the observed elevational variance in species richness.

The boundaries between distinct mountain ecosystems are characterised by ecotones; transitional zones where different habitats converge, which in themselves can be of great interest to the functioning and behaviour of ecosystems 'on the edge'. These ecotones create diverse microhabitats and ecological gradients, fostering species interactions and further facilitating biological diversity. Edge effects, resulting from the juxtaposition of different habitats, can further enhance biodiversity and adaptation by increasing habitat heterogeneity and niche diversity.

Plate 3.1 Trees in temperate latitudes have a relatively short growing season and survive winter through deciduousness or senescence, such as these birch trees in Norway. *(Photograph Rob Marchant).*

Due to more energy for growth in the tropics, species have evolved narrower physiological tolerances. As such, tropical mountains can tightly pack species in and hence harbour a disproportionate amount of Earth's diversity compared to adjacent lowlands and temperate regions (Suissa et al., 2021). This latitudinal gradient of mountain diversity was first pointed out by Janzen (1967) in a seminal paper, which summarised this pattern as 'species living on mountains closer to the equator have relatively smaller seasonal temperature ranges than species at comparable elevations at higher latitude'. As such, plants growing on a mountain situated on the equator can grow all year round rather than growing in the 'summer' and surviving the 'winter' through deciduousness or senescence (Plate 3.1). Although not all tropical mountain regions are unusually diverse, some are exceptionally so (Rahbek et al., 2019b). Both the turnover and the distinctness of mountain climates across elevational gradients are pronounced, particularly in the tropics. The biodiversity of individual mountain regions varies markedly across the world (Fig. 3.1). In the arctic and temperate zones, mountain regions typically have low species diversity, barely exceeding the richness of adjacent lowlands, and often do not host endemic species.

To some extent, this variation in biomes across latitudinal gradients mimics elevational gradients, but elevational gradients present fewer barriers than those found horizontally with latitude across the continents (Vetaas, 2021). It must be emphasised that such 'global rules' do not apply everywhere, with the biological richness of a particular mountain being dependent on multiple environmental dimensions, landscape history, microclimates, and biotic interactions (Dufour et al., 2006). Consequently, finding the most relevant geodiversity variables for explaining biological richness in a location is always context dependent (Määttänen et al., 2024). The importance of context should not be underestimated, both in terms of understanding the diversity and subsequently applying this knowledge.

The role of thermal regime, and how this changes along elevational gradients, needs to be explored in more detail (Griffiths et al., 2021). Throughout elevational and latitudinal gradients, temperature is regarded as a significant predictor of both plant and animal species diversity (Körner & Paulsen, 2004). However, this is not the only driver – a significant decrease in the amount of land area at high altitudes is also important. Under Rapoport's rule (Dewan & Acharya, 2024), where the area of land is positively correlated with species richness, a decreasing amount of land with increasing altitude results in fewer species (Körner, 2007, 2012). Low elevations, however, have the highest level of anthropogenic activities and a reduction in the diversity of habitat that can lead to less species richness.

Empirical data have shown that the elevational gradient of species richness can take many shapes, most often a hump-shaped mid-elevational peak in diversity (Di Biase et al., 2021; Lovett et al., 2006; Rahbek et al., 2019a) or a monotonic decline with increasing elevation. The decline with altitude is a consequence of increasingly harsher environmental conditions; lower productivity, less available area (due to the conical shape of the mountains), the nested distribution of species, and species at higher elevations that tend to be subsamples of the more tolerant species already occurring at lower elevations. The hump-shaped pattern of the distribution of diversity with altitude is probably the most widespread one, and the monotonic decrease may be due to a reduction in the available area with elevation (Colwell et al., 2004). The hump-shaped relationship between species richness and elevation arises from the occurrences of species with geographical ranges that overlap between two ecologically unfavourable boundaries, i.e., the lowlands and highlands

(Brehm et al., 2007; Fattorini, 2014), combined with the impact of decreasing space.

In addition to contemporary environmental-ecosystem diversity relationships, mountain regions can act as 'refugia' for some species as others move between higher and lower elevations as climates change (Chapter 4). The importance of mountains is especially so during periods of rapid climate change, as species only need to move shorter distances along the gradients to find their niche compared to lowlands (Perrigo et al., 2020). Specific landforms, such as kettle holes from retreating glacier river valleys, create specific microrefugia with unique morphological, hydrological, climatic, lithological, and soil characteristics. Geomorphic diversity, therefore, creates the basis for greater niche and habitat diversity. Geodiverse areas within mountain systems create a mosaic of microclimatic refugia that can be buffered from regional climate change, and therefore provide species with a range of suitable microclimate conditions (Gillson et al., 2013). Moreover, geodiverse areas can provide environmental gradients across which species may be able to track suitable conditions as climate changes (Hannah et al., 2014), and enable some species to persist on mountains for a geologically long time (Githumbi et al., 2025). Most analyses of biodiversity do not consider these legacy impacts of temporal changes and influences on biodiversity, despite their importance (Lausch et al., 2020). In summary, the diversity of the mountain system is driven by the diversity of its origins, the geology, its environment, and ensuing environmental dynamics through time and space (Fig. 3.4).

Despite the significance of mountain systems for biodiversity and the wider importance of this for humanity, there are massive deficiencies in our understanding of what species are found where, and although this continues to develop, there are multiple voids that need to be filled (Hu et al., 2020). Biodiversity data are often subject to several sources of error, such as incomplete/skewed data on the distribution of different taxa, or including spatial observation biases (Marshall et al., 2012). For example, biodiverse areas attract funding and further work, and in turn become more biodiverse (Ahrends et al., 2011), leaving other areas to become disproportionately biodiverse. A comprehensive synthesis of elevational patterns and drivers in biodiversity and ecosystem functions is still lacking for microorganisms across mountainous ecosystems. Microbes, which are abundant and diverse in mountain regions, play an essential role in virtually all biogeochemical cycling processes and are an important component of ecosystem response to global change (Wang et al., 2022). Given their high dispersal ability

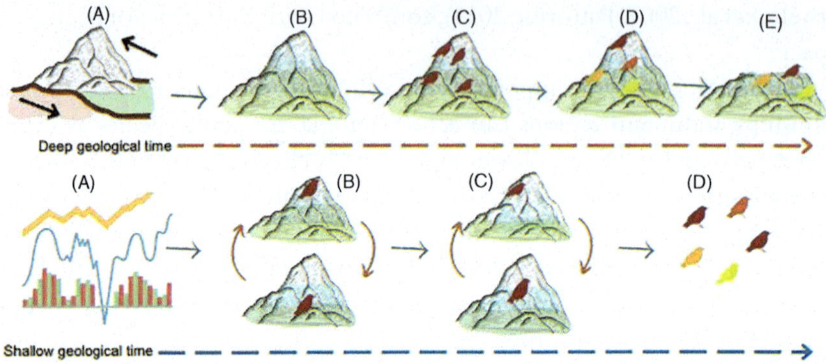

Figure 3.4 Simple depiction of creating mountain diversity over deep (top) and shallow (bottom) geological time. Top sequence: uplift and erosion form new, isolated mountain habitats (A), new immigrant populations arrive on the new mountain (B) and rapidly colonise new niches (C) where they go through a process of adaptation and evolve into new isolated populations forming new species (D). Habitat erosion and continued change drive some species to extinction, whereas other species adapt and survive (E). Bottom sequence depicts dynamic changes in connectivity driven by recent (Quaternary) climate changes, where speciation is driven by mountain climate cycling. In the Quaternary, the climate has oscillated between glacial and interglacial periods. Habitats periodically switch between being isolated and connected. Mountains provide refuge during interglacial periods, while times of glacial connectivity promote gene exchange and population viability. Secondary contact during periods of connectivity drives character displacement and speciation.

and high population densities, microorganisms tend to show more complex elevational diversity patterns than macro-organisms (Wang et al., 2022). It is hard to single out a single area of focus, but addressing the uncertainty surrounding the number of species in mountains, particularly in the Global South, and particularly in under-researched areas, should be a key priority area for future research (Wambulwa et al., 2021).

3.3 Mountain biodiversity hotspots

Mountains are globally significant centres of biodiversity, hosting 25 of the world's 34 biodiversity hotspots: areas characterised by high species diversity, high levels of endemism, high genetic diversity (Jump et al., 2014), and, importantly, high levels of threat (Myers et al., 2000). The geographical location and orientation of mountain ranges also play a role, along with the topography. Mountain systems such as the Eastern Arc Mountains, the Himalayas, and the Andes are positioned perpendicularly to atmospheric circulation patterns, creating complex climatic and biological patterns and an element of stability. With such a high amount of the world's storehouse of biodiversity and high levels of endemism, mountains are especially vulnerable to global change drivers, particularly land-use changes and invasive species. Mountain biodiversity hotspots for both large and small organisms play key roles in regulating global carbon, nitrogen, and water cycles and are thus fundamental for sustaining life on Earth (Myers et al., 2000). The Andes are amongst the most biodiverse regions on the planet (Myers et al., 2000). Spanning over 9000 km in length (following the mountain ridge) and reaching well over 6000 m in elevation, they are the longest and second-highest terrestrial mountain range on Earth after the Himalayas. The Andes are home to an estimated 40,000 plant species and thousands of vertebrate species with exceptionally high levels of species endemism (Tovar et al., 2022). Exceptionally high species diversity is also recorded in the Tibeto-Himalayan region, the New Caledonia archipelago, and the Eastern Arc Mountains (Myers et al., 2000); all areas that face a significant risk of habitat loss (Marchese, 2015). For example, it is estimated that the Himalayas support more than 10,000 species of vascular plants, with almost a third (31.6%) being endemic (Mittermeier et al., 2004), although the precise number of species remains controversial, and many new plant species continue to be described from the region (e.g., Borah et al., 2020). Conversely, the lowest species numbers were found in the temperate mountain ranges of Europe (Muellner-Riehl et al., 2019; Myers et al., 2000).

3.4 Mountain genetic diversity

Mountain biodiversity is multidimensional, including measures such as species richness, endemism, and phylogenetic diversity (Wang et al., 2022; Yu et al., 2021). Mountains often act as islands of biodiversity, isolated from surrounding areas by geographic barriers such as valleys, rivers, or deserts. This geographical isolation can promote allopatric speciation, where populations become genetically distinct, leading to the evolution of new endemic species (Wang et al., 2022). For example, birds in mountainous areas often have limited dispersal abilities, with isolation preventing gene flow between populations, leading to increased genetic divergence, speciation, fascinating ecological adaptation, and resultant endemic species. However, it is not just their isolation, but a combination of this geospatial character layered with the impacts of past climate changes that have been the fuel for the engine of diversification, evolution, and genetic diversity. As climates are inherently dynamic, over space and time, it is probable that only a certain proportion of any population would be impacted at any one time. Such a situation allows for the development of balanced polymorphisms where heterogeneous metapopulations can alter their gene frequencies as climatic conditions oscillate (Crawford, 2005). It is this climatic, geodiversity, and topographical diversity that make mountains evolutionary hotspots, where rapid speciation and adaptive radiation occur due to ecological opportunities created by environmental gradients (Scaldaferro et al., 2023; Wang et al., 2022).

Many recent phylogeographic studies have explored the roles that intermountain divergence, gene flow, and hybridisation have played in the current structuring of genetic diversity (Jump et al., 2014). There are, however, striking differences among species, suggesting a prominent role of stochasticity connected to rare long-distance dispersals between mountains (Peters et al., 2019). It is also noticeable that montane forest animal and plant species may have spread via forest corridors that formed between some mountains during previous climate regimes (Hemp et al., 2017). For example, *Rhododendron* speciation was associated with a 'species pump' driven by the dramatic orogenesis of the Hengduan Mountain, China (Wang et al., 2022).

Quantifying biodiversity has become more possible due to the increasing application of DNA barcoding. Molecular data are now available to calculate phylogenetic diversity, phylogenetic endemism, and biogeographically weighted evolutionary distinctiveness (Yu et al., 2021). With such a high amount of the world's storehouse of biodiversity and high levels of endemism, mountains are especially vulnerable to global change drivers, particularly land-use changes and invasive species (Chapter 6) (Plate 3.2). It is not just plants and animals that have been genetically shaped by mountains; Tibetans have developed remarkable physiological adaptations to high-altitude living, and have a unique genetic adaptation that allows them to have more efficient use of oxygen and lower concentrations of haemoglobin where oxygen levels are significantly lower.

Defining global hotspots of alpine plant diversity, including patterns of endemism, functional variation, and phylogenetic diversity, is still needed. This type of information, together with dynamic regional diversity models accounting for spatio-temporal connectivity and isolation, will provide a better understanding of the patterns described here. Such a development could become a tool for effective conservation of mountain biodiversity in response to climate change (Testolin et al., 2021). Mountains are indeed priority areas for conservation because of their high genetic diversity and

Plate 3.2 Invasive species can soon become a management challenge, such as these *Lupinus* (*left*) and Himalayan Balsam (*Impatiens glandulifera*) originating from the Americas and Asia, respectively, as ornamental plants have escaped and are becoming challenging invasive plants to control. *(Photographs Rob Marchant).*

unique evolutionary history – issues that will be explored in Chapters 6 and 7.

3.5 Mountain geobiodiversity

The amazing environmental diversity on mountains is shaped by geomorphic traits such as slope, aspect, curvature, variables of morphometry, lighting, and visibility. Hydrological factors, such as soil moisture, channel characteristics, drainage networks, flow directions, or valley depths, are also of fundamental importance to shaping geodiversity. As such, geodiversity is often seen as the foundation of biodiversity (Tukiainen et al., 2023) (Fig. 3.5). Geodiversity and biodiversity are positively connected and mark the increased union of biological and geological sciences (Muellner-Riehl et al., 2019), and also the need for a novel holistic approach that links biodiversity, geodiversity (in the broad sense), and their historical evolution at a range of temporal scales (Fig. 3.4). The inclusion of historical processes is crucial because past geological events, climatic fluctuations, biotic radiations, and

Figure 3.5 Illustration of geodiversity and its different elements that form the foundation for biodiversity. The abiotic setting is foundational to underpin drainage, productivity, and nutrient availability – geology and hence geodiversity tends to change rapidly on mountains. *Modified from Tukiainen et al. (2023).*

dispersal/migration events have had a major impact on the present-day distribution of biodiversity (Chapter 4).

By supporting and maintaining a variety of ecosystem functions and processes, geobiodiversity contributes to ecosystem service provision and human well-being (Gray, 2022). Hitherto, the focus has been on understanding geodiversity connections to biodiversity, often in the context of nature conservation (Tukiainen et al., 2023). In the face of the growing human pressure and land-use change, human disturbance can influence these geodiversity-biodiversity relationships. For instance, on the Loss Plateau in China, the geodiversity index shows links between topographic textural measures and landscape metrics derived from satellite images (Ren et al., 2021). Positive connections were found, especially in human-dominated regions. Exploring local-scale geodiversity can provide a better grasp of the underlying processes and mechanisms (Tukiainen et al., 2023). Both global and local explorations are needed to reveal and understand the possible interconnections across scales and to improve biodiversity modelling and conservation.

3.6 Mountains as sky islands

Mountains often act as islands of biodiversity, isolated from surrounding lowland areas by geographic barriers such as valleys, rivers, oceans, drylands, or deserts. The spectacular flora occurring on the tops of the mountains, which are isolated from the adjacent highland areas, has led to the term

Plate 3.3 Like an island, a mountain region may be viewed as a distinct biogeographical unit isolated from surrounding mountains by a 'sea' of cloud and lowlands, such as this view of Mount Meru in northern Tanzania, separate from the adjacent Ngorongoro Highlands and Mount Kilimanjaro. *(Photograph Rob Marchant).*

'sky islands'. This also includes those islands that are characterised by a mountain (Plate 3.3). Like an island, a mountain region may be viewed as a biogeographical unit in itself (Rahbek et al., 2019b). The isolated uniqueness of sky islands has attracted the attention of generations of biogeographers because of the extreme fragmentation and iconic examples of gigantism and convergent evolution (Brochmann et al., 2022).

The level of endemism is particularly high in the afroalpine flora (Plate 3.4): 67% of the 520 native species that regularly occur above 3200 m are endemic to these sky islands (Gehrke & Linder, 2014). Alpine areas are also more isolated from each other, forming fragmented systems of 'sky islands' surrounded by lowland environments that limit species dispersal (Testolin et al., 2021). These spectacular sky islands are under increasing threat stemming from the accelerating and destructive impact of human activities (Brochmann et al., 2022), which is making these islands even more isolated and potentially increasing their fragility.

3.7 Physiological adaptation to life on a mountain

Environmental conditions on mountains can be harsh with climatic stresses such as drought, flooding, ice-encasement, cryoturbation, and prolonged burial under snow or ice being common with changing severity and

Plate 3.4 The Afroalpine flagship genus *Dendrosenecio* shows conspicuous and intricate variation in growth form and morphology, resulting from parallel altitude/habitat adaptation on different mountains. Endemism is particularly high in the Afroalpine flora, such as these giant *Scenecios* on Mount Kilimanjaro. *(Photograph Rob Marchant).*

duration (Crawford, 2005). The unique mountain flora and fauna are adapted to extreme cold, UV light levels, often dry or unattainable moisture, and short growing seasons, temperature, precipitation, atmospheric pressure, and oxygen levels, all of which vary with altitude (Martin, 2001). For example, the ambient partial pressure for oxygen decreases with increasing altitude; at 4000 m, the PO_2 of inspired air is approximately 60% of that of sea level (Beall, 2006). Limits to plant survival arise, either from a failure to grow or from an inability to reproduce and continue to the next generation. Alpine species, which are adapted to extreme environmental conditions of high altitudes, are relatively few (Rolando et al., 2007), with the numbers of plants and animals dropping off with high altitude. As elevation increases, trees tend to become shorter and more stunted. Near the upper edge of the subalpine zone, the trees may appear as krummholz ('crooked wood'), where they are dwarfed, twisted, and contorted by the wind and snow. Krummholz formations often grow close to the ground to minimise exposure to the elements (Plate 3.5). Trees often develop extensive root systems to anchor themselves on steep or exposed slopes (Plate 3.6). These roots help prevent soil erosion, enhance biological weathering of the thin parent material, and contribute to the overall stability of the forest ecosystem.

Trees often have thick bark and needle-like leaves, which minimise water loss and resist freezing. Many conifers grow in a *spire-like shape,* with a narrow, pointed crown that helps shed snow, while low-growing plants can survive

Plate 3.5 As elevation increases, trees tend to become shorter and more stunted, known as krummholz growth form. *(Photograph Rob Marchant)*.

Plate 3.6 Trees in montane forests often develop extensive root systems to anchor themselves on steep slopes and are shaped by wind in exposure locations, such as this *Olea africana* on Mount Marsabit. *(Photograph Rob Marchant)*.

beneath the snowpack, insulated from the coldest temperatures (Plate 3.7). The branches are usually short and dense, reducing the surface area exposed to cold winds. Some species can photosynthesise at low temperatures, allowing them to take advantage of brief periods of warmth.

Mountains can be safely called cradles of diversity, with adaptive radiation at high elevation and the evolution of a particular morphological,

Plate 3.7 Many plants are adapted to survive under deep snow for much of the year. The conical shape of trees helps shed snow, while low-growing plants can survive beneath the snowpack, insulated from the coldest temperatures. Multiple Arctic-alpine species prefer acidic substrates and are often found on scree fields, snowbed sites, and in the vicinity of meltwaters. *(Photographs Rob Marchant).*

physiological, or behavioural trait to allow the species to better exploit montane niches in 'island-like' environments (Perrigo et al., 2020). There is a risk that some adaptations, although increasing fitness, may prove disadvantageous in the long-term. Adaptation increases habitat specialisation and therefore makes a species more vulnerable to environmental change and potential extinction risk (Crawford, 2005). For example, many animals might have dense fur to survive the colder conditions; in an increasingly warm world, such adaptations can become a hindrance. Most adaptations have both positive and negative aspects in relation to survival potential. Drought-tolerant species are usually not flood-tolerant (Crawford, 2005). In contrast to animals, plants are not automatically faced with extinction when sexual reproduction fails. Plants also differ from animals in their reproductive responses to climatic oscillations by not needing to grow to survive, as whole populations can lie dormant for decades, or even centuries, buried in the soil seed bank.

3.8 Mountains: Barriers, stepping stones, or bridges?

For many biogeographers, mountains are regarded for their role as barriers or bridges to curtail, or aid, species dispersal. Which of these roles that mountains play depends on the ecological and physiological requirements of the species, as well as their dispersal ability (Perrigo et al., 2020). Some ranges are connected and allow flow for ecosystem components across mountains and sometimes whole ranges, whereas other mountains are isolated. For example, the extensive mountain ranges of the Ethiopian highlands are more connected than the scattered free-standing volcanic mountains of East Africa that support extremely fragmented faunas and floras. The role of mountains as bridges that form dispersal corridors is clear from the widespread distribution of Northern Hemisphere taxa in Southern Hemisphere mountains, such as the European-centred plant genera *Carex* and *Ranunculus*, as well as the genus *Alchemilla* in Africa (Gehrke & Linder, 2009). Conversely, mountains acting as barriers can be seen in Amazonian freshwater fishes, where the separation of populations is congruent with the uplift of the Eastern Cordillera (Andes) (Lundberg et al., 1998). One significant question that remains is when, or whether, mountains can become saturated with diversity, and what are the 'brakes on diversity?' In the Himalayas, it has been suggested that the diversification rate of songbirds slowed as ecological niches were filled (Price et al., 2014). The shifting balance between speciation and extinction, in which mountains may act as 'cradles' (areas of especially rapid species origination), 'museums' (areas of especially long-term persistence of species), and 'graves' (areas with especially high rates of extinction) for biodiversity (Rahbek et al., 2019a) does seem to be a precarious one, particularly in a rapidly changing world (Chapter 6).

3.9 High-altitude ecosystems and plant distributions

From the densely forested foothills to alpine meadows and snow-covered peaks, mountains showcase a stunning diversity of plant life. Mountain ecosystems can typically be divided into three major vertical zones (Hedberg, 1961): species-rich montane forest associations, a sub-alpine ericaceous belt, and the sparse alpine zone. Like any classification system, there are multiple ecological ways to 'slice the mountain pie', whether into these three broad zones or many more, with the adjectives wet, moist, dry, and xeric being used to further differentiate ecosystems. The lower elevations often support lush forests, while mid to high altitudes are characterised by diverse shrublands and meadows. The vegetation zones on a typical mountain are stratified primarily by elevation, which affects temperature, moisture levels, and the amount of sunlight that different parts of the mountain receive. The position of these vegetation zones varies dramatically depending on the mountain's location (latitude, proximity to the ocean, regional landscape).

Although ecologists like to group vegetation into distinct ecosystem types, all species have individual limits to their distribution and an environmental tolerance they must live within – the so-called 'niche' (Fig. 3.6). The niche is a very important concept when understanding what controls the distribution of any organism. There's a multitude of ecophysiological and environmental factors that control individual growth, reproduction, and survival (e.g., Cain, 1944) that have been translated into the concept of realised environmental niche. This can be defined as the envelope of suitable conditions in multi-dimensional environmental space, within which populations of a species maintain positive growth; this is, of course, further constrained by biotic (e.g., competitive) interactions (Buri et al., 2017). Physiological tolerance to climatic factors is clearly a primary driver of plant distributions at all scales (Woodward, 1987). The niche can consist of a multitude of factors such as temperature, annual precipitation, soil type, dry spell intensity, UV tolerance, and frost-free days, to name but a few (Platts et al., 2008). A vegetation belt falls within environmental boundaries and margins and is composed of multiple niches. Relating physical climatic factors to species occurrence requires a knowledge of the many interactions between adaptive physiology and genetics of populations as they approach their environmental tolerances to survival.

3.9.1 Composition and distribution of lower montane forests

Montane forests deserve a special mention in this book, partly because of their name and partly because it is important to understand how those

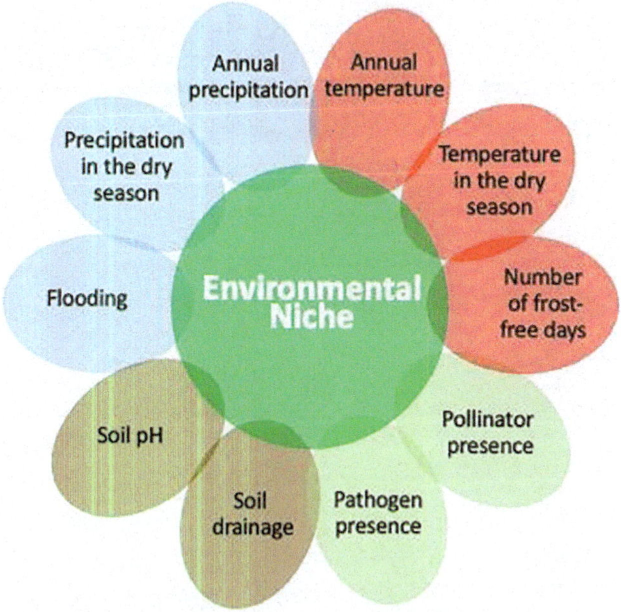

Figure 3.6 Environmental niche: The growth of any organism is governed by a whole series of overlapping variables – in this stylised example, a combination of temperature, precipitation, soil, and biotic factors interacts to determine the environmental niche.

ecosystems located at higher and lower altitudes function. The lowermost vegetation type is typically forests characterised by large emergent trees, a substratum of trees, and a rich understory of shrubs and herbs. In tropical regions, rainforests with species like mahogany, teak, and various tree ferns are found at the lowest altitudes. In temperate lowland regions, deciduous forests with oak, maple, and beech trees are common (Plate 3.8). In terms of the observable structure of montane forests, it must be stressed that the structure, composition, and certainly the distribution, have been impacted by a long history of human forest interaction, often leading to the loss of forest cover and removal of high-value large emergent timber trees.

Montane forests are complex and dynamic ecosystems that vary with elevation, latitude, and local conditions; they are characterised by a varied composition of trees, shrubs, and ground cover, with a multi-layered structure that supports a diversity of life; for example, there are some 700 montane forest trees on the Eastern Arc Mountains. Montane forests are often dominated by quite different trees depending on where the mountain system is located. *Piñon-Juniper* woodlands cover extensive areas in the southwestern US and Great Basin, characterised on the Colorado Plateau

Plate 3.8 Mountain tropical (*bottom*) or temperate forests (*top*) are characterised by large emergent trees, a substratum of trees, and a rich understory of shrubs and herbs. Montane forests can be lush with ferns, mosses, bromeliads, and orchids. *(Photographs Rob Marchant).*

by *Pinus edulis* and *Juniperus osteosperma* (Rocca et al., 2014). Lower montane forests in the Rocky Mountains are dominated by *Pinus ponderosa*. In the southern part of the region in southeastern Utah, lower montane forests usually occur as either relatively pure stands of ponderosa pine with some combination of species found in lower-elevation *Piñon-Juniper* woodlands or higher-elevation upper montane forests, especially *Pseudotsuga menziesii* with *Quercus gambellii* a common understory species. Lower montane forests in Idaho and Montana consist mostly of ponderosa pine or are co-dominated by Douglas-fir or *Larix occidentalis*. In temperate latitudes, coniferous trees dominate, especially pines (*Pinus* spp.), firs (*Abies* spp.), spruces (*Picea* spp.), and hemlocks (*Tsuga* spp.). These trees are well-adapted to colder climates and can withstand heavy snow and wind (Plate 3.7). At lower elevations, deciduous trees may be more abundant, along with species like oaks (*Quercus* spp.), maples (*Acer* spp.), birches (*Betula* spp.), and beech (*Fagus* spp.) (Plate 3.8). This mix of both deciduous and coniferous trees can be common, especially at intermediate elevations, and creates a diverse canopy structure supporting a wide variety of plant and animal life. The understory in montane forests can be dense with shrubs, including *Rhododendron* spp., *Azaleas*, and various berry bushes like *Vaccinium* spp. (Wang et al., 2022).

Montane forests are often characterised by a multi-layered canopy. The upper canopy is formed by the tallest trees, which can range from 15 to 40 m depending on elevation and species, but can be 100 m plus (Hemp et al., 2017). Below these emergent trees, there will be a mid-story of smaller trees and large shrubs that can vary considerably. In lower montane forests, the canopy might be denser with broad-leaved trees, while at higher elevations it becomes more open, with widely spaced conifers allowing lighter conditions to penetrate to the forest floor. The forest floor in montane forests is typically covered with a thick layer of leaf litter, needles, fallen branches, and decomposing wood. This layer is crucial for nutrient recycling and provides habitat for various fungi, insects, and microorganisms (Plate 3.9).

Tropical montane forests are characterised by unique species such as the *Podocarps* and evergreen broadleaf species in the Lauraceae (Plate 3.8). In tropical montane forests, epiphytic plants are common, creating a whole ecosystem within a tree. These orchids, ferns, mosses, and bromeliads thrive in the cool, humid conditions typical of montane environments. Ferns are a particularly diverse group of plants, ranging from tree ferns to small epiphytic ferns (Plate 3.10), and although the origin of the fern lineage dates to the Devonian, much of the extant diversity accrued relatively recently (Suissa et al., 2021).

3.9.2 Composition and distribution of upper montane forests

Upper montane forests reflect cooler temperatures with more precipitation, often in the form of snow during winter, due to an increase in altitude. Upper Montane forests are typically dominated by coniferous or ericaceous trees, which are well adapted to cold and snowy conditions. Like their low-altitude cousins, the composition of these forests varies greatly with location. For example, on temperate mountains, subalpine forests are found just below the tree line, resulting in a unique ecosystem that supports specialised plant life.

Plate 3.9 The forest floor in montane forests is typically covered with a thick layer of leaf litter, needles, fallen branches, and decomposing wood, often rich in mushrooms and toadstools. *(Photographs Rob Marchant)*.

The forest floor in subalpine forests is usually covered with a layer of needle litter, fallen branches, and decomposing wood. This layer is thinner than in lower elevation forests due to the slower rate of decomposition caused by the cold temperatures. The terrain in subalpine forests is often rocky and uneven, with exposed bedrock, boulders, and talus slopes with patches of bare ground (Plate 3.11). Soil is usually thin and nutrient-poor, further limiting the growth of plants.

Plate 3.10 Epiphytic plants such as ferns, mosses, and bromeliads are common, creating a whole ecosystem: Ferns ranging from tree ferns to small epiphytic ferns. *(Photographs Rob Marchant)*.

Upper montane forests in the Rocky Mountain region typically include a diverse mix of several tree species, including *Abies* spp., *Populus tremuloides*, and *Tsuga heterophylla*. Understories in the upper montane forest can be quite diverse, ranging from a near continuous mix of grasses, shrubs, and herbs to very sparse cover under relatively dense closed forest canopies (Rocca et al., 2014). Common species include *Abies lasiocarpa, Picea engelmannii, Pinus contorta, Pinus albicaulis*, and *Tsuga mertensiana*. The understory of upper montane forest is usually sparse, with low-growing shrubs such as *Vaccinium* spp. and *Salix* spp. The canopy is generally open, with trees spaced relatively far apart compared to lower elevation forests. This open structure allows more light to reach the forest floor, supporting the growth of understory vegetation. Tree density in upper montane forests can vary widely, depending on factors such as elevation, aspect, and local microclimate. In some areas, trees may form dense stands, while in others they may be more widely scattered or even isolated. Trees in subalpine forests are typically shorter than those in lower montane forests, reflecting the harsher growing conditions. Heights may range from 10 to 20 m in more sheltered areas, but in exposed locations, trees may be much shorter or stunted.

Plate 3.11 Upper montane forest around 3700 m on Kilimanjaro is dominated by *Erica arborea* following a recent fire. Subalpine forests are found just below the tree line on mountain slopes where the terrain is uneven, with thin soil patches of bare ground, and disturbance can be frequent and tree growth stunted. *(Photograph Rob Marchant).*

Herbaceous plants are also common in the upper montane understory, including various species of wildflowers, grasses, and sedges. These plants have adapted to survive the short growing season and may emerge quickly in the spring to take advantage of the brief period of warmth and sunlight. The ground cover in upper montane forests is generally sparse, reflecting the limited growing season and poor soil conditions. However, in some areas, the ground may be carpeted with mosses, low shrubs, or mats of wildflowers. In areas where snow persists late into the growing season, unique plant communities known as snowbed communities can develop. These areas are typically dominated by cold-tolerant, low-growing plants like *Carex spp.*, *Geum spp.*, and various alpine grasses. The forest floor and tree trunks in subalpine forests are often covered with a thick layer of moss and lichen, which thrive in the cool, moist environment (Plate 3.12).

Upper montane forests are highly specialised ecosystems that exist at the edge of tree growth on mountains. They are characterised by a dominance of hardy coniferous trees, a sparse understory, and a range of microhabitats created by the open canopy and rugged terrain. The structure and composition of these forests are shaped by the extreme conditions of cold, wind, and snow, and they play a crucial ecological role in water regulation, climate buffering, and providing habitat for a range of specially adapted species.

Plate 3.12 Subalpine forests are often covered with a thick layer of mosses and lichens. *(Photographs Rob Marchant).*

3.9.3 Treeline formation

The transition from forest to grassland can be one of the most visually striking vegetation boundaries (Plate 3.13). The tree line, also known as the timberline, is the highest elevation on a mountain where trees can grow. Altitudinal treelines are easily recognisable boundaries separating montane forest and alpine plant communities in mountain ranges. Beyond this line, environmental conditions become too harsh for tree survival, resulting in a landscape dominated by low-growing shrubs, grasses, and other non-woody plants, or simply bare rock and snow. The tree line is not always sharply defined; it can be a gradual transition, where trees become increasingly stunted and sparse until they disappear altogether. The nature of the habitat (oceanic or continental), the plant communities at the interface (wetland or dryland), and respiratory efficiency are affected by short-term temperature variations during the growing season, which play a role in reducing tree survival. Thus, although the treeline is a widespread and common phenomenon across the world's mountains, the factors controlling its position and character will vary greatly from one region to another (Crawford, 2005). Superimposed on

Plate 3.13 Treelines can be sharp or diffuse; sometimes reflecting the change in environment, other times a disturbance regime such as fire or grazing. Treelines from Jakobshorn, Switzerland (*top left*), Stubaai, Austria (*bottom left)*, Teide, Tenerife (*top right*), Sambuco, Italy (*bottom right*). *(Photographs Rob Marchant).*

this global diversity, the treeline is very rarely uniform around a mountain. South-facing slopes (in the Northern Hemisphere) receive more sunlight and warmth, often resulting in a higher treeline when compared to north-facing slopes. Local microclimates, such as sheltered valleys, can also support tree growth at slightly higher elevations. The treeline provides unique habitats for specialised plant and animal species, which have adapted to survive in these marginal conditions, and is often highly biodiverse as it supports species from both forested and alpine ecosystems (Rolando et al., 2007). Just below the tree line, trees often take on a dwarfed, twisted, and contorted growth form known as 'Krummholz' (crooked wood), or start to look increasingly stressed as they get to the edge of their niche. This growth form is an adaptation to the harsh conditions near the tree line, including strong winds, thin and nutrient-poor soils, cold temperatures, and heavy snow loads. Scattered but upright trees, with the degree of scatter changing across the treeline interface, is normal, rather than a line of trees. It is probable that climatic variation, together with the occurrence of extreme events, combine to create, over a period, a zone above which the probability of tree survival is much reduced.

The position of the tree line is determined by a combination of climatic and environmental factors, although temperature is primary in determining its location (Holtmeier, 2009 ;Liang et al., 2021). Trees require a minimum temperature during the growing season to photosynthesise, grow, and reproduce. When temperatures consistently fall below this threshold, tree growth becomes impossible. The freezing and thawing of snow and ice can damage tree roots and bark, further limiting tree survival close to the tree line. The growing season at higher elevations is often very short, varying from just a few weeks to a couple of months; if the growing season is too short, trees cannot produce enough energy to survive. Winds are generally stronger at higher elevations, and that can physically damage trees, especially young saplings, by breaking branches, desiccating needles, and stripping away soil around roots. Constant exposure to strong winds can prevent trees from growing tall and lead to the krummholz growth form (Plate 3.5). At high elevations, soils are often thin, rocky, and nutrient-poor. The limited availability of nutrients and water in such soils makes it difficult for trees to establish roots and grow. In some high-elevation areas, permafrost exists below the surface at varying frozen states and depths. This frozen layer can limit root growth and prevent trees from accessing water and nutrients in deeper soil layers. Higher elevations receive more intense solar radiation, including harmful ultraviolet (UV) rays. This can damage plant tissues and further limit the ability of trees to grow and survive near the tree line. Despite this factor-complex that controls the growth of trees, the treeline positions are still commonly associated with a growing season soil temperature of 6.7°C (Körner & Paulsen, 2004). In addition to the global temperature effects, other regional and local climate factors, such as moisture availability, winter snow cover, and wind damage (Lu et al., 2021), can co-determine the treeline position. Many studies have attempted to relate this sometimes very definite line to a range of temperature measurements (Körner & Kèorner, 1999). However, the treeline is more likely to be related to the length of the growing season that is needed for the production and development of new tissues (Körner & Kèorner, 1999). For example, increased spring snowfall may have negative effects on tree growth through a delayed snowmelt and thus a later onset of cambial activity in temperate regions (Lu et al., 2021). Despite the convenience of the concept of a generalised temperature limit for tree growth, it is important to take note of the fact that mean temperatures do not exist in nature!

Climate change will shift treelines, and warmer temperatures are already driving tree species upslope (Chapter 6). The consequences for species already adapted to the colder conditions near mountaintops are uncertain. However, climate warming alone cannot explain the observed range shifts of alpine plant species on European mountains (Grytnes et al., 2014). In addition to the global temperature effects, other regional to local climate factors, such as moisture availability (Elliott et al., 2020; Sigdel et al., 2018), winter snow cover, and wind damage (Kirdyanov et al., 2012), have all been shown to co-determine the treeline position. Multiple lines of evidence demonstrate that global climate change is re-shaping the distribution of many terrestrial organisms and ecosystems, including treelines (Lu et al., 2021). In addition to being an interesting biogeographical gradient, their dynamics epitomise the impacts of climate change on terrestrial ecosystems, calling for a better understanding of the underlying drivers and processes (Lu et al., 2021). However, a consensus about which climate factors ultimately determine treeline shift rates at hemispheric to global scales is still lacking.

Treeline shifts can have both positive and negative feedback on mountain biomes by affecting local and regional climate, soil carbon storage, nutrient cycling, hydrological processes, and biodiversity. A first quantitative synthesis of treeline shift for the Northern Hemisphere shows that an upward shift has been detected at 88.8% of sites, but the observed rates of treeline shift were relatively slow (mean treeline shift rate was quite varied, with global studies ranging from a low of 0.354 m $year^{-1}$ up to 1.1 m $year^{-1}$ despite rapid climate warming (Lu et al., 2021).

3.9.4 Trim line formation

The trim line on mountains is a distinct boundary, or visible line, on the landscape that marks the highest point reached by a glacier or ice sheet in the past. This line typically separates areas of different vegetation or surface characteristics and is often identified by a stark contrast between vegetated lower slopes and barren upper slopes. The trim line is formed by the erosive and depositional actions of glaciers. When glaciers advance, they erode the underlying rock and soil, removing vegetation. When they retreat, they leave behind a clear demarcation between the scoured, vegetation-free area

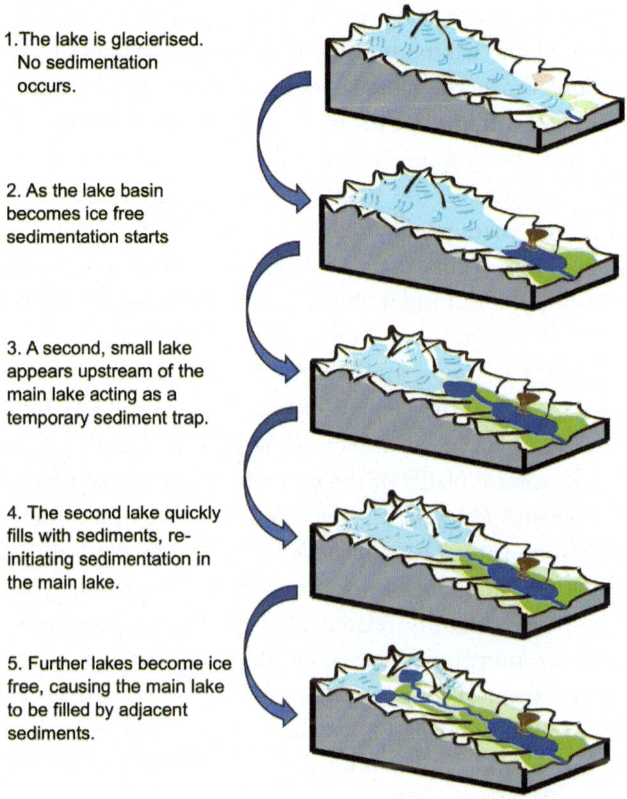

Figure 3.7 The creation of new land following deglaciation linking into the creation and infilling of lakes from sedimentation with rapid changes in both terms, and sediment supply and sediment transport.

above the trim line and the undisturbed, vegetated area below. Trimlines are becoming more common as the rate of global warming continues and glaciers retreat (Chapter 6). The creation of new land following deglaciation (Fig. 3.7) allows for a whole suite of new plant associations (Chapter 7).

3.9.5 Composition and distribution of the alpine herbaceous zone

The composition of the herbaceous vegetation on mountains varies greatly with location, but there is remarkable similarity in structure, and diversity is significantly reduced compared to lower altitudes. High levels of species endemism reflect the need to be specialised and adaptable to the extreme environment: in the Afroalpine region, more than half of the 191 genera in the area are represented by a single species, and just 14% of those are represented by more than five species (Gehrke & Linder, 2014). The uppermost, afroalpine habitat is home to some 520 species of vascular plants that can endure the harsh climate regime where days can be very warm and the nights are often cold throughout the year (Hedberg, 1961). Low-lying plants, tussock-forming shrubs, and grasses are highly adapted to a life of extremes. For example, the herbaceous areas of East African mountains are renowned for their peculiar giant rosette plants (Plate 3.4) and high number of endemic plants that are adapted to high UV, warm days, and frosty nights (Gehrke & Linder, 2014). The impressive plants of *Lobelia* and *Dendrosenecio* can rise to 10 m above the otherwise low-growing alpine vegetation (Plate 3.4). Quite similarly, the Andean mountains are dominated by ground rosettes from the genus *Espeletia* (Hedberg, 1964). Like many was fascinated by these classic examples of parallel evolution provided by the enigmatic giant rosette plants of the giant *Lobelia* and giant *Dendrosenecio* in Africa and *Puya* and *Espeletia* in South America. The giant lobelias arrived in East Africa from eastern Asia in the Miocene, and spread from there across the Congo basin in the early Pliocene to West and South America (Perez & Yu, 2021). These distantly related plants have independently developed strikingly similar morphological, physiological, and life history traits in response to the tropical alpine climate. These mountain systems provide an excellent example of a highly isolated system, where adaptive radiation between historical and contemporary processes has formed a flora rich in local endemics (Brochmann et al., 2022).

Similarly, the high-altitude tussock grasses and shrubs found on Teide are adapted to extreme diurnal variation in temperature, and the need to conserve water while also withstanding high levels of UV. There is a similarity in growth forms across many mountain areas as evolutionary processes have adapted and shaped a number of species to similar environmental stresses. Colonies of plant growth are common where the inner part of the plants' community is sheltered from the extremes by their neighbour (Plate 3.14), often forming extensive 'cushions'.

3.9.6 Composition and distribution of alpine grasslands

Grasslands cover about 25% of the total global land area and are the main type of terrestrial ecosystems, playing an essential role in climate regulation, soil, water, and biodiversity conservation (Phelps et al., 2025). Mountain grasslands are unique ecosystems that have been shaped over millions of years (Du et al., 2022), recognised by their physical and biological characteristics,

Plate 3.14 Colonies of plants growing together form 'cushion' plants, where each plant is protected by its neighbours. The form of growth is often similar on different mountains, such as these cushion forms from the Drakensberg (*left*), the Alps (*centre*), and the Chilean Andes (*right*). *(Photographs Rob Marchant).*

particularly the dominance of grasses, cold and wet climatic conditions, high soil organic matter content, and the large capacity of soils to store water (Phelps et al., 2025). For example, the Qilian Mountains are important for maintaining the ecological balance of the wider Qinghai-Tibet Plateau and safeguarding the recharge of inland river runoff in the Hexi corridor (Du et al., 2022). Grasslands are the most widespread ecosystem type in the Qilian Mountains, and have various ecological functions such as wind control and soil and water conservation. However, due to climate warming and extremes, the alpine grasslands in this region are degraded, and biodiversity is declining (Xu et al., 2009). The Qilian Mountains are a very important pastoral area, where population growth and overgrazing have led to a reduction in vegetation cover and biodiversity at lower elevations, increased soil erosion and desertification, and damage to the integrity of grassland ecosystems, making them more sensitive and less resilient to climate change (Li et al., 2019; Vitousek & Farrington, 1997), particularly at middle elevations (Du et al., 2022). Páramos are unique grassland ecosystems located in the upper parts of the Andes extending from northern Peru to Venezuela and occurring between the tree line and glaciers (Cárdenas et al., 2017). Montane grasslands and shrublands are home to distinct plant communities, dominated by tussock grasses and cushion plants with forest patches (*Polylepis* sp.) growing along water streams (Berrones et al., 2022). The hydrological

Plate 3.15 Montane grasslands are diverse and endemic systems that can have a long history, such as the Drakensberg mountains of South Africa. *(Photograph Rob Marchant).*

role of fog and cloud water in these environments is very important as it is the source of water for downstream communities (Berrones et al., 2022).

Grasslands are commonly present across many African montane systems (Plate 3.15). For example, Madagascar's grassy biomes cover 80% of the island and are central to Malagasy history, culture, and ecology (Phelps et al., 2025). The research community continues to debate the extent to which these biomes are ancient ecosystems or products of human-driven deforestation. This is a wide-ranging question with massive implications for ecosystem management. Grasslands are one of the most important ecological characteristics of the Drakensberg mountains across South Africa (Dlamini et al., 2025). Grasslands are located from the lower altitudes all the way to the summits and are extensive in other regions of southern Africa, such as the KwaZulu-Natal midlands. Management of Afromontane grassland in South Africa is highly relevant to current policies that advocate tree planting, even in regions with substantial pre-existing soil organic stocks, based on the assumption that this can enhance net ecosystem carbon storage and aid climate change mitigation. A growing number of studies document the significant impact of afforestation, and subsequent land degradation, on carbon dynamics in Afromontane grasslands (Dlamini et al., 2025).

In some regions, particularly where the snow cover is less permanent, cushion plants may be found. These are small, compact, dome-shaped plants that grow low to the ground, minimising exposure to wind and cold. Cushion plants include species like *Saxifraga* and *Draba*, but they are generally

Plate 3.16 Rocky and uneven terrain with exposed bedrock, boulders, and talus slopes is often an area where Lichens can dominate. *(Photograph Rob Marchant).*

found at the transition between the alpine and nival zones, rather than in the highest parts of the nival zone, where it becomes too harsh for even these specialists to survive.

3.9.7 Composition and distribution of the nival zone ecosystems

The nival zone is the highest elevation zone on a mountain, typically found above the alpine zone and extending to the mountain's summit. It is characterised by extreme environmental conditions, including cold temperatures, high winds, and perpetual snow and ice cover. These conditions limit the presence of vegetation, resulting in a landscape dominated by lichens, mosses, bare rock, glaciers, snowfields, and ice (Plate 3.16). The nival zone represents one of the most inhospitable environments on a mountain, and its composition and structure reflect the challenges of survival in such an extreme setting. Vegetation in the nival zone is extremely sparse, or often absent, due to the harsh environmental conditions. The growing season is very short, if it exists at all, and the persistent snow and ice cover prevent most plants from establishing roots. The few species that do survive are either highly specialised or have managed to inhabit crevices and sheltered locations. Due to the relatively restricted spatial area, they are also extremely rare. Lichens are one of the few types of organisms capable of surviving in the nival zone (Plate 3.16). They are symbiotic organisms composed of fungi and algae (or cyanobacteria) that can tolerate extreme cold, intense

UV radiation, and desiccation. Lichens can be found growing on exposed rock surfaces, often in cracks and crevices where they are slightly protected from the elements. Certain species of algae and cyanobacteria can grow in the nival zone, often within or beneath the snow and ice. Snow algae, for instance, are microscopic organisms that give patches of snow a red, green, or orange tint. These organisms are capable of photosynthesis even in cold temperatures and can survive in thin films of water between snow crystals.

3.10 Mountain animal diversity

Similar to plants, cold, dry climates, high levels of UV, wind exposure, and high seasonal variability are important environmental factors that restrict the distribution of animals on mountains. In addition, hypoxia and increasingly low levels of oxygen with altitude are important controls. Animals living at altitude have evolved a range of adaptive mechanisms during their evolutionary history to cope with these harsh environmental stresses (Zhou et al., 2014). Although mountains provide essential habitat for a wide variety of wildlife, including keystone species that are specially adapted to the mountain environment, such as snowshoe hares, lynx, mountain lion, gorillas, and panada, the following section will explore the distribution of two important animal groups – birds and insects.

3.10.1 Montane bird communities

Mountains, with their unique environmental gradients and complex structures, host diverse bird communities. Bird communities in montane areas, in common with plants, are typically stratified by elevation (Fig. 3.8). Unique topography, climate, available resources, and vegetation of mountains create distinct habitats that shape the presence and abundance of bird species.

Montane forests often exhibit high bird species richness, with many species being endemic to specific elevations or mountain ranges. For example, in the Andes, species like the Andean cock-of-the-rock (*Rupicola peruvianus*) and several species of Tanagers are found only in montane forests. Similarly, the Himalayas host species like the Satyr Tragopan (*Tragopan satyra*) and the Himalayan monal (*Lophophorus impejanus*). Birds like hummingbirds are common in montane forests, particularly in the Andes (Plate 3.17). These birds feed on the nectar of flowering plants, many of which have been specially adapted to be pollinated by specific hummingbird species.

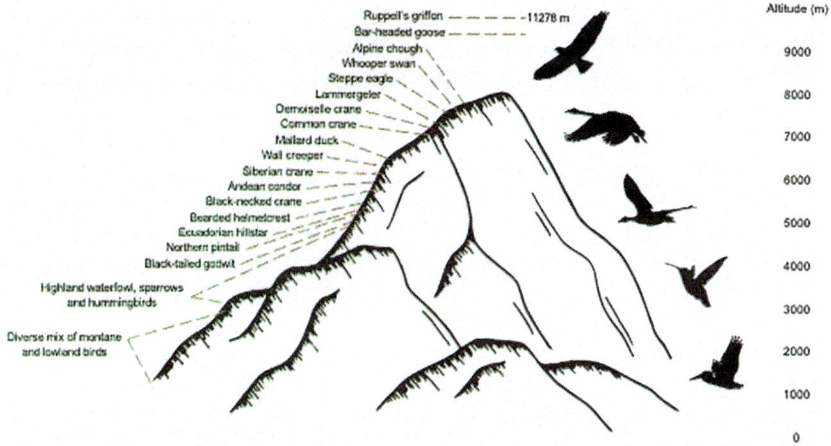

Figure 3.8 Bird species in montane areas are typically stratified by elevation. Although most birds live and fly at relatively low altitudes, species from several avian orders live, migrate, or occasionally ascend much higher. These include multiple species of raptor, waterfowl, and crane. *Reports from various sources in the literature are shown here (Eastwood & Rider, 1965; Faraci, 1986, 1991; Kanai, 2000; McCracken et al., 2009; Scott, 2011).*

Plate 3.17 Hummingbirds are common in montane forests, particularly in the Andes. *(Photographs Rob Marchant).*

Plate 3.18 Predatory and foraging birds, including the Drakensberg Vulture (*bottom left*), bee-eater (*top left*), hornbills (*top right*), and kingfisher (*bottom right*). (*Photographs Rob Marchant*).

Many montane forest birds are frugivores, feeding primarily on fruits, such as the various species of toucans in the Neotropics and hornbills in Asian montane forests. These birds play crucial roles in seed dispersal. Insectivorous birds, such as woodpeckers, flycatchers, and warblers, are also abundant in montane forests, exhibiting specialised foraging behaviours, such as bark-gleaning or aerial hawking. Predatory birds, including owls, buzzards, hawks, and eagles, are also present in montane forests. These birds are often top predators within their ecosystems. They can be indicators of environmental health and are often threatened and rare, such as the Drakensberg Vulture (Plate 3.18). Jays, crows, and some species of tanagers and thrushes have a varied omnivorous diet that includes fruits, insects, and small vertebrates. In addition to this classification by feeding habitat, it is also possible to separate by strata; birds in the canopy feed on fruits, nectar, and insects, and the midstory layer is mainly occupied by birds that forage on insects, fruits, and flowers in the lower branches of trees and large shrubs. The understory, characterised by dense shrubbery and young trees, supports a diverse community of birds that feed on insects, seeds, and small fruits. The ground layer is home to species that forage on the forest floor, searching for insects, seeds, and fallen fruits. In the complex structure of montane forests, birds often exhibit niche partitioning where different species use specific parts of the forest or have specialised feeding behaviours to reduce competition.

3.10.1.1 Migratory bird communities

The composition and structure of bird communities on mountains can vary seasonally. In some regions, migratory species may dominate during certain seasons, while resident species may shift their ranges or behaviour in response to seasonal changes in food availability and climate. Migratory species may use mountain forests as breeding grounds, wintering areas, or stopover sites during migration. For example, the black-and-white warbler (*Mniotilta varia*) is a migratory species that breeds in North American montane forests (Mann et al., 2021). Many migratory birds pass through high mountain areas during their migration, using these as migration routes (Hirschhofer et al., 2024). Some mountain ranges serve as important migration routes for birds travelling between breeding and wintering grounds. Mountain passes and ridges can be crucial for migratory species navigating through challenging terrain, and are increasingly being realised as important locations to support bird conservation (Hirschhofer et al., 2024).

3.10.1.2 Adaptation and behaviour of birds to a montane habitat

Birds in mountains are specifically adapted to colder temperatures and may have specialised behaviours for thermoregulation, particularly at higher altitudes (Scott, 2011). To adapt to hypoxia, every step in aerobic respiration in high-altitude birds has increased efficiency through the process of natural selection (Scott & Milsom, 2007). For adaptation to high-altitude hypoxic conditions, birds often exhibit physiological adaptations, such as increased lung capacity and enhanced oxygen utilisation, to cope with lower oxygen levels at higher altitudes. Birds, particularly those at high altitudes, have thick plumage for insulation, strong flying capabilities for dealing with high winds, and specialised feeding behaviours to exploit scarce food resources. Some bird adaptations can be both intra- and inter-specific. For example, (Lu et al., 2009) found that snowfinches (*Montifringilla* spp.) living at higher altitudes had longer wings as an adaptation to lower air density. However, cause and effect are difficult to establish, as strong winds may also be a driver of this morphological adaptation (Martin, 2001). The ability of birds to fly at high altitudes is critically dependent on the effective transport of oxygen from hypoxic air reaching all of the tissues (Scott, 2011). At least two factors account for this difference: (1) the tight mesh of capillaries surrounding avian muscle fibres due to a high degree of branching between longitudinal vessels, and (2) the smaller aerobic fibres of birds compared with similar-sized mammals (Ellis et al., 1990). Some species also show very high blood oxygen affinity (e.g., Andean Goose *Chloephagamelanoptera* (Jensen et al., 1991) and the White-tailed Ptarmigan (Martin, 2001) being particularly renowned for having blood with high oxygen exchange. In addition to enhancing oxygen uptake and transport, it is important to improve the efficiency of oxygen usage (Zhou et al., 2014). There is significant evidence for positive selection of genes (ND2, ND4, and ATP6) in high-altitude birds, which strongly suggests that, in addition to other physiological changes, positive selection in mitochondrial genes can explain how high-altitude birds attain sufficient energy despite hypoxia (Zhou et al., 2014). This necessity to adapt to high altitudes is likely to be an impediment to the colonisation of higher altitudes regardless of climatic changes.

3.10.2 High-altitude insect diversity and distribution

Similar to birds, the distribution of insects on the world's mountains is determined by a complex interplay of altitude, climate, vegetation, and evolutionary adaptations. Given their short life cycles, insects are remarkably adaptable, with species occupying even the most extreme environments, found at high altitudes. Mountain ecosystems host a unique array of insect species that have evolved to thrive in these challenging landscapes, and understanding the distribution and adaptations of these insects is crucial not only for their conservation but for the conservation of the main higher tropic levels that feed on them, such as the charismatic chameleons (Plate 3.19). These cascading linkages across food webs and chains are particularly important in the face of climate change and habitat loss (Chapter 6), which threatens these fragile mountain food webs and ecosystems. These impacts are not restricted to insects *per se,* but insects provide vast amounts of protein to support mountain food webs and provide the connection between the plant and the animal kingdoms (Hodkinson, 2005). Species such as the Cascades Frog (*Rana cascadae*), with a diet primarily consisting of mayflies and caddisflies, are reliant on insects.

Similar to plants and birds, insect distribution is typically stratified by altitude, with diversity decreasing with increasing altitude. Lower montane zones typically support a higher diversity of insects; in the Andean cloud forests of Ecuador, the diversity and abundance of insect species, such

Plate 3.19 Magnificent chameleons! from the Virunga volcanoes (*top middle*), the Drakensberg (*middle left*), and Kilimanjaro (*right*): like many animals are heavily reliant on a diet of insects. *(All Photographs Rob Marchant).*

as the glasswing butterflies (*Ithomiini*), vary significantly between slopes with different exposures. Large, iridescent butterflies are found at various elevations in the Andes, with different species occupying specific altitudinal zones (Plate 3.20). *Morpho sulkowskyi*, for instance, is commonly found in the montane cloud forests between 1000 and 2500 m, where it feeds on the nectar of flowers and plays a role in pollination (Lamas, 2004). The Apollo butterfly (*Parnassius apollo*) is found in the Himalayas at altitudes ranging from 2400 to 4000 m (Das et al., 2018). This butterfly is well adapted to the high-altitude environment, with a life cycle synchronised to the short growing season of alpine plants. In the Rocky Mountains, insect diversity is highest from about 1800 to 2400 m, where forests of pine, spruce, and fir provide a rich habitat for beetles, butterflies, and ants. Higher up, in the subalpine and alpine zones, species such as the Rocky Mountain Apollo butterfly (*Parnassius smintheus*) and the alpine tiger beetle (*Cicindela longilabris*) are suitably adapted to the harsher conditions (Das et al., 2018). *Parnassius smintheus* is adapted to the short growing season and cold temperatures of the region, and its larvae feed on stonecrop (*Sedum* species), a plant commonly found in alpine environments. The Appalachian tiger swallowtail (*Papilio appalachiensis*) is a butterfly species endemic to the Appalachian region. It is typically found in the montane and submontane zones, where it prefers the cooler, moist conditions of hardwood forests (Pavulaan &Wright, 2002).

Although insects are one of the most diverse biological groups, with large numbers of widely distributed species, there are relatively few studies

Plate 3.20 Mountain butterflies. *(Photographs Rob Marchant).*

on the large-scale diversity patterns of insects (Economo et al., 2018). Among insects, Hemiptera is the largest non-complete metamorphosis order of insects (Li et al., 2021). Mountains are home to extraordinarily abundant Hemiptera species, demonstrating the important role of mountains in preserving biodiversity (Li et al., 2021). It is estimated that about 103,590 Hemiptera species are recorded in the world (Stork, 2018). Most Hemipteran insects are phytophagous, and many species are important agricultural and forestry pests (Li et al., 2021). In common with plant diversity, in ascending a mountain, the number of species tends to decline, with only a few specialised species found at the highest altitudes. For example, in the Andes, insect diversity is highest in the cloud forests (around 1500–3000 m). At higher altitudes, such as the páramo, the diversity of insects decreases and those present are highly specialised, such as the giant Andean grasshopper (*Tettigonia viridissima*), which is adapted to the cold, windy conditions with a slow life cycle (Lamas, 2004). The Sierra Nevada Blue butterfly (*Plebejus podarce*) is a species found primarily in the subalpine and alpine zones of the Sierra Nevada, at elevations between 2500 and 3500 m. This butterfly is adapted to the cold, windy conditions of these high-altitude environments and is closely associated with the *Lupinus* plants that its larvae feed on.

The distribution of insects is closely tied to the availability of food resources, particularly from plants (Hodkinson, 2005). Different insect species are associated with specific types of vegetation, and changes in the plant communities with altitude influence the composition of insect communities. For example, the distribution of the Edith's checkerspot butterfly (*Euphydryas editha*) in the Sierra Nevada is closely linked to the availability of its host plants, such as Indian paintbrush (*Castilleja*) and plantain (*Plantago*). This butterfly is typically found in meadows and open areas where these plants are abundant. Another example is the Andean bear beetle (*Mecistophylla thalia*), which is closely associated with *Polylepis* trees that dominate the Peruvian high-altitude montane forests (Lamas, 2004).

The distribution of insects on mountains is also influenced by microclimates created by variations in slope aspect, vegetation cover, and proximity to water sources (Hodkinson, 2005). For example, on Mount Kilimanjaro in Tanzania, different insect communities are found on the drier, warmer southern slopes compared to the wetter northern slopes. In the Sierra Nevada, the diversity and abundance of insect species, such as the Sierra Nevada Blue butterfly (*Plebejus podarce*), vary significantly between the drier, sunnier eastern slopes and the wetter, more vegetated western slopes.

Insect declines have been reported in many large-scale studies, though these are mainly from Western Europe and North America (Wagner, 2020). In Europe, the longest time series existing for insects is butterfly monitoring data from the United Kingdom. Butterflies are good insect models as they are easy to identify, popular, and they react quickly to environmental changes (Thomas, 2005). The main drivers of change, in line with global changes elsewhere, are habitat loss/degradation, climate change, and chemical pollution (Warren et al., 2021). The impacts caused by the abandonment of traditional management on Southern European mountain landscapes, harbouring a high diversity of both habitats and species, are not fully understood, though abandonment is leading to a loss of habitats dominated by sparse vegetation, thereby giving rise to a successional progression towards forest habitats with consequent changes in butterfly species composition (Mora et al., 2023).

3.10.3 Endemic insects

Due to short and metamorphic life cycles, insects are highly adapted to life on a mountain and have a high degree of endemism, particularly on isolated mountains such as the Sierra Nevada de Santa Marta in northern Colombia; separate from the Andes, it is one of the most biodiverse mountains in the world (Fortier et al., 2025), with high levels of endemism, particularly among insects. For example, the Santa Marta montane butterfly (*Parides iphidamas*) is endemic to this region and closely linked to the variety of plant species, particularly those that serve as larval host plants. The Tepuis, or table-top mountains, of Venezuela and Guyana are another unique mountain system in Latin America, characterised by their isolated summits, distinct insect fauna, and many endemic species such as the Roraima grasshopper (*Apteropeoedes roraimae*), only found on the summit of Mount Roraima (Lamas, 2004). This grasshopper is adapted to the cool, moist conditions of the Tepui summit and feeds on the sparse vegetation. The isolation of the Tepuis has led to the evolution of a unique insect fauna, with many species found on no other mountain. The Central American Volcanic Arc, stretching from southern Mexico through Guatemala, El Salvador, Honduras, and Nicaragua, is a region of exceptionally high insect diversity (Kohlmann et al., 2024). The mountainous terrain and volcanic soils support a wide variety of plants, which in turn support diverse insect communities. For example, the high-altitude forests of Guatemala's Sierra de los Cuchumatanes are home to the rare Guatemalan silk moth (*Rothschildia orizaba*). This moth is adapted to the cooler temperatures and high humidity of the montane forests, where its larvae feed on the leaves of various hardwood trees found primarily in undisturbed forests. The Alpine longhorn beetle (*Rosalia alpina*), found in the Alps at higher altitudes, in old-growth forests where it lays its eggs in decaying wood, is an indicator species of well-preserved alpine ecosystems and is protected under European conservation laws.

3.10.4 Insect adaptations and survival strategies

Similar to birds, insects at high altitudes have evolved various physiological adaptations to cope with low temperatures and low oxygen levels. These adaptations include antifreeze proteins that prevent ice formation in tissues, increased metabolic rates that generate body heat, and changes in body size that reduce heat loss. For example, the high-altitude beetle *Pterostichus latemarginatus*, found in the Peruvian Andes, has evolved to survive in temperatures that regularly drop below freezing at night (Lamas, 2004). This beetle produces glycerol, an antifreeze, in its body fluids to prevent freezing during cold nights. Some species also have a reduced metabolic rate to conserve energy in cold environments (Golding et al., 2023). The freeze-tolerant beetles of the genus *Pterostichus*, found in the Arctic and subarctic mountains, produce glycerol and other cryoprotectants that lower the freezing point of their bodily fluids, allowing them to survive in temperatures as low as −20°C (Sømme et al., 1979).

Insects, being one of the most diverse groups of organisms, have colonised almost every habitat within mountainous regions, where they have evolved unique adaptations to survive (Hodkinson, 2005). Behavioural adaptation, such as basking in the sun to regulate body temperature, is commonly used. For example, the Himalayan bumblebee (*Bombus himalayanus*) is adapted to life at altitudes of up to 5600 m, making it one of the highest-flying insects in the world (Singh et al., 2024). This species has a dense coat of hair to insulate against the cold and prevent moisture loss, and a metabolic rate that allows it to remain active in low temperatures. Its ability to forage at such high altitudes is critical for the pollination of alpine flowers and is another amazing example of co-adaptation. Similarly, the Alpine bumblebee (*Bombus alpinus*), found in the European Alps, often basks in the sun to raise its body temperature before foraging. Many mountain insects will migrate to lower altitudes during colder periods or seek shelter in crevices to avoid the harshest environmental conditions (Hodkinson, 2005).

The life cycles of many mountain insects are synchronised with the short growing seasons typical of high-altitude environments (Harmsen, 1989). These insects often have extended larval stages and undergo a diapause (a period of suspended development), or have rapid development cycles to take advantage of the brief periods of favourable conditions. For example, the Andean pierid butterfly *Catasticta incerta incerta*, found in the high-altitude grasslands of the southern Andes, has a life cycle that is tightly synchronised with the short summer season. Its larvae develop quickly during the brief summer months, feeding on grasses before pupating and emerging as adults in time to reproduce before the onset of winter (Padrón & Vélez, 2020). The larvae of the Himalayan butterfly *Polyommatus coridon* overwinter in the soil and emerge as adults in the brief summer, coinciding with the flowering of their host plants. This synchronisation ensures that the butterflies can complete their life cycle within the short growing season of the high Himalayas (Mehra et al., 2017).

3.11 Species distribution modelling

As emphasised earlier, the concept of niche is fundamental to understanding the relationship between environment and species distributions. Species distribution modelling (SDM) is a technique used to explore the relationship between geographical occurrences of species and corresponding environmental variables (Wang et al., 2022). Since both abiotic environments and biotic interactions ultimately drive species distributions, joint species distribution models (JSDMs) have emerged as extensions of SDMs to simultaneously explore interspecific interactions and the response to an abundance of environmental variables (Abrego et al., 2020). Owing to the increasing demands for models in biodiversity assessments on mountains, the big challenge for projecting species responses to future global change scenarios is to incorporate models with more necessary local- and landscape-level predictors that also capture biotic interspecific interactions.

The vast majority of SDM work is correlative, based on fitting statistical models to datasets of species occurrences and making predictions from these models. This approach is commonly used to predict new situations (space, time, and environments), but the ability of these models to be transferred to new conditions is rarely tested. Correlative models can fail to transfer owing to invalid assumptions that distributions represent niches (e.g., biotic interactions, dispersal limitation, and non-equilibrium distributions), and also due to modelling failures (e.g., chance correlations and extrapolation) (Vesk et al., 2021). Plant functional traits can be used as predictors of species distribution along environmental gradients to provide a route for generalised ecological inference – how traits influence species occurrence. However, SDM performance is hampered by spatial inaccuracy of occurrence data along with low numbers of records, spatial biases of where records are from, and a lack of environmental data at the right resolution (Vesk et al., 2021).

Species distribution models need to be calibrated with environmental variables at appropriate spatial 'resolution' or 'grain size' for sound predictive outputs (Platts et al., 2014). This resolution needs to be consistent with the resolution or coordinate accuracy of the species distribution data, as well as capturing the relevant biological scale of the system and species studied. Spatially-explicit climatic layers are commonly available at coarse resolution (e.g., 1 km or greater) over large extents by interpolating measurements from meteorological stations. However, these layers do not adequately capture the heterogeneous microenvironments of mountain landscapes and do not accurately represent microclimatic conditions for plants (Platts et al., 2010). Moreover, the meteorological stations on which they rely are often found at low or mid elevations, and relatively far from each other, with very few recordings at high elevations. In Switzerland, these climatic layers can be found at a 25 m resolution, but are based on interpolations at the national scale (Buri et al., 2017). Although soil is an important component of the environment for plants, only a few studies that focused on tree or shrub species have successfully incorporated soil variables into plant SDMs to date (Buri et al., 2017).

Differences in multifaceted biodiversity distributions found between finer and coarser resolutions indicate strong consequences for ecological studies implementing SDMs; too coarse a resolution may lead to misleading predicted species distributions. For example, grains $\geq$ 1 km in SDM conservation studies of sessile species are relatively common (Chauvier et al., 2022). The 'wrong' or inappropriate grain size misleads related decision-making; e.g., species reintroduction, conservation planning, or ecosystem services protection (Chauvier et al., 2022). Realistic SDMs should normally include additional key local-scale ecological factors, biotic interactions, and dispersal limitations, being the most crucial ones (Chauvier et al., 2022), whilst accounting for dispersal is increasingly feasible through supplementary modelling processes, such as those dealt with in eco-evolutionary mechanistic models (Hagen et al., 2021).

The role of climate in constraining plant species distributions is a foundational ecological concept that has taken on new urgency in light of anthropogenic climate change (Rödder et al., 2021; Stark & Fridley, 2022). Many SDMs are currently being used to explore the impacts of multiple issues, such as climate change, conservation planning and strategies, invasive alien species, impacts of climate change and ecological problems (Qazi et al., 2022). The correlative approach was the most common approach, with MaxEnt one of the most frequently used models, but the majority of studies failed to present uncertainties (Qazi et al., 2022). It has long been documented that mountain biota will move upwards as a result of global climate warming (e.g., Braun-Blanquet, 1957). Individual species will move at different speeds and possess species-specific potentials to use migration paths (Gottfried et al., 1999; Rödder et al., 2021). The most important features of migration paths in high mountains are relief and surface texture. The manner in which organisms can adapt to follow climatic changes depends intrinsically on their ability to move, i.e., to overcome topographical migration barriers. 'Winners' and 'losers' of the upward migration effect will emerge (Gottfried et al., 1999). Accurately monitoring rare plant populations is critical in identifying threats to occurrences and establishing long-term population trends (Reckling et al., 2021).

In such a mountainous environment, most improvements to plant SDMs have resulted from the addition of edaphic predictor parameters. However, the right balance must be found between the time spent creating new predictors and the predictive improvement in plant SDMs. Accurately modelling precipitation at very high resolution in mountainous landscapes has yet to be resolved, hence its exclusion from the SDM (Buri et al., 2017). The lack of consideration for the intricacies of mountain habitats in the modelling procedures – especially the choice of predictor variables – often obscures the ecological value of their findings (Oke & Thompson, 2015; Platts et al., 2010). Mountain niches are influenced by microtopography; small differences in microtopography will have significant effects on soil temperature, freeze-thaw cycles, snow drifting, and wind, all of which affect mountain niche properties (Oke & Thompson, 2015). Thus, geographic variables likely serve as surrogates for elusive climatic factors. For this reason, it is crucial to utilise a well-constructed set of predictor variables that are relevant to the biology of species that occupy unique habitats like those of mountain regions (Platts et al., 2008). With the increasing availability of data sets on species occurrence and environmental variables (e.g., from remote sensing), it is now possible to build models over wide geographic areas at a biologically meaningful resolution that matches the scale at which species perceive habitat and respond to environmental circumstances. These more accurate, finer-grained models are crucial to avoid the caveat of overestimating the effect of large-scale forces like climate, and to better integrate local-driving forces such as functional species–habitat relationships into projections (Barras et al., 2021). With the combination of results from different disciplines (i.e., ecology, vegetation science, biogeography, environmental sciences, and geosciences), we can better differentiate, disentangle, and understand underlying factors controlling current distribution, species ranges, and compositions to derive forecasts under climate change scenarios (Bobrowski et al., 2021).

For example, *Pinus armandii* (White pine) populations at the rear edge of the distribution range face extirpation, such as in the Tsinling Mountains and Daba Mountains, and some are gradually approaching the upper limit. SDMs have become a major focus of large-scale ecology and biogeography studies, and these models are widely applied to replicate the habitat ranges of plants and insects, assess the effects of global environmental change on species and ecosystems, evaluate the risk of species invasion and spread, and manage and plan the habitats for species, communities, and ecosystems (Ning et al., 2021). The consequences of climate change may result in three possible fates for forest tree populations: to persist through migration, adapt to new conditions, or go locally extinct. Therefore, all three scenarios may come true in the response of *P. armandii* to climate change. For instance, in the Tsinling Mountains and Daba Mountains in China, the range of *P. armandii* is projected to persist through a shift by tracking and adapting to new ecological niches, while populations will likely face extirpation in the Hengduan Mountains (Ning et al., 2021).

Modelling ecological niches and species distribution across vast areas in remote, high mountain regions like the Himalayas constitutes a challenging task. The main obstacles arise from large-scale sampling of species occurrence (and in rare cases absences) and acquisition of sufficiently high-quality, fine-scale environmental predictors. Sampling deficits can be primarily attributed to the difficult accessibility of the terrain and the geographical extent of the Himalayan mountain system. However, the quality and quantity of occurrence data determine the significance of the model. The basic assumption of the model is based on uniform or random sampling of the study area. The selection of an appropriate predictor variable set depends not only on the availability of high-quality environmental data, but also on the spatial scale, the autecology of the species, and the degree of complexity of the ecological niche (Bobrowski et al., 2021; Peterson et al., 2011). Often, the acquisition of sufficient numbers of species' occurrences, the evaluation of environmental data quality, and the ability to account for the species ecological requirements are neglected. This, in turn, could lead to flawed predictions, misinterpretations, and biased implications, especially when nature conservation strategies and environmental management are in focus.

3.12 Conclusions on mountain plant and animal diversity

Mountain systems are diverse at all ecological levels. The diverse climatic, geological landscape, and fire regimes – both past and present – all contribute to their ecological diversity. Understanding these mountain ecosystem foundations is key to the more applied aspects of mountain systems explored in Chapters 7 and 8. Indeed, understanding the underlying biological foundations is essential for identifying both the challenges facing mountain ecosystems and potential solutions – we have to know the mountain ecosystem to work with it, rather than against it. There are many knowledge gaps and areas of uncertainty about species' response to current and future climate change. In particular, we know very little about the details of how biotic interactions (e.g., between plants and their competitors, herbivores, pollinators, pathogens, and soil microbes) will be affected by altered climates, and how these interactions will influence individual species and communities. This type of information, together with dynamic models, can be used to explore how, at what rate, and where species may change in the future (Chapter 7). Accounting for the spatio-temporal connectivity of mountain biota to mountain environments will provide a better understanding of the patterns we have and a basis for the effective conservation of mountain biodiversity in response to climate, social, and economic changes.

References

Abrego, N., Roslin, T., Huotari, T., Tack, A. J. M., Lindahl, B. D., Tikhonov, G., Somervuo, P., Schmidt, N. M., & Ovaskainen, O. (2020). Accounting for environmental variation in co-occurrence modelling reveals the importance of positive interactions in root-associated fungal communities. *Molecular Ecology, 29*(14), 2736–2746. https://doi.org/10.1111/mec.15516.

Ahrends, A., Rahbek, C., Bulling, M. T., Burgess, N. D., Platts, P. J., Lovett, J. C., Kindemba, V. W., Owen, N., Sallu, A. N., Marshall, A. R., Mhoro, B. E., Fanning, E., & Marchant, R. (2011). Conservation and the botanist effect. *Biological Conservation, 144*(1), 131–140. https://doi.org/10.1016/j.biocon.2010.08.008.

Antonelli, A., Kissling, W. D., Flantua, S. G. A., Bermúdez, M. A., Mulch, A., Muellner-Riehl, A. N., Kreft, H., Linder, H. P., Badgley, C., Fjeldså, J., Fritz, S. A., Rahbek, C., Herman, F., Hooghiemstra, H., & Hoorn, C. (2018). Geological and climatic influences on mountain biodiversity. *Nature Geoscience, 11*(10), 718–725. https://doi.org/10.1038/s41561-018-0236-z.

Ausilio, G., Sand, H., Månsson, J., Mathisen, K. M., & Wikenros, C. (2021). Ecological effects of wolves in anthropogenic landscapes: The potential for trophic cascades is context-dependent. *Frontiers in Ecology and Evolution, 8*. https://doi.org/10.3389/fevo.2020.577963.

Barras, A. G., Braunisch, V., & Arlettaz, R. (2021). Predictive models of distribution and abundance of a threatened mountain species show that impacts of climate change overrule those of land use change. *Diversity and Distributions, 27*(6), 989–1004. https://doi.org/10.1111/ddi.13247.

Beall, C. M., & Andean, T. (2006). Ethiopian patterns of adaptation to high-altitude hypoxia. *Integrative and Comparative Biology, 46*(1), 18–24. https://doi.org/10.1093/icb/icj004.

Berrones, G., Crespo, P., Ochoa-Sánchez, A., Wilcox, B. P., & Célleri, R. (2022). Importance of fog and cloud water contributions to soil moisture in the Andean Páramo. *Hydrology, 9*(4), 54. https://doi.org/10.3390/hydrology9040054.

Birks, H. J. B. (2021). High-elevation limits and the ecology of high-elevation vascular plants: Legacies from Alexander von Humboldt1. *Frontiers of Biogeography, 13*(3), 1–18. https://doi.org/10.21425/F5FBG53226.

Bobrowski, M., Weidinger, J., Schwab, N., & Schickhoff, U. (2021). Searching for ecology in species distribution models in the Himalayas. *Ecological Modelling, 458*, 109693. https://doi.org/10.1016/j.ecolmodel.2021.109693.

Borah, D., Gap, N., & Singh, R. K. (2020). Pedicularis khoiyangii (Orobanchaceae), a new species from the Eastern Himalaya, India. *Phytotaxa, 430*(4), 287–293. https://doi.org/10.11646/phytotaxa.430.4.3.

Braun-Blanquet, J. (1957). Ein Jahrhundert Florenwandel am Piz Linard (3414 m). In *Extrait du Bulletin Jardin Botanique Etat Bruxelles, volumes jubilaire Walter Robyns* (pp. 221–232). Station Internationale de Geobotanique Mediterraneenne et Alpine.

Brehm, G., Colwell, R. K., & Kluge, J. (2007). The role of environment and mid-domain effect on moth species richness along a tropical elevational gradient. *Global Ecology and Biogeography, 16*(2), 205–219. https://doi.org/10.1111/j.1466-8238.2006.00281.x.

Brochmann, C., Gizaw, A., Chala, D., Kandziora, M., Eilu, G., Popp, M., Pirie, M. D., & Gehrke, B. (2022). History and evolution of the afroalpine flora: In the footsteps of Olov Hedberg. *Alpine Botany, 132*(1), 65–87. https://doi.org/10.1007/s00035-021-00256-9.

Buri, A., Cianfrani, C., Pinto-Figueroa, E., Yashiro, E., Spangenberg, J. E., Adatte, T., Verrecchia, E., Guisan, A., & Pradervand, J. N. (2017). Soil factors improve predictions of plant species distribution in a mountain environment. *Progress in Physical Geography, 41*(6), 703–722. https://doi.org/10.1177/0309133317738162.

Cain, S. A. (1944). *Foundations of plant geography*. Harper and Brothers.

Cárdenas, M. F., Tobón, C., & Buytaert, W. (2017). Contribution of occult precipitation to the water balance of páramo ecosystems in the Colombian Andes. *Hydrological Processes, 31*(24), 4440–4449. https://doi.org/10.1002/hyp.11374.

Chauvier, Y., Descombes, P., Guéguen, M., Boulangeat, L., Thuiller, W., & Zimmermann, N. E. (2022). Resolution in species distribution models shapes spatial patterns of plant multifaceted diversity. *Ecography, 2022*(10). https://doi.org/10.1111/ecog.05973.

Colwell, R. K., Rahbek, C., & Gotelli, N. J. (2004). The mid-domain effect and species richness patterns: What have we learned so far? *The American Naturalist, 163*(3). https://doi.org/10.1086/382056.

Crawford, R. M. M. (2005). Mountain ecosystems. In *Peripheral plant population survival in polar regions* (pp. 43–76). Springer Berlin Heidelberg. https://doi.org/10.1007/3-540-27365-4_2.

Das, G. N., Gayen, S., Ali, M., Jaiswal, R. K., Lenin, E. A., & Chandra, K. (2018). *Insecta: Lepidoptera (Butterflies)* (pp. 611–650). Faunal Diversity of Indian Himalaya.

Dewan, S., & Acharya, B. K. (2024). Rapoport's rule explains the range size distribution of butterflies along the Eastern Himalayan elevation gradient. *Biotropica, 56*(3). https://doi.org/10.1111/btp.13311.

Di Biase, L., Pace, L., Mantoni, C., & Fattorini, S. (2021). Variations in plant richness, biogeographical composition, and life forms along an elevational gradient in a mediterranean mountain. *Plants, 10*(10), 2090. https://doi.org/10.3390/plants10102090.

Dlamini, L. X., Lévêque, J., Thevenot, M., Feig, G. T., Mathieu, O., & Kotzé, E. (2025). Afforestation and wildfire-induced grassland degradation: Impact on soil carbon dynamics in temperate grasslands, Drakensberg, South Africa. *Catena, 252*. https://doi.org/10.1016/j.catena.2025.108886.

Du, Q., Sun, Y., Guan, Q., Pan, N., Wang, Q., Ma, Y., Li, H., & Liang, L. (2022). Vulnerability of grassland ecosystems to climate change in the Qilian Mountains, northwest China. *Journal of Hydrology, 612*, 128305. https://doi.org/10.1016/j.jhydrol.2022.128305.

Dufour, A., Gadallah, F., Wagner, H. H., Guisan, A., & Buttler, A. (2006). Plant species richness and environmental heterogeneity in a mountain landscape: Effects of variability and spatial configuration. *Ecography, 29*(4), 573–584. https://doi.org/10.1111/j.0906-7590.2006.04605.x.

Eastwood, E. R. I. C., & Rider, G. C. (1965). Some radar measurements of the altitude of bird flight. *British Birds, 58*(10), 393–426.

Economo, E. P., Narula, N., Friedman, N. R., Weiser, M. D., & Guénard, B. (2018). Macroecology and macroevolution of the latitudinal diversity gradient in ants. *Nature Communications, 9*(1). https://doi.org/10.1038/s41467-018-04218-4.

Elliott, G. P., Bailey, S. N., & Cardinal, S. J. (2020). Hotter drought as a disturbance at upper treeline in the southern rocky mountains. *Annals of the American Association of Geographers, 111*(3), 756–770. https://doi.org/10.1080/24694452.2020.1805292.

Ellis, C. G., Mathieu-Costello, O., Potter, R. F., MacDonald, I. C., & Groom, A. C. (1990). Effect of sarcomere length on total capillary length in skeletal muscle: In vivo evidence for longitudinal stretching of capillaries. *Microvascular Research, 40*(1), 63–72. https://doi.org/10.1016/0026-2862(90)90008-f.

Esquerré, D., Brennan, I. G., Catullo, R. A., Torres-Pérez, F., & Keogh, J. S. (2019). How mountains shape biodiversity: The role of the Andes in biogeography, diversification, and reproductive biology in South America's most species-rich lizard radiation (Squamata: Liolaemidae). *Evolution, 73*(2), 214–230. https://doi.org/10.1111/evo.13657.

Faraci, F. M. (1991). Adaptations to hypoxia in birds: How to fly high. *Annual Review of Physiology, 53*(1), 59–70. https://doi.org/10.1146/annurev.ph.53.030191.000423.

Faraci, F. M. (1986). Circulation during hypoxia in birds. *Comparative Biochemistry and Physiology – Part A: Physiology, 85*(4), 613–620. https://doi.org/10.1016/0300-9629(86)90270-7.

Fattorini, S. (2014). Disentangling the effects of available area, mid-domain constraints, and species environmental tolerance on the altitudinal distribution of tenebrionid beetles in a Mediterranean area. *Biodiversity and Conservation, 23*(10), 2545–2560. https://doi.org/10.1007/s10531-014-0738-y.

Fedyń, I., Sobociński, W., Czyżowicz, S. Ł, Wyka, J., & Ciach, M. Ł (2024). Ecosystem engineers cause biodiversity spill-over: Beavers are associated with breeding bird assemblages on both wetlands and adjacent terrestrial habitats. *Science of The Total Environment, 950*, 175166. https://doi.org/10.1016/j.scitotenv.2024.175166.

Fortier, R. P., Kullberg, A. T., Aragón, L., Bernal-Escobar, M., Contreras, M. P., Hurtado, C. P., Lautenschlager, L., & Feeley, K. J. (2025). Plant diversity and endemism of the Sierra Nevada de Santa Marta, Colombia, with floristic comparisons to surrounding mountains. *Annals of the Missouri Botanical Garden, 110*, 71–87. https://doi.org/10.3417/2025921.

Gehrke, B., & Linder, H. P. (2014). Species richness, endemism and species composition in the tropical Afroalpine flora. *Alpine Botany, 124*(2), 165–177. https://doi.org/10.1007/s00035-014-0132-0.

Gehrke, B., & Linder, H. P. (2009). The scramble for Africa: Pan-temperate elements on the African high mountains. *Proceedings of the Royal Society B: Biological Sciences, 276*(1667), 2657–2665. https://doi.org/10.1098/rspb.2009.0334.

Gillson, L., Dawson, T. P., Jack, S., & McGeoch, M. A. (2013). Accommodating climate change contingencies in conservation strategy. *Trends in Ecology and Evolution, 28*(3), 135–142. https://doi.org/10.1016/j.tree.2012.10.008.

Githumbi, E. N., Finch, J., Kinyanjui, R. N., Courtney-Mustaphi, C., Musili, P., Rucina, S., Lejju, J., Liljenberg, P., & Marchant, R. (2025). Late quaternary montane forest dynamics from equatorial East Africa: A biome perspective. *Journal of Biogeography, 52*(9). https://doi.org/10.1111/jbi.15173.

Golding, D., Rupp, K. L., Sustar, A., Pratt, B., & Tuthill, J. C. (2023). Snow flies self-amputate freezing limbs to sustain behavior at sub-zero temperatures. *Current Biology, 33*(21), 4549. https://doi.org/10.1016/j.cub.2023.09.002.

Gottfried, M., Pauli, H., Reiter, K., & Grabherr, G. (1999). A fine-scaled predictive model for changes in species distribution patterns of high mountain plants induced by climate warming. *Diversity and Distributions, 5*(6), 241–251. https://doi.org/10.1046/j.1472-4642.1999.00058.x.

Gray, M. (2022). Geodiversity and the ecosystem approach. *Parks Stewardship Forum, 38*(1). https://doi.org/10.5070/p538156117.

Griffiths, A. R., Silman, M. R., Farfán Rios, W., Feeley, K. J., García Cabrera, K., Meir, P., Salinas, N., & Dexter, K. G. (2021). Evolutionary heritage shapes tree distributions along an Amazon-to-Andes elevation gradient. *Biotropica, 53*(1), 38–50. https://doi.org/10.1111/btp.12843.

Grytnes, J. A., Kapfer, J., Jurasinski, G., Birks, H. H., Henriksen, H., Klanderud, K., Odland, A., Ohlson, M., Wipf, S., & Birks, H. J. B. (2014). Identifying the driving factors behind observed elevational range shifts on European mountains. *Global Ecology and Biogeography, 23*(8), 876–884. https://doi.org/10.1111/geb.12170.

Hagen, O., Flück, B., Fopp, F., Cabral, J. S., Hartig, F., Pontarp, M., Rangel, T. F., Pellissier, L. oïc, & Tanentzap, A. J. (2021). gen3sis: A general engine for eco-evolutionary simulations of the processes that shape Earth's biodiversity. *PLOS Biology, 19*(7), e3001340. https://doi.org/10.1371/journal.pbio.3001340.

Hannah, L., Flint, L., Syphard, A. D., Moritz, M. A., Buckley, L. B., & McCullough, I. M. (2014). Fine-grain modeling of species' response to climate change: Holdouts, stepping-stones, and microrefugia. *Trends in Ecology and Evolution, 29*(7), 390–397. https://doi.org/10.1016/j.tree.2014.04.006.

Harmsen, R. (1989). Recent evolution and dispersal of insects on the East African mountains. In *Quaternary and environmental research on east african mountains* (pp. 245–256). CRC Press.

Hedberg, O. (1961). The phytogeographical position of the afroalpine flora. *Recent Advances in Botany, 1*, 914–919.

Hedberg, O. (1964). Features of afroalpine plant ecology. *Acta Phytogeographica Suecica, 49*, 1–144.

Hemp, A., Zimmermann, R., Remmele, S., Pommer, U., Berauer, B., Hemp, C., & Fischer, M. (2017). Africa's highest mountain harbours Africa's tallest trees. *Biodiversity and Conservation, 26*(1), 103–113. https://doi.org/10.1007/s10531-016-1226-3.

Hirschhofer, S., Liechti, F., Ranacher, P., Weibel, R., & Schmid, B. (2024). High-intensity bird migration along Alpine valleys calls for protective measures against anthropogenically induced avian mortality. *Remote Sensing in Ecology and Conservation, 10*(3), 360–373. https://doi.org/10.1002/rse2.377.

Hodkinson, I. D. (2005). Terrestrial insects along elevation gradients: Species and community responses to altitude. *Biological Reviews, 80*(3), 489–513. https://doi.org/10.1017/s1464793105006767.

Holtmeier, F. K. (2009). In *Mountain timberlines: Ecology, patchiness, and dynamics. Advances in global change research: 36* (pp. 1–419). Springer Dordrecht. https://www.springer.com/series/5588.

Hu, A., Wang, J., Sun, H., Niu, B., Si, G., Wang, J., Yeh, C. F., Zhu, X., Lu, X., Zhou, J., Yang, Y., Ren, M., Hu, Y., Dong, H., & Zhang, G. (2020). Mountain biodiversity and ecosystem functions: Interplay between geology and contemporary environments. *ISME Journal, 14*(4), 931–944. https://doi.org/10.1038/s41396-019-0574-x.

Janzen, D. H. (1967). Why mountain passes are higher in the tropics. *The American Naturalist, 101*(919), 233–249. https://doi.org/10.1086/282487.

Jensen, T. H., Weber, R. E., Fermi, G., & Tame, J. (1991). Adaptation of bird heamoglobins to high altitudes: demonstration of molecular mechanism by protein engineering. *Proceedings of the National Academy of Sciences, 88*(15), 6519–6522.

Jump, A. S., Carr, M., Ahrends, A., & Marchant, R. (2014). Genetic divergence during long-term isolation in highly diverse populations of tropical trees across the Eastern Arc mountains of Tanzania. *Biotropica, 46*(5), 565–574. https://doi.org/10.1111/btp.12139.

Kanai, Y. (2000). Migration of demoiselle cranes in Asia based on satellite tracking and fieldwork. *Global Environmental Research, 4*, 143–153.

Kirdyanov, A. V., Hagedorn, F., Knorre, A. A., Fedotova, E. V., Vaganov, E. A., Naurzbaev, M. M., Moiseev, P. A., & Rigling, A. (2012). 20th century tree-line advance and vegetation changes along an altitudinal transect in the Putorana Mountains, northern Siberia. *Boreas, 41*(1), 56–67. https://doi.org/10.1111/j.1502-3885.2011.00214.x.

Kohlmann, B., Salomão, R. P., & Solís, Á. (2024). New World dung beetle (Coleoptera: Scarabaeidae: Scarabaeinae) colonization of a recent Miocene insular territory: The case of Costa Rica. *Ecology and Evolution, 14*(6). https://doi.org/10.1002/ece3.11436.

Körner, C. (2007). The use of 'altitude' in ecological research. *Trends in Ecology and Evolution, 22*(11), 569–574. https://doi.org/10.1016/j.tree.2007.09.006.

Körner, C., & Paulsen, J. (2004). A world-wide study of high altitude treeline temperatures. *Journal of Biogeography, 31*(5), 713–732. https://doi.org/10.1111/j.1365-2699.2003.01043.x.

Körner, C. (2012). *Alpine Treelines: Functional Ecology of the Global High Elevation Tree Limits*. Springer.

Körner, C., & Kèorner, C. (1999). *Alpine plant life: Functional plant ecology of high mountain ecosystems*. Springer.

Lamas, G. (Ed.). (2004). *Atlas of neotropical lepidoptera: Checklist: Part 4A, Hesperioidea-Papilionoidea*. Association for Tropical Lepidoptera.

Lausch, A., Schaepman, M. E., Skidmore, A. K., Truckenbrodt, S. C., Hacker, J. M., Baade, J., Bannehr, L., Borg, E., Bumberger, J., Dietrich, P., Gläßer, C., Haase, D., Heurich, M., Jagdhuber, T., Jany, S., Krönert, R., Möller, M., Mollenhauer, H., Montzka, C., & ..., Thiel, C. (2020). Linking the remote sensing of geodiversity and traits relevant to biodiversity—Part II: Geomorphology, terrain and surfaces. *Remote Sensing, 12*(22), 1–61. https://doi.org/10.3390/rs12223690.

Li, J., Li, Q., Wu, Y., Ye, L., Liu, H., Wei, J., & Huang, X. (2021). Mountains act as museums and cradles for hemipteran insects in China: Evidence from patterns of richness and phylogenetic structure. *Global Ecology and Biogeography, 30*(5), 1070–1085. https://doi.org/10.1111/geb.13276.

Li, L., Zhang, Y., Wu, J., Li, S., Zhang, B., Zu, J., Zhang, H., Ding, M., & Paudel, B. (2019). Increasing sensitivity of alpine grasslands to climate variability along an elevational gradient on the Qinghai-Tibet Plateau. *Science of The Total Environment, 678*, 21–29. https://doi.org/10.1016/j.scitotenv.2019.04.399.

Liang, E., Wang, Y., Piao, S., Lu, X., Camarero, J. J., Zhu, H., Zhu, L., Ellison, A. M., Lu, X., Liang, E., Wang, Y., Babst, F., & Camarero, J. J. (2021). Mountain treelines climb slowly despite rapid climate warming. *Global Ecology and Biogeography, 30*(1), 305–315.

Linnaeus, C. (1781). On the increase of the habitable earth. *Amoenitates Academicae, 2*, 17–27.

Lovett, J. C., Marshall, A. R., & Carr, J. (2006). Changes in tropical forest vegetation along an altitudinal gradient in the Udzungwa Mountains National Park, Tanzania. *African Journal of Ecology, 44*(4), 478–490. https://doi.org/10.1111/j.1365-2028.2006.00660.x.

Lu, L. M., Mao, L. F., Yang, T., Ye, J. F., Liu, B., Li, H. L., Sun, M., Miller, J. T., Mathews, S., Hu, H. H., Niu, Y. T., Peng, D. X., Chen, Y. H., Smith, S. A., Chen, M., Xiang, K. L., Le, C. T., Dang, V. C., Lu, A. M., & ..., Chen, Z. D. (2018). Evolutionary history of the angiosperm flora of China. *Nature, 554*(7691), 234–238. https://doi.org/10.1038/nature25485.

Lu, X., Ke, D. H., Zeng, X. H., & Yu, T. L. (2009). Reproductive ecology of two sympatric Tibetan snowfinch species at the edge of their altitudinal range: Response to more stressful environments. *Journal of Arid Environments, 73*(12), 1103–1108. https://doi.org/10.1016/j.jaridenv.2009.06.011.

Lu, X., Liang, E., Wang, Y., Babst, F., & Camarero, J. J. (2021). Mountain treelines climb slowly despite rapid climate warming. *Global Ecology and Biogeography, 30*(1), 305–315. https://doi.org/10.1111/geb.13214.

Lundberg, J. G., Marshall, L. G., Guerrero, J., Horton, B., Malabarba, M., & Wesselingh, F. (1998). The stage for Neotropical fish diversification: A history of tropical South American rivers. *Phylogeny and Classification of Neotropical Fishes,*, 27, 13–48.

Määttänen, A. M., Kemppinen, J., & Hällfors, M. (2024). *Geodiversity is an inseparable but underutilized part of connectivity in nature conservation amid climate change.* Authorea https://doi.org/10.22541/au.172977173.33452815/v1.

Kumar, M. (2019). Macroecological patterns and drivers of Himalayan plant species diversity and distribution through the Ages. *Frontiers of Biogeography, 11*(2). https://doi.org/10.21425/f5fbg38754.

Mann, J. T., Mann, E. A., & Clark, A. B. (2021). Social information affects prospecting, but not settlement, by Black-and-white Warblers (Mniotilta varia). *Ibis, 163*(3), 798–811. https://doi.org/10.1111/ibi.12921.

Marchese, C. (2015). Biodiversity hotspots: A shortcut for a more complicated concept. *Global Ecology and Conservation, 3*, 297–309. https://doi.org/10.1016/j.gecco.2014.12.008.

Marshall, A. R., Platts, P. J., Gereau, R. E., Kindeketa, W., Kang'ethe, S., & Marchant, R. (2012). The genus Acacia (Fabaceae) in East Africa: Distribution, diversity and the protected area network. *Plant Ecology and Evolution, 145*(3), 289–301. https://doi.org/10.5091/plecevo.2012.597.

Martin, K. (2001). Wildlife in alpine and sub-alpine habitats. In D. H. Johnson, & T. A. O'Neil (Eds.), *Wildlife-habitat relationships in Oregon and Washington* (pp. 285–310). Oregon State University Press. http://www.forestry.ubc.ca/alpine/docs/alpecol.pdf.

McCracken, K. G., Bulgarella, M., Johnson, K. P., Kuhner, M. K., Trucco, J., Valqui, T. H., Wilson, R. E., & Peters, J. L. (2009). Gene flow in the face of countervailing selection: Adaptation to high-altitude hypoxia in the a hemoglobin subunit of yellow-billed pintails in the Andes. *Molecular Biology and Evolution, 26*(4), 815–827. https://doi.org/10.1093/molbev/msp007.

Mehra, D., Kirti, J. S., & Sidhu, A. K. (2017). Biodiversity and conservation status of Butterflies in Western Himalaya, India: An appraisal. *Journal of Entomology and Zoology Studies, 5*(5), 1171–1177.

Mittermeier, R. A., Gil, P. R., Hoffmann, M., Pilgrim, J., Brooks, T., Mittermeier, C. G., & Fonseca, G. A. B. (2004). *Hotspots revisited: Earth's biologically richest and most endangered ecoregions.* CEMEX.

Mora, A., Wilby, A., & Menéndez, R. (2023). South European mountain butterflies at a high risk from land abandonment and amplified effects of climate change. *Insect Conservation and Diversity, 16*(6), 838–852. https://doi.org/10.1111/icad.12676.

Muellner-Riehl, A. N., Schnitzler, J., Kissling, W. D., Mosbrugger, V., Rijsdijk, K. F., Seijmonsbergen, A. C., Versteegh, H., & Favre, A. (2019). Origins of global mountain plant biodiversity: Testing the 'mountain-geobiodiversity hypothesis. *Journal of Biogeography, 46*(12), 2826–2838. https://doi.org/10.1111/jbi.13715.

Myers, N., Mittermeler, R. A., Mittermeler, C. G., Da Fonseca, G. A. B., & Kent, J. (2000). Biodiversity hotspots for conservation priorities. *Nature, 403*(6772), 853–858. https://doi.org/10.1038/35002501.

Ning, H., Ling, L., Sun, X., Kang, X., & Chen, H. (2021). Predicting the future redistribution of Chinese white pine Pinus armandii Franch. Under climate change scenarios in China using species distribution models. *Global Ecology and Conservation, 25*, e01420. https://doi.org/10.1016/j.gecco.2020.e01420.

Oke, O. A., & Thompson, K. A. (2015). Distribution models for mountain plant species: The value of elevation. *Ecological Modelling, 301*, 72–77. https://doi.org/10.1016/j.ecolmodel.2015.01.019.

Padrón, P. S., & Vélez, A. (2020). *Description of the immature stages of the high Andean pierid butterfly Catasticta incerta incerta (Dognin, 1888)(Lepidoptera: Pieridae)* (pp. 65–71). Tropical Lepidoptera Research.

Pavulaan, H. and Wright, D. M., 2002. Pterourus appalachiensis (Papilionidae: Papilioninae), a new swallowtail butterfly from the Appalachian region of the United States.

Perez, M. P., Yu, W.-B. (2021) Pleistocene origin and colonization history of *Lobelia columnaris* Hook. f.(Campanulaceae: Lobelioideae) across sky islands of West Central Africa. Authorea. https://doi.org/10.22541/au.161201731.10223625/v1.

Perrigo, A., Hoorn, C., & Antonelli, A. (2020). Why mountains matter for biodiversity. *Journal of Biogeography, 47*(2), 315–325. https://doi.org/10.1111/jbi.13731.

Peters, M. K., Hemp, A., Appelhans, T., Becker, J. N., Behler, C., Classen, A., Detsch, F., Ensslin, A., Ferger, S. W., Frederiksen, S. B., Gebert, F., Gerschlauer, F., Gütlein, A., Helbig-Bonitz, M., Hemp, C., Kindeketa, W. J., Kühnel, A., Mayr, A. V., Mwangomo, E., & ..., Steffan-Dewenter, I. (2019). Climate–land-use interactions shape tropical mountain biodiversity and ecosystem functions. *Nature, 568*(7750), 88–92. https://doi.org/10.1038/s41586-019-1048-z.

Peterson, A. T., Soberón, J., Pearson, R. G., Anderson, R. P., Martínez-Meyer, E., Nakamura, M., & Araújo, M. B. (2011). *Ecological niches and geographic distributions ecological niches and geographic distributions* (pp. 1–316). Princeton University Press. https://press.princeton.edu/books/paperback/9780691136882/ecological-niches-and-geographic-distributions.

Phelps, L. N., Razanatsoa, E., Davis, D. S., Hackel, J., Rasolondrainy, T., Tiley, G. P., Burney, D., Cox, R., Godfrey, L., Hempson, G. P., Hixon, S., Andermann, T., Andriambololonera, S., Andriamiarisoa, L. R., Antonelli, A., Besnard, G., Birkinshaw, C., Bond, W., Chikhi, L., & ..., Lehmann, C. E. R. (2025). Advancing transdisciplinary research on Madagascar's grassy biomes to support resilience in ecosystems and livelihoods. *Ecological Monographs, 95*(2). https://doi.org/10.1002/ecm.70011.

Platts, P. J., Ahrends, A., Gereau, R. E., McClean, C. J., Lovett, J. C., Marshall, A. R., Pellikka, P. K. E., Mulligan, M., Fanning, E., & Marchant, R. (2010). Can distribution models help refine inventory-based estimates of conservation priority? A case study in the Eastern Arc forests of Tanzania and Kenya. *Diversity and Distributions, 16*(4), 628–642. https://doi.org/10.1111/j.1472-4642.2010.00668.x.

Platts, P. J., Garcia, R. A., Hof, C., Foden, W., Hansen, L. A., Rahbek, C., & Burgess, N. D. (2014). Conservation implications of omitting narrow-ranging taxa from species distribution models, now and in the future. *Diversity and Distributions, 20*(11), 1307–1320. https://doi.org/10.1111/ddi.12244.

Platts, P. J., McClean, C. J., Lovett, J. C., & Marchant, R. (2008). Predicting tree distributions in an East African biodiversity hotspot: Model selection, data bias and envelope uncertainty. *Ecological Modelling, 218*(1-2), 121–134. https://doi.org/10.1016/j.ecolmodel.2008.06.028.

Price, T. D., Hooper, D. M., Buchanan, C. D., Johansson, U. S., Tietze, D. T., Alström, P., Olsson, U., Ghosh-Harihar, M., Ishtiaq, F., Gupta, S. K., Martens, J., Harr, B., Singh, P., & Mohan, D. (2014). Niche filling slows the diversification of Himalayan songbirds. *Nature, 509*(7499), 222–225. https://doi.org/10.1038/nature13272.

Qazi, A. W., Saqib, Z., & Zaman-ul-Haq, M. (2022). Trends in species distribution modelling in context of rare and endemic plants: A systematic review. *Ecological Processes, 11*(1). https://doi.org/10.1186/s13717-022-00384-y.

Rahbek, C., Borregaard, M. K., Antonelli, A., Colwell, R. K., Holt, B. G., Nogues-Bravo, D., Rasmussen, C. M. Ø., Richardson, K., Rosing, M. T., Whittaker, R. J., & Fjeldså, J. (2019a). Building mountain biodiversity: Geological and evolutionary processes. *Science, 365*(6458), 1114–1119. https://doi.org/10.1126/science.aax0151.

Rahbek, C., Borregaard, M. K., Colwell, R. K., Dalsgaard, B., Holt, B. G., Morueta-Holme, N., Nogues-Bravo, D., Whittaker, R. J., & Fjeldså, J. (2019b). Humboldt's enigma: What causes global patterns of mountain biodiversity? *Science, 365*(6458), 1108–1113. https://doi.org/10.1126/science.aax0149.

Reckling, W., Mitasova, H., Wegmann, K., Kauffman, G., & Reid, R. (2021). Efficient drone-based rare plant monitoring using a species distribution model and AI-based object detection. *Drones, 5*(4), 110. https://doi.org/10.3390/drones5040110.

Ren, Y., Lü, Y., Hu, J., & Yin, L. (2021). Geodiversity underpins biodiversity but the relations can be complex: Implications from two biodiversity proxies. *Global Ecology and Conservation, 31*, e01830. https://doi.org/10.1016/j.gecco.2021.e01830.

Rocca, M. E., Brown, P. M., MacDonald, L. H., & Carrico, C. M. (2014). Climate change impacts on fire regimes and key ecosystem services in Rocky Mountain forests. *Forest Ecology and Management, 327*, 290–305. https://doi.org/10.1016/j.foreco.2014.04.005.

Rödder, D., Schmitt, T., Gros, P., Ulrich, W., & Habel, J. C. (2021). Climate change drives mountain butterflies towards the summits. *Scientific Reports, 11*(1). https://doi.org/10.1038/s41598-021-93826-0.

Rolando, A., Caprio, E., Rinaldi, E., & Ellena, I. (2007). The impact of high-altitude ski-runs on alpine grassland bird communities. *Journal of Applied Ecology, 44*(1), 210–219. https://doi.org/10.1111/j.1365-2664.2006.01253.x.

Scaldaferro, M. A., Srsic, A., Romanutti, A. A., & Acosta, M. C. (2023). Mountains as refugia in the Dry Chaco: Phylogeography and species distribution modelling of the southernmost chilli pepper, Capsicum chacoense Hunz. (Solanaceae). *Biological Journal of the Linnean Society, 140*(1), 130–148. https://doi.org/10.1093/biolinnean/blad045.

Scott, G. R. (2011). Elevated performance: The unique physiology of birds that fly at high altitudes. *Journal of Experimental Biology, 214*(15), 2455–2462. https://doi.org/10.1242/jeb.052548Canada.

Scott, G. R., & Milsom, W. K. (2007). Control of breathing and adaptation to high altitude in the bar-headed goose. *American Journal of Physiology - Regulatory Integrative and Comparative Physiology, 293*(1), R379. https://doi.org/10.1152/ajpregu.00161.2007.

Sigdel, S. R., Wang, Y., Camarero, J. J., Zhu, H., Liang, E., & Peñuelas, J. (2018). Moisture-mediated responsiveness of treeline shifts to global warming in the Himalayas. *Global Change Biology, 24*(11), 5549–5559. https://doi.org/10.1111/gcb.14428.

Singh, A. P., De, K., Uniyal, V. P., & Sathyakumar, S. (2024). Unveiling of climate change-driven decline of suitable habitat for Himalayan bumblebees. *Scientific Reports, 14*(1). https://doi.org/10.1038/s41598-024-52340-9.

Sømme, L., Conradi-Larsen, E.-M., & Somme, L. (1979). Frost resistance in alpine, adult melasoma collaris (coleoptera). *Oikos, 33*(1), 80. https://doi.org/10.2307/3544515.

Spehn, E. M., Rudmann-Maurer, K., & Körner, C. (2011). Mountain biodiversity. *Plant Ecology and Diversity, 4*(4), 301–302. https://doi.org/10.1080/17550874.2012.698660.

Stark, J. R., & Fridley, J. D. (2022). Microclimate-based species distribution models in complex forested terrain indicate widespread cryptic refugia under climate change. *Global Ecology and Biogeography, 31*(3), 562–575. https://doi.org/10.1111/geb.13447.

Stork, N. E. (2018). How Many Species of Insects and Other Terrestrial Arthropods Are There on Earth? *Annual Review of Entomology, 63*, 31–45. https://doi.org/10.1146/annurev-ento-020117-043348.

Suissa, J. S., Sundue, M. A., & Testo, W. L. (2021). Mountains, climate and niche heterogeneity explain global patterns of fern diversity. *Journal of Biogeography, 48*(6), 1296–1308. https://doi.org/10.1111/jbi.14076.

Testolin, R., Attorre, F., Borchardt, P., Brand, R. F., Bruelheide, H., Chytrý, M., De Sanctis, M., Dolezal, J., Finckh, M., Haider, S., Hemp, A., Jandt, U., Kessler, M., Korolyuk, A. Y., Lenoir, J., Makunina, N., Malanson, G. P., Montesinos-Tubée, D. B., Noroozi, J., & ..., Jiménez-Alfaro, B. (2021). Global patterns and drivers of alpine plant species richness. *Global Ecology and Biogeography, 30*(6), 1218–1231. https://doi.org/10.1111/geb.13297.

Thomas, J. A. (2005). Monitoring change in the abundance and distribution of insects using butterflies and other indicator groups. *Philosophical Transactions of the Royal Society B: Biological Sciences, 360*(1454), 339–357. https://doi.org/10.1098/rstb.2004.1585.

Tovar, C., Carril, A. F., Gutiérrez, A. G., Ahrends, A., Fita, L., Zaninelli, P., Flombaum, P., Abarzúa, A. M., Alarcón, D., Aschero, V., Báez, S., Barros, A., Carilla, J., Ferrero, M. E., Flantua, S. G. A., Gonzáles, P., Menéndez, C. G., Pérez-Escobar, O. A., Pauchard, A., & ...,

Hollingsworth, P. M. (2022). Understanding climate change impacts on biome and plant distributions in the Andes: Challenges and opportunities. *Journal of Biogeography, 49*(8), 1420–1442. https://doi.org/10.1111/jbi.14389.

Tukiainen, H., Toivanen, M., & Maliniemi, T. (2023). Geodiversity and Biodiversity. *Geological Society Special Publication, 530*(1), 31–47. https://doi.org/10.1144/SP530-2022-107.

Vesk, P. A., Morris, W. K., Neal, W. C., Mokany, K., & Pollock, L. J. (2021). Transferability of trait-based species distribution models. *Ecography, 44*(1), 134–147. https://doi.org/10.1111/ecog.05179.

Vetaas, O. R. (2021). Mountain biodiversity and elevational gradients1. *Frontiers of Biogeography, 13*(3), 1–4. https://doi.org/10.21425/F5FBG54146.

Vitousek, P. M., & Farrington, H. (1997). Nutrient limitation and soil development: Experimental test of a biogeochemical theory. *Biogeochemistry, 37*(1), 63–75. https://doi.org/10.1023/A:1005757218475.

Von Humboldt, A., & Bonpland, A. (1826). Voyage aux régions équinoxiales du nouveau continent, fait en 1799, 1800, 1801, 1802, 1803, et 1804. *Libraire grecque-latine-allemande, 12.*

Wagner, D. L. (2020). Insect declines in the anthropocene. *Annual Review of Entomology, 65*, 457–480. https://doi.org/10.1146/annurev-ento-011019-025151.

Wambulwa, M. C., Milne, R., Wu, Z. Y., Spicer, R. A., Provan, J., Luo, Y. H., Zhu, G. F., Wang, W. T., Wang, H., Gao, L. M., Li, D. Z., & Liu, J. (2021). Spatiotemporal maintenance of flora in the Himalaya biodiversity hotspot: Current knowledge and future perspectives. *Ecology and Evolution, 11*(16), 10794–10812. https://doi.org/10.1002/ece3.7906.

Wang, K., Zhou, X. H., Liu, D., Li, Y., Yao, Z., He, W. M., & Liu, Y. (2022). The uplift of the Hengduan Mountains contributed to the speciation of three Rhododendron species. *Global Ecology and Conservation, 35*. https://doi.org/10.1016/j.gecco.2022.e02085.

Warren, M. S., Maes, D., Swaay, C. A. M., Goffart, P., Dyck, H. V., Bourn, N. A. D., Wynhoff, I., Hoare, D., & Ellis, S. (2021). The decline of butterflies in Europe: Problems, significance, and possible solutions. *Proceedings of the National Academy of Sciences, 118*(2). https://doi.org/10.1073/pnas.2002551117.

Woodward, F. I. (1987). *Climate and plant distribution*. Cambridge University Press.

Xu, J., Grumbine, R. E., Shrestha, A., Eriksson, M., Yang, X., Wang, Y., & Wilkes, A. (2009). The melting Himalayas: Cascading effects of climate change on water, biodiversity, and livelihoods. *Conservation Biology, 23*(3), 520–530. https://doi.org/10.1111/j.1523-1739.2009.01237.x.

Yu, Y., Li, J., Zhou, Z., Ma, X., & Zhang, X. (2021). Response of multiple mountain ecosystem services on environmental gradients: How to respond, and where should be priority conservation? *Journal of Cleaner Production, 278*, 123264. https://doi.org/10.1016/j.jclepro.2020.123264.

Zhou, T., Shen, X., Irwin, D. M., Shen, Y., & Zhang, Y. (2014). Mitogenomic analyses propose positive selection in mitochondrial genes for high-altitude adaptation in galliform birds. *Mitochondrion, 18*, 70–75. https://doi.org/10.1016/j.mito.2014.07.012.

CHAPTER 4

High-altitude ecosystems and plant distributions

4.1 Introduction

Mountain environments are incredibly dynamic with diverse ecosystems that change through time: they have changed, they are changing, and they will continue to change into the future. Heraclitus may have been referring to the concept of environmental change when he declared in ancient Greece 'Change is the only constant'. Indeed, it is this constancy of change that still allows people to dismiss the challenges of predicted future climate change under the banner of 'climate change has always happened'. As we will see, current change, and that predicted to come, is so different from anything experienced in the past, that there is more than a passing cause for concern – the mountain alarm bells are ringing! The major difference of the current environmental emergency is that the driver is human, as opposed to natural cycles of climate variation. Anthropogenic change is occurring more rapidly, more intricately, and more uncontrollably for communities worldwide, each with its unique history and future aspirations (IPCC, 2022; Steffen, 2021).

As detailed in Chapter 3, geology, topography, and climate all matter for mountain biodiversity. But how do these relationships change as the surrounding climate transitions through time? Temporal variability is of two types: short-term variation occurring over the course of a day, a season, a year, and longer-term variation through climatic cycles that can extend to glacial-interglacial periods and geological time. Variation through time is important; the 'Time Theory' (Pianka, 1966) has been used to explain why communities that have had a longer uninterrupted climatological history (such as parts of the tropics, which have largely avoided the Pleistocene glaciations) will be more speciose for both ecological and evolutionary reasons. Ecologically, more species will have had time to populate the area, whilst evolutionarily, more species will have had time to evolve from existing populations (Parks & Mulligan, 2010). Relative stability will have led to the generation of complex, highly specialised species interactions, generating many species at low populations and leading to high diversity if these can be maintained

Mountain Diversity: Dynamics of High-Altitude Environments.
DOI: https://doi.org/10.1016/B978-0-443-22251-1.00004-6

in co-existence. These suggestions of long-term stability and unchanging tropical environments have been largely overturned by a plethora of palaeo data from the tropics that document species shifts, some as extreme as those in more temperate latitudes. However, the complete removal of species from the highest latitudes by mobile ice sheets was largely absent from the tropics.

4.2 Proxy tools for understanding the past

The instrumental record is limited by being short, sparse, incomplete, and imperfect, particularly in mountain regions where there is a chronic shortage of monitoring stations and research effort. One major issue is that the short and sparse instrumental record cannot capture low-frequency and long-term variation; it records a relatively narrow amplitude of environmental variation that neither fully characterises the past nor the likely near future. The dynamics of past climates cannot be fully understood by simply extending what has been observed during the instrumental period, because most of the climate signals that are reconstructed from paleoclimate archives were of greater amplitude, longer duration, larger spatial footprint, and lower frequency than any that have been observed in the instrumental period (Baker & Fritz, 2015; Birks, 2012). Thus, lessons learned from studying the palaeoclimate and palaeoecological record provide insight into both past and future mountain climate relationships.

Data that enables us to travel through time and understand the changes that have happened are derived from cores through accumulated material such as ice, deep-sea sediments, lake sediments, peats, soils, tree rings, speleothems, and even elephant tusks – the list of material that carries an imprint of the environment when it was growing or accumulating is quite endless. Lake sediments can provide very useful archives to reconstruct past environments (Kappenberg et al., 2021), as they accumulate material constantly over long periods of time. Peat deposits, characterised by the in-situ accumulation of plant detritus, can provide valuable archives of peatland development processes and their interactions with the surrounding environment. Additionally, they have great advantages in revealing regional ecological changes and climate-human-environment interactions in historical periods. In the past two decades, there has been a large increase in the quantity and quality of palaeoclimate records from mountains. Lacustrine sediment records and speleothem isotopic records are foremost among these incremental archives, whereas tree rings and data from rapidly accumulating peat bogs also offer excellent insights into the past. Each type of archive claims some advantages in dating accuracy, temporal resolution, fidelity of proxy, or length of record (Baker & Fritz, 2015). The reality is they all have their own strengths and weaknesses, but combined can be used to provide a greatly improved understanding of the spatio-temporal variation of past climates. These proxies of changing environments can be used to travel back through time and show clear patterns of temperature, greenhouse gas concentration, and precipitation changes. These accumulated materials will record micro fossils, chemical signatures, or DNA that can be used to reconstruct past ecosystems, environments, and act as the drivers of change. An extensive delve into the world of paleoscience is not possible within this book, but the author would urge people to explore the fascinating world of 'travelling through time' in any or all of the following books: Brechin and Lee (2024), Chakrabarty (2021), Chang and Turner (2019), Harris (2022), Lamb (2013), Mélières and Maréchal (2015).

To provide insight into the long term, we have to employ palaeoecology, the study of past ecology and environment. This provides insights into long-term change in key landscape parameters, including vegetation abundance and composition (land cover), fire regimes, biogeochemical cycles, water quality, pollution, and soil erosion (Gillson et al., 2025). Palaeoecology utilises 'proxies' – measurable variables that are biological, geochemical, and molecular in nature, contained within various environmental archives (Birks, 2012). Common proxies used in palaeoecological studies include fossil pollen, charcoal, stable isotopes, diatoms, and spores preserved in lake sediment records to reconstruct long-term changes in vegetation, fire history, climate, and herbivory (Birks, 2012). Such approaches are complemented by other time-series data covering the recent past, including, but not limited to, long-term monitoring, historical records, and satellite imagery.

One of the keys to unlocking the past is preserved fossils that get trapped in sediments as they accumulate. Diatoms (Bacillariophyta) are unicellular eukaryotic organisms (algae) responsible for much of the primary production on Earth, with the frustules being preserved as fossils in a wide range of aquatic environments, including peatlands and lakes (Battarbee et al., 2001). The pattern and shape of the siliceous cell walls allow them to be identified at the species level, which then allows the past ecosystem composition and the changing environment to be reconstructed. Diatoms are abundant and diverse in peat, and have considerable potential as biomonitors of long-term environmental variation (Battarbee et al., 2001). Diatom composition depends on a series of factors, such as moisture, water acidity, water table depth, salinity, temperature, nutrients, and oxygen availability

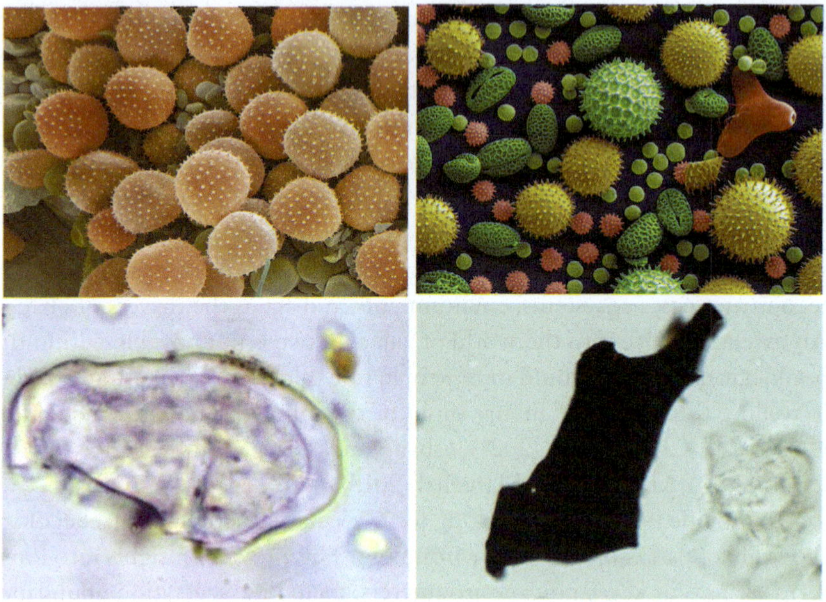

Plate 4.1 Pollen grains (*top right*), phytoliths (*bottom left*), charcoal (*bottom right*), and fungal spores (*top left*) preserve a direct record of the surrounding ecosystem composition and distribution and can be used to understand how this has changed over time.

(Battarbee et al., 2001). Another main group of microfossils is pollen grains (Plate 4.1), which preserve really well in sediments and record a fingerprint of the parent vegetation, as this changes in response to environmental change and/or human interaction, so does the composition of the pollen record. Numerous studies of biodiversity focus on the last few decades, but it is important to study longer trends in biodiversity (Karachurina et al., 2023). To reconstruct past alpha and beta diversity from pollen data, a standard set of statistical methods can be used, although with limitations, particularly because it is often not possible to identify pollen grains to species level (Berglund et al., 2008; Jantz et al., 2014; Marchant et al., 2018; Reitalu et al., 2019). A further problem arises when pollen analysis is dominated by taxa that produce high amounts of pollen (e.g., *Pinus, Picea,* and *Betula*) at the expense of entomophilous forbs and taxa that produce very little pollen (e.g., *Fagus, Larix,* and *Quercus*), leading to the underestimation of floristic richness (Karachurina et al., 2023). Indeed, to make the most interpretive value of records from the past, these proxies need to be calibrated, and an understanding of the pollen-vegetation, diatom-water character, and charcoal-fire type relationships established (Julier et al., 2021). Phytolith assemblages provide another dimension, rather than just a pollen record alone, for reconstructing changes in grassy ecosystems and their composition (Li et al., 2024). Pollen grains combined with phytoliths, with the addition of charcoal and fungal spores, can provide amazing insights into the interactions between fires, grazers, and ecosystem composition and distribution, and how this has evolved through decades, centuries, or millennia.

Geochemical characteristics can be used to reconstruct past environments and climates. Al and Ti are frequently used as indicators of chemical weathering due to their high immobility and insolubility (Liang et al., 2013, 2018). The geochemistry data records from the mountains of central China show the periodicity of freshwater discharges into the North Atlantic, the so-called Heinrich events, which lead to winter monsoon intensification and severe droughts across the Tibetan Plateau (Porter & An, 1995; Raduła et al., 2021). In the study by Song et al. (2023), these changes to monsoonal activity are characterised by abrupt changes in Al, Ti, and Zr/Rb ratios due to shifts in rainfalls, low weathering rates, and reduction of the wetland area. The results illustrate that the environmental changes from the mountains of central China are in sync with monsoon intensity deduced by orbital parameters, and the role of the North Atlantic circulation in driving abrupt climate and ensuing erosion events (Song et al., 2023). Such amazing insights into global teleconnections and understanding the spatial and temporal connections across our world's mountains underscore the importance of extending our instrumental record through proxy data.

Plant DNA from lake sediments is another interesting proxy of past change that enables us to establish, with high accuracy, the changing nature of the ecosystem composition. For example, the presence and dominance of Larch in the vegetation can be reconstructed better than by using pollen alone (Karachurina et al., 2023; Schulte et al., 2021). In the Altai Mountains of southern Russia and northern China, maximum plant diversity occurred between 7000 and 3700 years ago. This is quite a common situation, as it was before extensive human ecosystem interaction had an impact on most parts of the world. In the Altai Mountains, active algal blooms are inferred from a high abundance of the cyanobacteria *Aphanizomenon* and *Microcystis* over the last 500 years that document the increasing impact of human populations on mountain environments (Karachurina et al., 2023).

The reconstruction of palaeo-vegetation and palaeo-fire is essential for understanding climate history, as well as for identifying trends of future fire and vegetation regimes (Kappenberg et al., 2021). Forest fires generate a significant amount of pyrogenic material called black carbon through the incomplete combustion of terrestrial biomass. The analysis of lignin-derived phenols and black carbon expands the information from pollen records that already provide excellent perspectives on past ecosystem transitions. Combined with the additional fire proxies, fire and vegetation residues on a macro and molecular scale can be detected in soil and sediments, thus providing a more complete insight into the vegetation history (Kappenberg et al., 2021; Rahman et al., 2024). The presence of black carbon suggests that fire frequency and/or intensity were highest in the low mountain ranges in central Germany during warm and humid periods when forest was established, peaking around 70, 85, and 115 years ago (Kappenberg et al., 2021).

A multiproxy approach can help to identify the drivers of such transitions, including fire (reconstructed from charcoal abundance), herbivory pressure (reconstructed from dung fungal spores), water availability, climate, and nutrient status (from isotopes and diatoms) (Marchant et al., 2018). This combination of a diverse range of tools and techniques, each with its own merits and limitations, is the best approach to travelling through time and space. The right combination of these methodologies potentially allows for a much-needed temporal continuum spanning from millennia to months, thereby improving our understanding of landscape change, variability at local and regional spatial scales, and, importantly, the drivers of change. This nuanced understanding proves crucial for informing decision-making processes that consider whether and/or how far we have exceeded Holocene 'safe operating spaces' and are in the realms of adapting to no-analogue conditions. Knowledge of such variability could help in mapping out trajectories that could stabilise essential Earth system processes and the associated ecosystem services (Gillson et al., 2025), particularly vital for sustainable mountain social ecological systems (Chapter 7).

4.3 Climatic cyclicity of the Quaternary

The Quaternary period, spanning the last 2.58 million years to the present, has been characterised by significant geological, climatic, and ecological changes, particularly in mountains where glacial-interglacial cycles have shaped mountain topography and ecosystems. The primary driver of interglacial-glacial cycles is the Milankovitch cycles: changes in Earth's orbit around the sun and tilt of the Earth's axis relative to the sun, which affect how much solar radiation reaches the Earth. These changes in

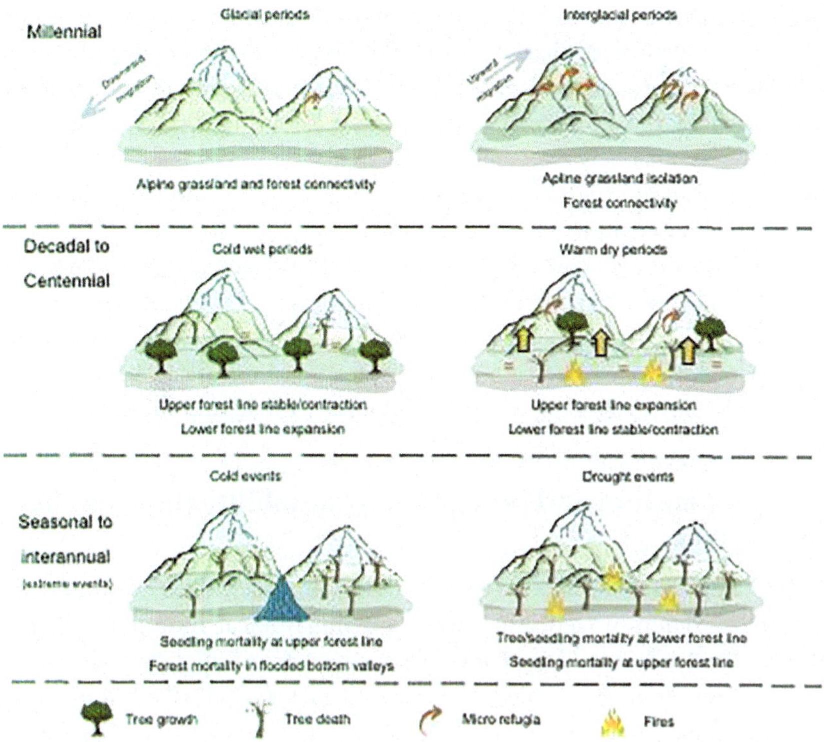

Figure 4.1 Schematic diagram of key changes that occur over different time scales to glacial-interglacial cycles through to seasonality extremes.

Earth–Sun relationships alter the amount of energy arriving to drive Earth's climate systems. This changed energy budget impacts on global temperature, precipitation, atmospheric, and oceanic circulation systems, leading to shifts in climate belts, ecosystem composition, and altered levels of carbon dioxide in the atmosphere (Boom et al., 2002). Changing greenhouse gas concentrations (CO_2, CH_4), ice-albedo feedback, atmospheric dust, and ocean circulation further amplify these changes (Peterson et al., 2024). It is these past Quaternary oscillations that have affected Earth's mountain landscapes and influenced species evolution and migration (including humans), that help us to understand current and future climate dynamics (Fig. 4.1). Interestingly, Earth has spent most of the Quaternary under a Glacial system with temperatures some 10°C lower than present, precipitation some 20% reduced, seasons shifted, and atmospheric CO_2 around 200 ppmV, compared to the c. 427 ppmV of today. Glacial periods tend to last around 85,000 years, with the interglacials being relatively short-lived events of 10,000–15,000 years. This means Earth has spent some 80%–90% of the Quaternary under a glacial environment – this is the norm. We are currently in an interglacial period, called the Holocene, which began about 11,700 years ago. The Holocene is relatively wet and warm, with ice sheets restricted to the polar areas and the highest mountain ranges. Herein is the main difference between the current phase of climate change and that of many of the past – it is the only time the planet has gone from a hot, wet, high CO_2 world to a hotter and more variable future with high levels of CO_2. Alarmingly, the current changes in climate are happening, when large tracts of land have been transformed, and the migration ability of plants and animals to accommodate a changing climate has been largely curtailed – uncharted times lie ahead!

4.4 Temperature and precipitation variability through the Quaternary

As climates changed, lake levels rose and fell, palaeolakes arose and disappeared, glaciers waxed and waned on the tropical mountains. Clearly observed in drill cores from Lake Titicaca is a record of glacial advances in the Eastern Cordillera of the Andes and rising lake water balance coincident with global glacial stages, whereas glacial retreat and major lake-level decline were coincident with interglacial periods (Baker & Fritz, 2015). A large number of pollen records from the tropical Andes depict patterns of late-Quaternary vegetation change through time, but the majority of records span less than 50,000 years (Marchant et al., 2018). As a recorder of past climate, ice cores have been used to derive long-term climate records, most notably from the polar ice sheets but also from high-altitude glaciers. For example, the Guliya ice core on the northern Tibetan Plateau provides detailed palaeoclimate information over the past 130,000 years (Thompson et al., 1997). Glacial advances and retreats also deposited moraines and carved out landforms (Chapter 2), marking both horizontal and vertical glacial extents of glaciers past (Yang et al., 2023). The Tibetan Plateau, known as the Third Pole, contains the largest glacier inventory (>100,000 km^2) outside of polar regions (Yang et al., 2023). This record of the climate's past is not just about describing and documenting change, but very much about navigating the future! Glaciers from the Tibetan Plateau feed major rivers and provide fresh water for over a billion people across South and East Asia (Immerzeel et al., 2020): it is vital that we know about how these relationships have changed, are changing, and could change in the future.

4.4.1 Temperature change

The interval of time spanning the Last Glacial Maximum (LGM; 21–18 thousand years ago, ka) to the pre-industrial era represents the most recent large-scale reorganisation of the climate system (Osman et al., 2021). Temperature ranges over a full LGM cycle could have reached 5°C–10°C below present in the high Northern and Central Andes above 2500 m (e.g., Hooghiemstra & Flantua, 2019), while at mid-latitudes of the Southern Andes, a decrease in mean summer temperatures of 6°C–8°C below modern values has been estimated (Marchant et al., 2009). In temperate latitudes, where the North American and Eurasian ice sheets are near their maximum extent, terrestrial cooling relative to the preindustrial of some 20°C across the glaciated high northern (>45°N) and southern (>45°S) latitudes is common (Osman et al., 2021). Temperatures on Mount Taibai were estimated to be 7.0°C–9.4°C lower than present at the LGM (Liu et al., 2022). Cooler conditions, up to 11°C colder than today, with moisture reduced by 20% was needed to sustain palaeoglaciers of Mount Karanfil, Turkey (Köse et al., 2022).

Thus, everywhere was substantially cooler but with high variability in the degree of temperature reduction. This global variability was linked to two primary climatic mechanisms: radiative forcing from albedo, such as ice sheets, and influences of ocean overturning circulation and seasonal insolation (Osman et al., 2021). For example, regions affected by the East Asian Monsoon, Indian Monsoon, and Westerlies saw greater cooling around the LGM, while areas blocked by high mountains are less affected by these monsoons, and Westerlies experienced less cooling (Liu et al., 2022). This interesting finding shows (1) the value of the palaeo record and (2) how important topographical feedback can be on the climate system – insights can only be achieved by digging into the palaeo record. By accessing the temporal record and applying a 'time for space' approach, we can document how ecosystems, environments, and human interactions have changed across multiple dimensions.

The importance of local feedback can be admirably demonstrated by some mountain forests. For example, forests on the Northern Andes show very little change in species composition over glacial-interglacial cycles, which is likely the effect of a continuous high habitat connectivity unaffected by the Quaternary climate fluctuations (Flantua et al., 2019; Hooghiemstra & Flantua, 2019). Other mountain ecosystems, particularly those close to oceanic influences, seem to have relatively little ecological change, particularly at the level of biomes (Finch et al., 2014; Marchant et al., 2009) due to 'climate stability' driven by the continued buffering of warm flows of moisture from the ocean to mountains. Overall, the LGM was characterised by global cooling and drying, but with large site-to-site variance that emphasises the importance of topography and other local factors in controlling the climate and ensuing vegetation cover (Davis et al., 2024).

Despite the relatively small magnitude of climate change in the Holocene compared to the deglaciation, the understanding of Holocene climate evolution still poses significant challenges (Bao et al., 2025). The Holocene continued to be a period of pronounced climate change with significant spatial variability in the direction and amount of change. The number of sedimentary records generally increases through the Holocene (Githumbi et al., 2025), and it is clear that this period was characterised by complex forcings and heterogeneous climates that make for a challenging test of climate model performance and application (Kaufman & Broadman, 2023). Glaciers generally retracted to their Holocene minima between 10,000 and 6000 years ago and advanced in the mountains of both hemispheres between 5000 and 3000 years ago, as evidenced by extensive proxy data and models (Kaufman & Broadman, 2023). This overall signal reflects global mean surface temperature peaking during the first half of the Holocene, at roughly 6.5 ka, as evidenced by extensive proxy data and models (Bao et al., 2025; Kaufman & Broadman, 2023).

The period around 4200 years ago is considered to be a phase of abrupt worldwide climate change, also known as the '4.2 ka BP event' (Bond et al., 2001; Marchant & Hooghiemstra, 2004), encompassing 200–300 years of mega-droughts, cold temperatures, and strong winds that affected large areas (Liu & Feng, 2012). The event is often linked to social collapses, such as in Mesopotamia and profound change across the low latitudes (Staubwasser & Weiss, 2006). North of the Alps, on the other hand, little convincing evidence has been found so far, and it is assumed that this event had more visible consequences south of the Mediterranean and the Middle East (Bevan et al., 2017).

Rates of climate warming in the world's mountains substantially exceed the global mean, with dramatic effects on the cryosphere, hydrosphere, and biosphere (Schickhoff et al., 2021). Human activity has induced global warming, in which extremely high-temperature events occur more frequently, especially on mountains and particularly leading to multiple threats (Chapter 6). Tree ring data show that temperatures have been increasing, particularly since 1850 and even more notably since 1950 (Sun et al., 2022).

Indeed, the world's mountains have been warming at a disproportionately fast rate, particularly in the last 40 years, since many changes have started to accelerate (Pepin et al., 2022).

4.4.2 Precipitation change

Alpine lakes are natural rain gauges, and reconstructing changes in their water level is key to understanding the regional hydrological environment, the climate change behind this, and the associated evolution of the climate systems. The Oxygen isotope composition of sediments can be used as a particularly good proxy of water level, hence evapotranspiration and rainfall (Bittner et al., 2022). As precipitation today is characterised by high spatial variability (Chapter 2), it is not surprising that precipitation variation in the past was also highly variable. Furthermore, there would have been differences in summer and winter regimes due to changes in convective processes. Precipitation in central China during the LGM is thought to be 60%–80% of modern levels on Mount Taibai (Liu et al., 2022). A reconstruction based on pollen data showed increases in winter precipitation around the eastern end of the Alps, admittedly with high variability, but a general reduction of precipitation relative to today, of around 15% (Davis et al., 2024). Palaeoecological records record monsoon-aridity regime fluctuations over Asia, characterised by typical monsoon over the east and south, contrasted to severe aridity in the interior (Sha et al., 2023). The climate in the Qinling Mountains in eastern China was also characterised by approximately 300 mm $year^{-1}$ less precipitation compared to the Holocene. Likewise, in upland lakes of South America, the LGM was characterised by a dry climate (Guimarães et al., 2023). Overall, the LGM was characterised by climatic drying but with large site-to-site variance that emphasises the importance of topography and other local factors in controlling the precipitation and vegetation (Davis et al., 2024). Precipitation patterns have been more variable across mountains (Tovar et al., 2022), which would have resulted in the relative stability of some areas.

Precipitation reached a maximum (940 mm $year^{-1}$) in the mid-Holocene (Zhang et al., 2025a). A similar situation is recorded from Lake Garba Guracha, Bale Mountains, Ethiopia, with precipitation increased at the beginning of the Holocene, leading to an overflowing lake between ca. 10,000 and ca. 8000 years ago (Bittner et al., 2022). A humid early Holocene was followed by a cooling and drying trend through the late Holocene with significant variability (Zhang et al., 2025a). For example, dry phases are recorded at Lake Garba Guracha, Bale Mountains, Ethiopia, at 10,800, 10,200, 3500, and 500 years ago (Bittner et al., 2022). Daye Lake, in the Qinling Mountains, in eastern China, reached its highest level 1500 years ago; subsequently, the lake fell to its lowest level around 1000 years ago (Chen et al., 2021). Based on *Polylepis* tree-ring reconstructions, an unprecedented decrease in precipitation over the past 700 years has been recorded over the arid Altiplano since the mid-20th century (Morales et al., 2012).

The Little Ice Age, over the period of 1400–1900 CE, is a globally cold event in the gradual warming Holocene. This cold climate caused glaciers to advance in many regions of the world, and the landforms left by the Little Ice Age glaciers, such as terminal-lateral moraines and trimlines, were preserved well due to weak weathering and movement of the landforms (Zhang et al., 2024). Since the Little Ice Age, glaciers have respectively lost 13.3%, 23.9%, and 42.8% of their area, retreated by 15.4%, 25.3%, and 39.6% of their length, and reduced by 20.8%, 39.0%, and 61.0% of their volume in the western, central, and eastern subregions of the Tanggula Mountains. This also reflects that temperate and continental-temperate transitional glaciers have experienced stronger ablation since the Little Ice Age than the continental glaciers in the interior of the Tibetan Plateau (Zhang et al., 2024). A similar trend is found in the humid eastern Peruvian Andes based on *Cedrela-Juglans* tree rings that date to 1817 (Humanes-Fuente et al., 2020). A 404-year precipitation reconstruction based on the tree rings of *Siberian larch* growing on the south slopes of the Altay Mountains demonstrated some severe drought events which could be found in the historical documents, such as the drought in the late Ming Dynasty (1640s) and the Ding-Wu Disaster (1870s). These precipitation variations in Central Asia seem related to the ENSO and North Atlantic Oscillation – Ocean circulation in the Pacific and Atlantic oceans, respectively (Pang et al., 2023). Tree-ring minimum earlywood density on *Picea schrenkiana* in the arid central Tianshan Mountains provides robust signals of drought, with far-reaching impacts on global water resources, agricultural productivity, ecological balance, and societal stability, particularly in arid and semi-arid mountain regions (Ma et al., 2025). Droughts in the Tianshan region, which often manifest as three consecutive years of aridity, were the most severe from 1916 to 1918, and the second driest period occurred from 1808 to 1810. Notably, a renowned drought event known as the 'Ding-Wu Great Famine' (1877–1878 CE) also occurred during this period (Ren et al., 2021). It has been suggested that this super drought event not only significantly impacted central and eastern China but also extended

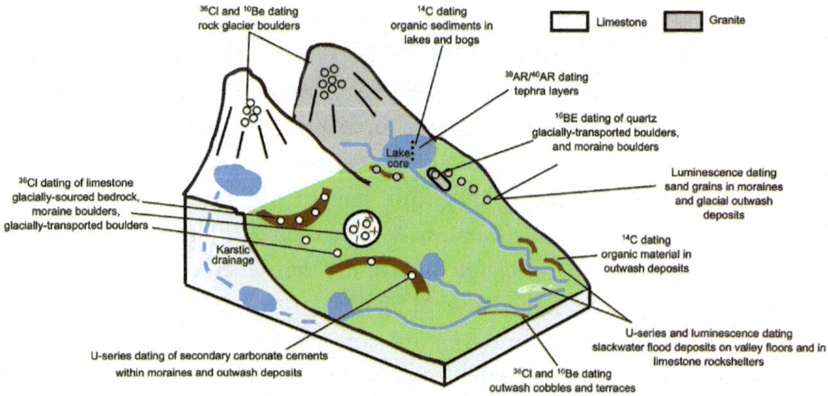

Figure 4.2 Schematic diagram of radiometric dating methods employed in glaciated catchments in the Mediterranean mountains.

westward, thereby influencing the Tianshan region of Central Asia, which was initiated by a typical El Nino-Southern Oscillation (ENSO) warm event (Ren et al., 2021). Synchronised notable drought periods were found more recently (1950–1952, 1984–1986, 2007–2009, and 2011–2013 CE) (Ma et al., 2025). As global warming continues, so does the change in the scale of atmospheric motion and frequency of extreme drought events. Again, these global teleconnections further demonstrate the importance of these long-term climate reconstructions to understand not only the magnitude of change, but also the nature of the driver of change and how these operate across space and through time.

4.5 Changes in mountain glaciers

Mountain glaciers are extremely sensitive to climate change, and their shape, area, and volume change with fluctuations in temperature and precipitation and the annual timing of this (Allard et al., 2021; Liu et al., 2022). As mountain glaciers change, they leave formed geomorphological features, such as erosional trimlines and depositional moraines (Köse et al., 2022; Zhang et al., 2024). Previous ice caps and glaciers have shaped many of the high mountain landscapes that we see today across the world's mountains (Allard et al., 2021) (Chapter 2). Reconstructing past glaciers can provide valuable insights into the climatic conditions beyond the instrumental climate records (Brooks et al., 2022; Cui et al., 2022). Radiometric dating, field mapping, and chronostratigraphy (Fig. 4.2) have established that the most extensive glaciations were during the Middle Pleistocene (Giraudi & Giaccio, 2017) (e.g., Allard et al., 2021). The Mediterranean mountains, although currently ice-free, were repeatedly glaciated during the Pleistocene, with the glacial archive providing a distinctive record of palaeotemperature and moisture conditions. Dated evidence is particularly compelling in the Pindus Mountains, Dinaric Alps, and Italian Apennines of the central Mediterranean Pleistocene (Giraudi & Giaccio, 2017) (e.g., Allard et al., 2021), the central Taurus Range in Turkey (Köse et al., 2022), and the Romanian Carpathians (Kłapyta et al., 2020).

The timing, extent, and nature of glaciation during the Late Pleistocene show considerable variability across the globe, with glaciers responding to precipitation supply as well as air temperatures (Allard et al., 2021; Hughes et al., 2013). Mountain glaciers in different parts of the world reached their maxima at very different times, and there were likely multiple waxings and wavings of the glaciers on an individual level. In the Late Pleistocene, glacier maxima have been identified at various times between 70,000 and 29,000 years, but with greater variability in the western Mediterranean (Allard et al., 2021). There is broad agreement throughout the Mediterranean that glaciers were at, or close to, their maximum Late Pleistocene extent around 40,000 to 30,000 years ago in response to a marked fall in solar radiation receipt (Berger & Loutre, 1991). The Tibetan Plateau has been affected by repeated glaciations throughout the Quaternary (Sun et al., 2020). Based on geomorphological surveys, the extension of the Eryehai Valley glacier on Mount Taibai comprised three lateral moraines dating to around 25,000, 22,000, and 19,000 years old (Liu et al., 2022).

Climate fluctuations during the late stages of the Last Glacial Maximum resulted in the complex evolution of mid-latitude alpine glaciers in both hemispheres (Clarke et al., 2019; Wang et al., 2024). Interrogating the glacial record helps us to better understand sediment and meltwater delivery to river systems, which is the dynamics of Mediterranean plant and animal refugia (Blischak et al., 2020). Glaciers and their meltwater streams were key elements of Mediterranean mountain cold-stage refugia and their runoff regimes. It is important to recognise that Mediterranean mountain ecosystems have evolved during the Pleistocene (Allard et al., 2021). An analysis of regional ages, based on a compilation of new and published ^{10}Be ages across the Shaluli Mountains, Sichuan, China, suggests that glacial termination was largely synchronous, and three phases of glacial fluctuations occurred across the region during the LGM. Glacial fluctuations are often linked to wider teleconnections across global climate circulation systems (Farinotti et al., 2020). For example, glacial fluctuations in the Shaluli Mountains were influenced by sea temperature variations in the Indo-Pacific and the discharge of freshwater and/or ice drift into the North Atlantic, which subsequently weakened Atlantic Meridional Overturning Circulation (Wang et al., 2024).

4.6 Climate change-induced biodiversity and ecosystem shifts

The recent glacial-interglacial cycles caused significant shifts in species distribution, with most plants and animals adapting and migrating to the shifting altitudinal distribution of climate. Although a few large mammals like mammoths and sabre-toothed cats became extinct (possibly due to interaction with early human populations), most species were able to migrate in response to climate shifts (Jump et al., 2014). These large-scale climate transformations altered the distribution of every plant and animal and hence the distribution, composition, and structure of mountain ecosystems (Chapter 3). Forests expanded during warm periods and retreated during cold periods, with many montane forest taxa moving to lower elevations to fill the void left by retracting tropical rainforests (Los et al., 2019). At higher altitudes, tundra and grasslands expand and migrate down the altitudinal gradient (Wambulwa et al., 2021). However, this was not just a case of ecosystems simply migrating up and down mountains on a climatic metronome. There was great variability and a differential degree of climate change across the world's mountains, even within a single mountain. As species today behave individually, the response to past climatic changes would have been individual.

Pollen analysis shows that vegetation patterns in the Erxianyan region in western Hubei have fluctuated significantly over the past 60,000 years (Song et al., 2023). Although these changes are specific to this mountain range, they can be used to typify ecosystem response to climate change across a glacial to interglacial cycle. From 60,000 to 28,000 years ago, the dense forest was dominated by *Quercus*, along with several temperate deciduous broad-leaved trees, some herbs, and ferns, mirroring the mixed subtropical evergreen and temperate deciduous broad-leaved trees that covered much of central China (Dodson et al., 2021; Song et al., 2023). The highest abundance of herb taxa occurred between 21,000 and 19,000 years ago, corresponding to the LGM. Charcoal values fluctuated frequently but remained relatively high, indicating an openness to the vegetation, likely induced by fires (Song et al., 2023). Central China contains many important pockets of forest refugia in its complex topography, allowing whole, but probably reduced, dense forest biomes to survive during the glacial period (Dodson et al., 2021; Song et al., 2023). Warming and more moisture in the early Holocene led to a rapid increase in tree cover. From 6000 to 4000 years ago, when tree cover was greatest, charcoal abundance was greatly reduced under a warm and humid climate (Song et al., 2023). From 2500 years ago, the increasing dominance of *Pinus, Alnus*, and *Artemisia* was accompanied by an expansion of grasses, which suggested a cooler and drier climate. The loss of temperate taxa (*Quercus, Juglans, Corylus*, and *Salix*) and ferns has been interpreted as responding to reduced moisture as the East Asia summer monsoon weakened in the late Holocene (Song et al., 2023). A rise in grasses probably resulted from human activity in forest management and crop establishment. Historical documents show that Xianfeng County experienced intensified human activity characterised during the late Holocene (Song et al., 2023).

In 1969, Jurgen Haffer published his seminal paper on the Amazon refugia hypothesis, which set in motion decades of research examining the nature of climate variation during the Quaternary and its impacts on tropical biota. Haffer (1969) proposed that during cold, dry intervals of the Quaternary, rainforest contracted into small isolated areas (refugia), where water availability was higher, separated by vast open areas of savanna, which promoted isolation and biotic differentiation and hence enhanced speciation in Amazon forest birds. This concept of refugia has been extrapolated to mountains, where the interaction of precipitation, topography, and winds allows some organisms to persist through intervals of variable regional climate, due to local stable conditions, leading to the development of fine-scale high biodiversity. Glacial climate cycles did not alter the distribution ranges of all species, further demonstrating the individual nature of species response to change.

Glacial-interglacial cycles certainly played a central role in the evolution and divergence of mountain species via range contraction and expansion (Wang et al., 2016). As glaciers expanded and there was ensuing isolation of populations, this led to speciation and endemism in mountain regions. Reconstructing changes in past distributions can offer valuable insights into the evolutionary patterns of vegetation types and their spatial distribution across mountainous landscapes (Li et al., 2024). Climate oscillations over the Quaternary can be viewed as the 'fuel for the engine of evolution' that drives mountain species diversity. Such a mechanism that has been

Plate 4.2 Inversion layer forms clouds at relatively low altitudes in the Stubai valley. *(Photograph Rob Marchant).*

convincingly demonstrated in the Himalayas for *Primula tibetica* identified four distinct lineages, each of which had experienced bottlenecks and multiple expansions during the Pleistocene (Ren et al., 2021). These accounts strengthen the suggestion that cycles of expansion and contraction, driven by orbital variations, could be one of the key drivers of montane biodiversity (Muellner-Riehl et al., 2019; Spicer et al., 2020). Diversification of montane plant species is not only driven by orogenesis, but also by a complex interaction of factors such as topography and climate (Wambulwa et al., 2021). There is a continuing debate about the relative contribution of changes in air temperature and lapse rate, precipitation regime, atmospheric CO_2 concentration, UV-B radiation flux, wildfire, and herbivory (Los et al., 2019). Given the complex nature of mountain geodiversity (Chapter 2), it is likely the response would have been quite mountain-specific.

Since the beginning of the Quaternary, mountains have acted as refugia for species during climatic oscillations because they have been able to support plants as they moved along these environmental gradients. The causes of why refuges form in some locations and not others are very location-dependent. In certain mountain regions, pockets of air get trapped in small isolated valleys high in the mountains (Plate 4.2), protected by an inversion layer, in which wind and cold air act like a cap to prevent air from circulating out of the valley. Such pockets can result in extremely stable local climatic conditions that might have favoured the survival of species during periods of past climatic change (Rahbek et al., 2019). These micro-environments lead to the splitting and isolation of species and promote the differentiation of populations and the creation of new species (Gavin et al., 2014). Studying past refugia can help us to identify the specific geographical locations of the current refugia and evaluate their conservation significance, analysing the evolutionary processes that buffer species against rapidly changing climate (Scaldaferro et al., 2023). This mosaic of responses by plant taxa to past climatic changes could have important lessons for the future – do places of past stability allow for stability in the future?

The ability of a species to persist in microrefugia may depend on the specific interaction between the species' traits and the climatic drivers that control its distribution (Hylander et al., 2015). Thus, while models based on thermal microclimate conditions may predict suitable habitat for more species than macroclimate conditions, the accuracy of these predictions will vary by species (Stark & Fridley, 2022). Given the importance of these refugial areas, there is a need to explicitly consider microclimatic conditions when projecting future climates to understand organismal responses (Barros-Souza & Borges, 2023; Stark & Fridley, 2022). Unfortunately, caution should be observed when generalising about how the past can be used to navigate the future, and the future is clearly very different from anything experienced in the past! One thing we do see is that different species have varying adaptive potential (Githumbi et al., 2025). Selecting priority regions by integrating these approaches with the multiple facets of biodiversity (e.g., species, genetic, phylogenetic, and functional diversity, and option values) represents one of the central challenges in studies and practices of conservation planning (Cai et al., 2021). One such example of a region sensitive to climate change with a rich assemblage of vegetation and high biodiversity is the mountains of central China (Song et al., 2023). The Hengduan Mountains contain about 9000 vascular plant species in an area of about 500,000 km^2 (Wu et al., 2005), ranking as one of the most biodiverse mountain ranges surrounding the Qinghai–Tibetan Plateau. The Hengduan Mountains may have also acted as a refuge for plants during the Quaternary and prevented species from extinction (Rahbek et al., 2019). Thus, these mountains could be a target for future conservation investment to preserve and maximise biodiversity. Similarly, the Eastern Arc Mountains of Tanzania are known to have been relatively stable to past climate shifts and are likely to be relatively stable in the future due to the strong and pervasive relationship with the continued moisture supply from the Indian Ocean (Mumbi et al., 2008).

Plate 4.3 Glaciers are retreating increasingly fast, as documented in the valleys above Obergurgel, which has been monitoring the retreat of the Ötztal glacier has retreated several kilometres in less than 170 years. The extent of this is marked with a series of plates that document the year of the terminus; the older ones are now colonised by lichens, while the most recent areas are bare rock. *(Photograph Rob Marchant)*.

4.7 Recent environmental change during the Holocene Interglacial

Global surface temperature has been rising significantly since the late nineteenth century, and the upward trend has been even more dramatic in recent decades, manifested as increasing climate variability, leading to droughts, floods, and fires (IPCC, 2022). The pronounced warming has exacerbated the instability of the climate system, causing problems such as water scarcity, rapid glacier retreat (Plate 4.3), permafrost melt, and deaths of trees, particularly across mountain systems (Cong et al., 2023; Yuan et al., 2024). This enhanced warming is expected to continue in the future and will have severe consequences for both human and natural systems. Placing this in the context of long-term climate variation is important to show how the rate and magnitude of warming are way outside the range of natural variability.

Tree rings, due to their continuous annual resolution, accurate dating, wide geographical distribution, and numerous replicates, are one of the best proxies for longer palaeoclimatic information from the recent past (Björklund et al., 2023; Cui et al., 2024; Ljungqvist et al., 2020). For example, tree rings have been used to create a robust 346-year chronology of *P. wilsonii* from the western Qinling Mountains: cold conditions from 1810 to 1820 CE, which were mainly due to the strong eruption of the Tambora Volcano in 1815. Distinct warm periods around 1670, 1690, and 1780 CE were bracketed by cold intervals from 1750 to 1770 and around 1790 CE (Yuan et al., 2024). Worryingly, the reconstruction also reveals unprecedented warming conditions in recent decades, especially after the 1980s. The consistency of the reconstructed series, with temperature reconstructions for Asia and the Northern Hemisphere, indicates that the thermo-dynamics of the western Qinling Mountains and wider Mongolia are symptomatic of large-scale climate systems linked to North Atlantic Ocean temperatures (Baldermann et al., 2021). Mirroring the long-term insights, these telecoupled influences are important drivers of multidecadal thermal variations in the western Qinling Mountains (Yuan et al., 2024). Moreover, it is possible to separate how variation in solar activity, volcanic eruptions, and greenhouse gas forcing have contributed to the temperature change in the region. Tree rings clearly demonstrate that the recent warming trend, particularly since the 1980s, has been a disconnection from nature's climate drivers. Such information can be useful for constraining predictions of future warming rates and applying these insights to issues such as forestry management (Yuan et al., 2024).

Juniperus saltuaria tree-ring width has been used to develop a 669-year record of spring-time relative humidity that provides a long-term spring atmospheric moisture record for the central Hengduan Mountain region (Cui et al., 2024). The reconstruction recorded two periods with the highest frequency of extreme spring moisture, corresponding to the peak period of the Little Ice Age and the current warm period. Before the current warm period, spring atmospheric moisture variations over the central Hengduan Mountains were modulated by the solar activity and sea-air interactions. However, with the increase in anthropogenic greenhouse gases, global warming has become the main factor influencing atmospheric moisture conditions in spring in the mountain area. This transformation of the climate system may be the main reason for the recent unprecedentedly high occurrence of spring atmospheric droughts (Cui et al., 2024).

4.8 Retreat of glaciers and the creation of ice-free areas

Glaciers are well-recognised indicators of a changing climate, and have shaped mountains as they expanded and contracted through glacial and interglacial cycles (Fig. 4.1). Glacier mass balance depends on snow accumulation and snow, firn, and ice melt during the cold and warm seasons, respectively, and is thus considered a reliable indicator of climate fluctuations

(Vidaller & Alonso-González, 2021). Although glaciers have been, in general, retreating since the end of the Little Ice Age, from the mid-1980s glaciers worldwide have undergone drastic retreat at a rate that has accelerated in recent decades (Dangles et al., 2025). Glaciers are retreating, shrinking, and disappearing worldwide due to global warming (Losapio et al., 2021). The retreat of glaciers is affecting landscape configuration, water availability, and natural hazards, impacting downstream ecosystems and society (Fell et al., 2017). In the European Alps, glaciers have lost 25%–30% of their surface area over the last 60 years, and the rate of glacier reduction is now 200%–300% faster than it was 20 years ago (Sommer et al., 2020). Depending on the climate scenario, the Alps may lose up to 94% of their 2020 ice volume by the end of the 21st century (Steffan et al., 2022; Zekollari et al., 2019), with important consequences for the appearance of new landscapes and ecosystems. The retreat of glaciers since the mid-19th century is attributable to anthropogenic factors that have increased in recent years (Bach et al., 2018; Ficetola et al., 2021). The southernmost glaciers in Europe are located in the Pyrenees, Apennines, and Balkan Peninsula at latitudes between 41°N and 43°N and elevations between 2000 and 3300 m (Grunewald & Scheithauer, 2010) and can all be classified as threatened – the glaciated area of the Pyrenees has been reduced from 294 to 229 ha (−23.2%) and three glaciers have disappeared (Vidaller & Alonso-González, 2021). There is no sign of slowdown in the shrinkage of Pyrenean glaciers, meaning we are moving towards an ice-free Pyrenees in the coming decades (Vidaller & Alonso-González, 2021), which adds to the other mountains rapidly losing their glaciers, such as the iconic Kilimanjaro.

4.8.1 Retreat of glaciers: New landscapes and new lakes

Long-lasting mountain lakes can provide indispensable cultural and ecological value (Dangles et al., 2025). However, retreating glaciers produce new lacustrine environments filled with sediment transported by them (Mölg et al., 2021). Such lakes pose both risks and opportunities (Dangles et al., 2025); for example, they impact mountain biodiversity, hold significant hydropower potential, might be attractive to tourists, serve as reservoirs for artificial snow production in ski areas, and provide irrigation reservoirs (Steffan et al., 2022). In these new lakes, high amounts of glacial flour absorb ultraviolet radiation, which is particularly strong in high mountains and acts as a habitat for bacteria. The expansion of the ice-free area releases substantial quantities of new potential habitats for species, but the decrease in meltwater levels increases the risk of desiccation in glacier forefields (Cauvy-Fraunié & Dangles, 2019). If glaciers disappear completely in the next few decades, more substantial changes in the biodiversity of glacier-fed systems will occur, with unclear consequences for ecosystem functioning and services to human populations (Cauvy-Fraunié & Dangles, 2019). With close to 700 newly detected glacier lakes in the Swiss Alps, this transformation can be significant across the landscape (Steffan et al., 2022). This large number of new lakes calls for reflection about their future use. Already, we are seeing conflicting interests, spanning from a strict preservation due to the potential importance of these emerging natural habitats, through to their exploitation for commercial use (Steffan et al., 2022). Most of the current debates are often centred on anthropogenic interests, but in our 'enlightened times' it should include the ecosystem services provided by lakes and also how these could be used as barometers of change. For example, ice cover can be derived from low spatial resolution MODIS data (Tom et al., 2022), whereas high-resolution multi-temporal digital aerial imagery provides opportunities for a more comprehensive and detailed analysis of climate change-induced glacier retreat and mass loss (Geissler et al., 2021).

4.8.2 Retreat of glaciers: New communities as an ecological experiment

As glaciers retreat, new glacier forelands are exposed to colonisation by living organisms, primarily plants, making space for species colonisation and primary succession (Cauvy-Fraunié & Dangles, 2019). Given the alarming rates of glacier retreat, specialist species inhabiting land close to glaciers are at risk of extinction (Losapio et al., 2021). Therefore, there are winners and losers from the rapidly changing glacial world. With the alteration of both taxonomic and interaction diversity following glacier retreat, discussions on glacier foreland protection and management are needed (Zimmer et al., 2022). The development of new ecosystems is rapid following this retreat (Fig. 4.3). We are reaching a point where it is necessary to take actions to manage glacier forelands in order to protect the unique biodiversity close to glacier margins and promote alpine grasslands to enhance species interactions, ecological networks, and ecosystem functioning.

How communities change or replace one another over time through ecological succession has been a cornerstone of ecology since its inception. Where a glacier once flowed, the severe disturbance created areas lacking lifeforms, a given species or a set of species colonising a newly exposed

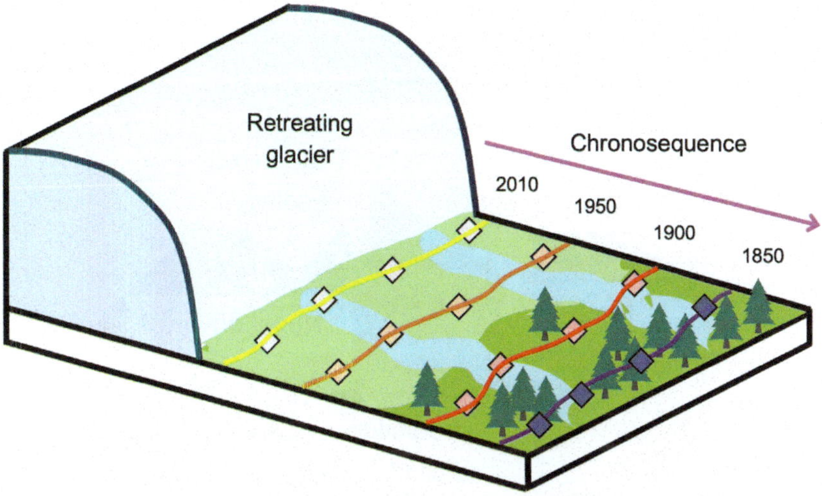

Figure 4.3 Plants can be sampled using a systematic series of plots across the time series of glacial retreat. A chronosequence is a sequence of dated sites for which time after retreat has been measured using a wide range of approaches (after Ficetola et al., 2021).

surface, which would then be further colonised by other species to reach complex communities (Cantera et al., 2024). Studying the environmental changes occurring after the retreat of glaciers provides a great opportunity to comprehend the formation of communities and an understanding of the temporal dynamics of ecosystem composition and functioning (Ficetola et al., 2021). Ecosystems can be sampled using a systematic series of plots across the time series of glacial retreat in a chronosequence (Fig. 4.4) to make them ideal for ecological succession studies (Bosson et al., 2023) and to understand how plant communities respond to glacier recession; crucial insights for projecting biodiversity dynamics and planning sustainable ecosystem management (Erschbamer & Caccianiga, 2016). Studies of deglaciated areas carried out since the beginning of the twentieth century were a milestone in the development of ecological succession theory, with a strong emphasis on plants and invertebrates (Ficetola et al., 2021).

Rather than describing relatively simple models of succession, by using community network and ecosystem functional approaches, it is possible to go beyond a mere description of biodiversity patterns and to tease apart the role of different processes such as habitat filtering, dispersal, and biotic interactions (Cortés et al., 2018). Integrating data from multiple trophic levels, such as producers, consumers, parasites, and decomposers, with a

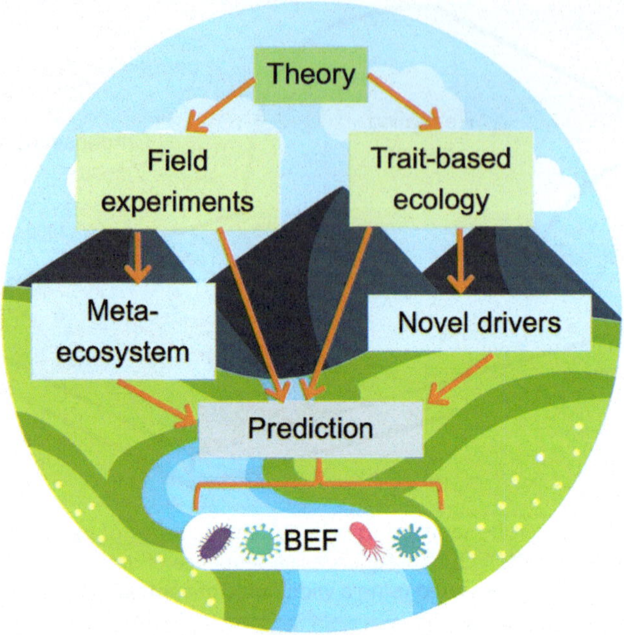

Figure 4.4 Integrating data from multiple trophic levels, such as producers, consumers, parasites, and decomposers, with a combination of field-based, modelling, and trait-based ecology.

combination of field-based, modelling, and trait-based ecology (Fig. 4.4) provides a more complete understanding of the functioning of these new ecosystems. Colonisation can be surprisingly fast, as the rate at which different taxa colonise glacier forelands is not constant, and there are clear differences between both taxonomic groups and regions of the world. Factors triggering differences in the rate of biotic colonisation include (1) successional stage (e.g., early versus mid and late-successional stage), (2) traits of the surrounding species pool (e.g., dispersal ability, diet), and (3) abiotic factors, such as climate, geomorphic stability of surfaces, and altitudinal gradients along glacier forelands (Hågvar et al., 2020; Rosero et al., 2021), and are shaped by both the addition and the replacement of taxa (Cantera et al., 2024).

Time is one factor driving colonisation, with the rates influenced by multiple stochastic and environmental factors, such as the availability of potential propagules, the surrounding species pool, microclimates, topography, solar radiation, bedrock type, and soil properties. Understanding how

Plate 4.4 Nitrogen-fixing plants such as this *Lotus campylocadus* at Arenas Negras, Tenerife, can help speed up the colonisation of new land, as clearly shown with the older lava flow on the left vs the new 16th-century lava on the right. *(Photographs Rob Marchant)*.

environmental factors shape community changes and functions is pivotal for predicting the effects of ongoing environmental changes (Bjorkman et al., 2018; Ficetola et al., 2021). Biotic interactions have always been considered important for the evolution of communities, as certain taxa can facilitate and hamper colonisation by other functional components. In the early stages of succession, for example, nitrogen-fixing organisms (e.g., actinorhizal plants) dominated due to the initial lack of available nitrogen (Koffel et al., 2018) (Plate 4.4). In short, these new lands can provide key insight into the mechanisms that enable the advancement of a more complete theory of ecosystem dynamics (Ficetola et al., 2021) (Fig. 4.5).

Global change scenarios forecast an acceleration of glacier retreat during the next few decades (Hock et al. 2019); therefore, deglaciated terrains will constitute a growing component of high-mountain ecosystems. Sound ecological knowledge is crucial for understanding the consequences of ongoing climate change and for defining effective strategies of management and adaptation in these highly dynamic environments. The effects of climate warming produce 'winners' and 'losers' as the response of plants to increasing temperature varies among species and communities (Losapio et al., 2021). To fully understand the changing nature of the communities, we need more data; indeed, the value of community survey data may help give a better understanding, conserving, managing, and restoring ecosystems with scarce (economic) resources. To do so, addressing both which and how species will interact with each other while being impacted by global warming is of paramount importance (Losapio et al., 2015).

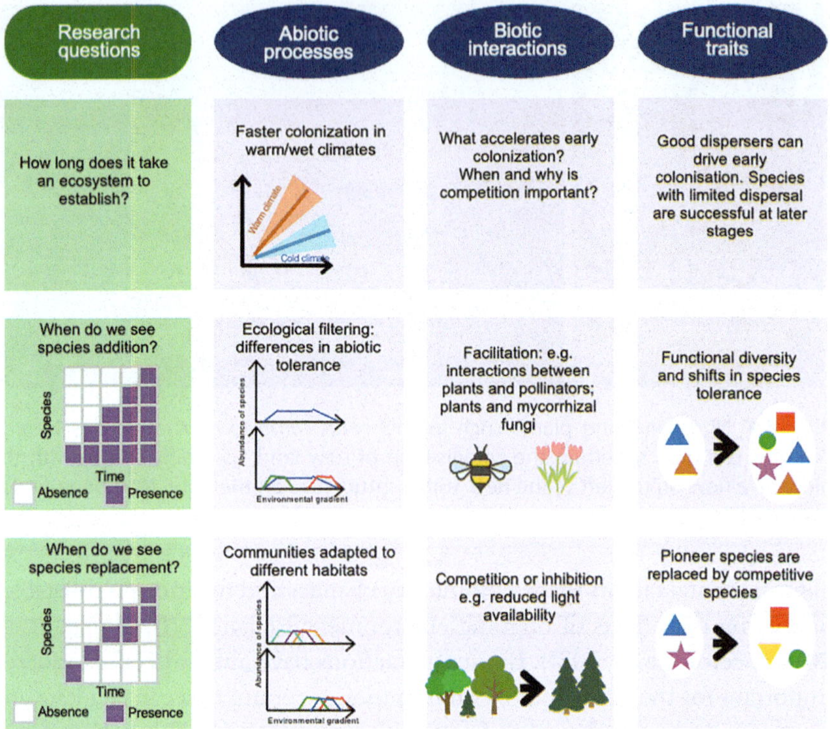

Figure 4.5 A non-exhaustive list of open research issues on the dynamics of ecological communities following glacier retreat. These include questions related to the pattern of community change, the identification of drivers of community change, the analysis of multitrophic linkages and functional diversity, and the identification of general patterns from case studies. Each question can benefit from multiple traditional and emerging tools (after Ficetola et al., 2021).

4.8.3 Retreat of glaciers: Glacier impact on water resources

Climate-driven modifications of ecosystems will continue under 21st-century global environmental change, further threatening biodiversity and ecosystem services (Pörtner et al., 2019). Alpine river biodiversity around the world is also under threat from glacier retreat driven by rapid warming, yet our ability to predict the future distributions of specialist cold-water species is currently limited. As Alpine regions are warming at a faster rate than the global average and have high levels of endemism, Alpine biodiversity is particularly vulnerable to climate change. By linking glacial models, hydrological models, and SDMs, it is possible to locate potential refuge areas for aquatic alpine invertebrate species responding to glacier retreat, at the scale of an entire mountain range. A major benefit that a coupled modelling approach provides is identifying future refuge locations relative to current protected zones, such as national parks or conservation areas.

Glacier water resources play an important role in maintaining economic and social sustainable development and resilience in the middle and downstream reaches (Roberts et al., 2018). However, the lack of long-term glacier measurements due to the difficulties and dangers in extremely high-altitude areas limits our understanding of glacier variations and their impact on water resources (Bazai et al., 2021). Glacier retreat caused by global warming alters the hydrological regime and poses far-reaching challenges to water resources and nature conservation of the headwater of the Yangtze River and its vast, densely populated downstream regions (Gao et al., 2024). The Dongkemadi Glacier is projected to continue to retreat, with glacial runoff increasing and reaching peak water around 2060 to 2085. After this tipping point, water resources will likely decrease (Gao et al., 2024). To understand these linkages across the landscape and how they will change under predicted climate change modelling frameworks (Fig. 4.6) allows us to systematically quantify hydrological processes in previously glaciated basins, which sheds light on further large-scale studies (Gao et al., 2024).

4.9 Evolution of mountain-human interactions

Globally, mountains are centres of food production and tourism income, and providers of energy, timber, and non-timber products, with more than half of humanity relying on the 60%–80% of the world's freshwater that is accumulated, stored, and distributed from mountains (Liniger & Weingartner, 2000). Using insights from historical documentary sources, repeat photographs, and archaeology, we can reconstruct human-environment interactions and entanglements over the entirety of our own species' history on Earth (Beneš et al., 2023). Archaeological records can be used to explore, at both site and landscape scale, how human activity has influenced changes in ecosystems and vice versa (Millhauser & Earle, 2022). Archaeologically informed long-term studies of human-environment interactions have a key role to play in identifying land and resource use strategies that have proved to be ecologically sustainable over hundreds of years (Ferro-Vázquez & Stump, 2023; Marchant et al., 2018). There are numerous examples of where and when human activities have enhanced regional biodiversity over the long term (Roberts et al., 2017), and how human communities have 'weathered' severe climate change in the past (Petek-Sargeant & Lane, 2021).

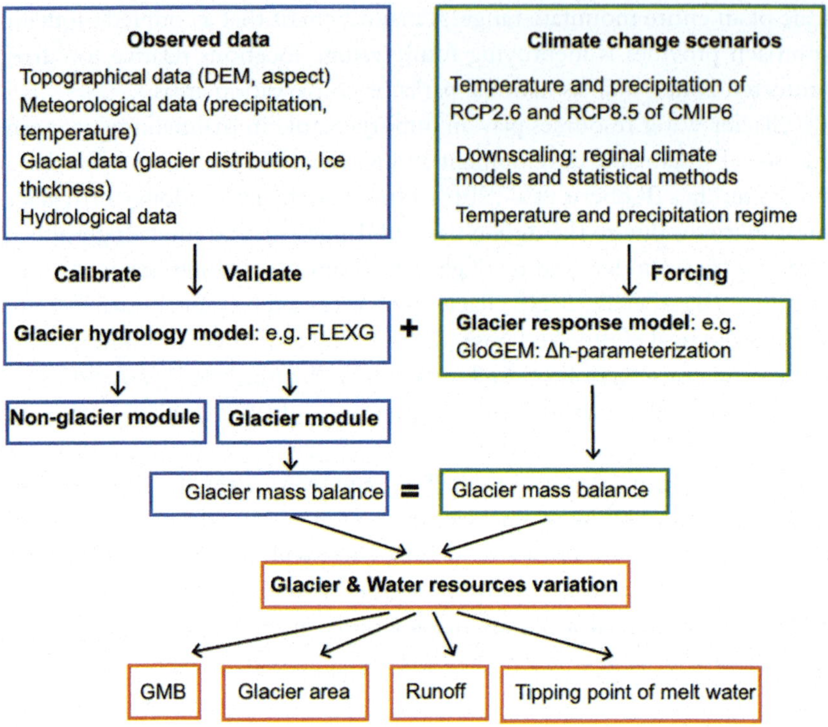

Figure 4.6 The FLEXG-Δh modelling framework, and the workflow of model calibration, validation, and prediction of future climate change on the glacier area and melting water.

These insights can be used to explore the drivers and human dimensions of land cover change and interaction with environmental change over decadal to millennial timescales (Gillson & Marchant, 2014), and to understand the changing relationship between people and mountains. For example, the Altai Mountains are a pivotal location for the ancient nomadic civilisation in arid Central Asia and a key origin of nomadic cultures in north China (Li et al., 2024b). Archaeological findings indicate that the exchange of crops and livestock between the Eastern and Western cultures in the Altai Mountains dates back to the Palaeolithic period (Wilkin et al., 2020). Although it is clear that an intensification of human land use occurred through the late Holocene, detailed insight into the intricate and changing relationships between human activities, ecosystems, and climate change in Mountain systems during the late Holocene remains frustratingly limited.

The mid to late-Holocene was a time of profound cultural transition: settled agrarian societies emerged and were established across much

Plate 4.5 Exotic tree species, such as *Eucalyptus,* have been extensively planted across African mountains, such as the Jimma Highlands, Ethiopia. *(Photograph Rob Marchant).*

of the world's mountains. A broad range of species were domesticated, and their distribution ranges dramatically altered; metallurgy and new technologies were developed, and various complex societies emerged and collapsed (Marchant et al., 2018). Since the Holocene, land use and cover change, primarily driven by anthropogenic activities such as cultivation and grazing, have profoundly influenced mountain ecosystems and the physical, ecological, and biological functioning of Earth's planetary systems (Wende et al., 2021). For example, expansion of iron-using early farming communities between c. 3500–2000 years BP across west-central and eastern Africa apparently correlates with widespread forest loss (Marchant et al., 2018). These early transitions were a precursor for more recent, extensive changes in forest composition, distribution, and structure associated with industrial-scale forestry over the colonial and post-independence periods. For some mountain areas, this long history of human-ecosystem interaction has resulted in total forest clearance or 'selective' deforestation, where all the large and valuable timber trees have been removed. In some cases (e.g., parts of the Eastern Arc Mountains), the post-World War II period was also characterised by significant deforestation and afforestation projects (von Hellermann, 2020), with these planting regimes often driven, as elsewhere (Fairhead & Leach, 1996), by crisis narratives and characterised by exotic species (Plate 4.5). Indeed, afforestation with alien species, especially in ancient grasslands or heathlands as we are increasingly seeing in southern African mountains, could damage the resilience of mountains (Vorontsova et al., 2016).

Plate 4.6 Many montane forests are in various successional phases of regeneration, with large open forests clearly representing a previously extensive deforestation, even though currently the forest is protected, such as the Ngorongoro Crater here. *(Photograph Rob Marchant).*

Even currently protected mountain ecosystems, such as Bwindi-Impenetrable Forest, Ngorongoro Highlands, and the Eastern Arc Mountains, are in a successional phase of regeneration with large open forests clearly representing a previously deforested ecosystem (Plate 4.6). Although details of when, how, and what tree species were removed are unknown, today's informed restoration decisions are highly challenging (Grainger et al., 2019). In other areas, high altitude grasslands have been particularly vulnerable to changed fire regimes or afforestation with exotic plantations, especially *Pinus* and *Eucalyptus*, that are known to impact water provision, and affect water and soil quality and agricultural productivity (Plate 4.5). Again, however, there are misconceptions and misunderstandings about the history with a lack of knowledge on ancient montane grassland antiquity and biodiversity (Vontrosova et al 2016). The impact of these transformative changes to montane ecosystems on water, food, biodiversity, and energy is largely unquantified, although there are significant impacts on biodiversity (Ahrends et al., 2011, 2021), hydrology (Cuni-Sanchez et al., 2021), carbon storage and sequestration (Cuni-Sanchez et al., 2018), and nutrient flows leading to reduced crop yields (Chanie et al., 2020). The shifts in human-environment-climate interactions, and the legacies of these today on ecosystem composition and distribution, are often assumed to be profound but have rarely been quantified.

Evidence from recent decades shows that growing populations, changes in food and energy consumption strategies, habitat destruction, invasive species, pollution, and climate change have contributed to the loss of natural ecosystems and declining biodiversity (IPCC, 2022). Much of this focused on mountain areas. The increase in human population and human activities across the world's mountains has driven the current rate of species extinction to c. 100 to 1000 times that of the background extinction rate (Cai et al., 2021; Steffen et al., 2015). However, there is still a great deal of uncertainty on how human activities and climatic changes have impacted the hydrological dynamics of mountain systems. For example, a 3300-year peat-based diatom record, combined with plant macrofossils and palynological and geochemical proxies, documents hydrological changes and vegetation succession during the late Holocene in the Greater Khingan Mountains, Northeast China. The most dramatic change during the last 3300 years occurred around 500 years ago, when the diatom community shifted from planktonic to benthic dominance, concurrent with vegetation shift from an *Eriophorum*-dominated rich fen to a *Sphagnum*-and Ericaceae-dominated fen following a decrease in water level driven by human disturbance (Han et al., 2023). Other regulatory services of the mountains have also seen marked transitions. For example, the long-term carbon accumulation in the Altai Mountains was highest in the middle Holocene, and then decreased in the late Holocene due to shifting local hydrological conditions, human interaction, and temperatures (Zhang et al., 2025b).

Thus, it is clear that humans began to inhabit and modify mountain environments over the Late Quaternary, introducing agriculture, deforestation, and infrastructure development (Li et al., 2024b). Mountains became centres of human culture, spirituality, and adaptation. Although there are multiple case studies to draw on, the northeast margin of the Qinghai Tibet Plateau has many attributes that make it typical of many mountain transformations: a long history of human interaction leading to a densely populated and extensively cultivated area. In relation to the rest of the plateau, it receives more precipitation, and in common with all mountain systems, the diversity of environments leads to a diversity of resources. Hunter-gatherers occupied this part of the Tibet Plateau from the late Pleistocene (Zhang et al., 2016), although the extent to which early human populations transformed the landscape is thought to be minimal. Agricultural activities have been a driving force for humans to adapt and modify the environment, as charted by palynological and archaeological evidence used to reconstruct past environments and cultural interactions on the Qinghai Tibet Plateau ecosystems. Charred seeds of millet were discovered in the Majiayao cultural sites (Jia et al., 2013), providing direct evidence for the onset of cultivation in this area. The firing of pottery and utilisation of wood resources accelerated deforestation; pottery-making technology developed during the Majiayao culture, with this culture renowned for its exquisite pottery that required a temperature of 800°C for the firing (Cui et al., 2015). During the mid-late Bronze Age (3600–2400 years ago), the northeast Qinghai Tibet Plateau was occupied by people of the Xindian and Kayue cultures. There was a change in the farming practice and a shift from the monoculture of millet to growing multiple crops. Barley, as a cold-resistant crop, became a staple in the highlands with sheep, as cold-tolerant livestock, were introduced into this region in the early Bronze Age, most likely via cultural exchange, and grazing livestock was broadly adopted (Dodson et al., 2013) with the growing importance of grazing in the Kayue culture. The development of pastoralism allowed farming activities to expand from low altitude to high altitude areas, despite the dry and cold climate, and for people to settle permanently in areas with elevations above 3000 m (Wende et al., 2021). Grazing by domesticated livestock had an impact on mountain grassland communities, such as an increase of *Stellera chamaejasme* across the landscape (Guo et al., 2021). This spread of an overgrazing indicator documents the degradation of a previously *Artemisia-dominated* steppe. As a result of the deep-rooted expansions of mountain communities, the pristine forests on the Qinghai Tibet Plateau were largely destroyed, and the grassland cover was disturbed (Wende et al., 2021). Further impacts would follow as populations expanded and moved into increasingly marginal areas. Soil erosion and landslides occurred more frequently following the deterioration of the natural vegetation as the natural forest became increasingly patchy and scattered across the hills (Wende et al., 2021). These past transformations have produced a patchwork of legacies and challenges that will be explored in Chapter 6; they also offer potential solutions and insights into what land use management strategies worked in the past and could work in the future to be explored in Chapter 7.

Human migrations into the Americas are unique in the context of the global expansions of our species from Africa (Shipton et al., 2018). Not only were the Americas the last continental landmasses to be occupied by *Homo sapiens*, but they are also characterised by a compression of time between the first human arrival and the emergence of agriculture relative to other parts of the world (Solís-Torres et al., 2021). Ancient and modern genetic data indicate that human expansion into the Americas was rapid, with numerous migratory waves along the mountain chains linking North to

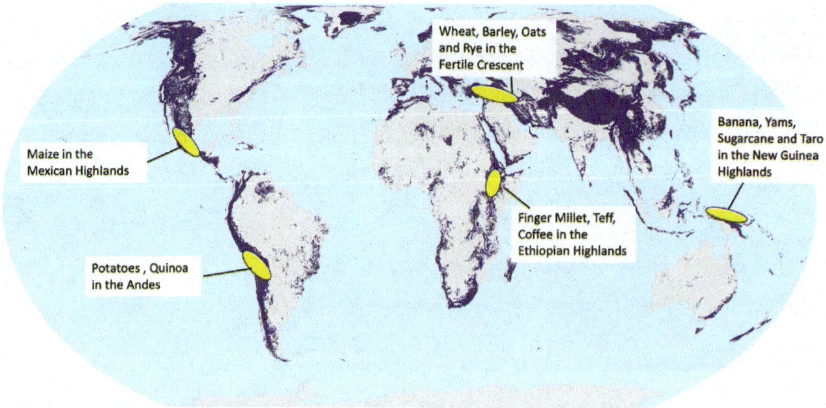

Figure 4.7 Centre of global crop domestication associated with mountains.

South America (Ardelean et al., 2020). Climatic and environmental change had a major impact on early human dispersals and resource use (Zarillo & Kooyman, 2006). For example, the route into North America would have been blocked by the Laurentide Ice sheet until it underwent contraction after c. 21,000 years ago. Once populated, the spread and adoption of new technology and settlement were rapid, as was the evolution of dietary shifts as documented by stable isotope analyses (Pezo-Lanfranco & Colonese, 2024). Increasingly intense use of tropical plants led to the domestication of teosinte (wild relatives of maize), cacao, tomato, potatoes, and different species of squash (Piperno & Smith, 2012). The seasonal river valleys of the Santa Marta mountains in Mexico Valley have become the centre of one of the world's most relied upon crops today, i.e., maize (Piperno, 2011). The ecotonal tropical forest and grassland environments of montane southern Mexico provided a productive setting for the growing and increasingly varied human societies (Acosta et al., 2018). Hunting and gathering strategies adopted by the occupants of rock shelters in Santa Marta implied that there was development of new technologies for the exploitation of tropical environments and the emergence of agriculture and urban societies in the Late Holocene (Solís-Torres et al., 2021) (Fig. 4.7).

The Central Andes was considered a second centre of domestication for maize, but currently is recognised as an 'improvement centre' where local adaptations of earlier domesticated maize occurred (Kistler et al., 2018). The management and cultivation of semi-domesticated and domesticated plants on the Andes dates back to 8350 years ago, although the Andean Altiplano was settled approximately 6000 years ago when hunter-gatherers

Plate 4.7 Salt production from mountain systems such as in Nan, northern Thailand (*left*) and the Salineras de Maras, Peru (*right*) would have been transported to lowland regions. Often, food from the lowlands would have been transported back to altitude, as clear from the marine shell in the Atacama surrounded by pottery (*centre*). *(Photographs by Rob Marchant, apart from the right Photograph taken by Alexandrine Massot).*

began to cultivate crops (Dillehay et al., 2005). Shortly thereafter, llama and alpaca herders played an important role in developing crop cultivation strategies, with potatoes being uniquely adapted to a mobile pastoral lifestyle (Langlie, 2021; Pezo-Lanfranco & Colonese, 2024). Stable isotope data reveal that fish, terrestrial fauna, and cultivated plants variably contributed to the diet of prehistoric Andean populations, indicative of significant trade from coastal, mid-valley, and high-altitude settlements (Marchant et al., 2004) (Plate 4.7). Plant cultivation supported the development of the earliest complex societies during the Formative Period (from 3000 years ago). Similarly, in the highlands, the societies that built ceremonial centres are based on a largely plant-based economy (Pezo-Lanfranco & Colonese, 2024). These food production systems were complemented by the development of irrigation technologies as early as 4550 years ago (Pezo-Lanfranco & Colonese, 2024), leading to big economic and social changes as water and productivity became managed. These early economies would have been spatially connected to lowland areas, particularly through key resources such as salt production from mountain systems that would have been transported to lowland regions (Plate 4.7). Often, food from the lowlands would have been transported back to mountain communities (Plate 4.7) with deep-rooted trade routes evolving that demonstrate the early connected nature of mountains to lowlands.

Fire can be the predominant natural disturbance in many mountain ecosystems; humans have historically employed fire for purposes such as hunting, forest clearing, resource management, and defence (Marlon, 2020).

Understanding the roles humans have in shaping landscape history through the use of fire is crucial for developing effective conservation and land management strategies (Li et al., 2024b). Comparing four sequences involving vegetation, fire, nomadic pastoralism, and agricultural activities at different elevations of the Altai Mountains, it is clear there are complex interactions between the natural climate-vegetation-fire dynamics and changes in anthropogenic burning over the late Holocene (Li et al., 2024b). Fire activities before ~2000 years ago were influenced by temperature at higher elevations and by limited steppe cover at lower elevations. Over the past 2000 years, human activities associated with agriculture and pastoralism have increasingly influenced fire dynamics, although with a climate influence underpinning this dynamic. For example, during the Medieval Warm Period, fire activities were notably high due to warm climate conditions and intensified pasture activities at higher elevations, while lower fire activities may be linked to population migration at lower elevations (Li et al., 2024b).

Using a unique network of palaeofire records spanning the past 2000 years to the 21st century, it is clear that climate change has led to unprecedented fire activity in the Rocky Mountain subalpine forests (Higuera et al., 2021). These high-elevation forests are useful sentinels to monitor the impacts of climate change because their typically cool, moist climate prevents frequent fires, and they have also experienced less historical land-use change and fire suppression than lower-elevation forests. The extensive fire activity during the 21st century is unprecedented in the past two millennia, with rates of burning now nearly twice the late-Holocene average (Higuera et al., 2021). Elsewhere, the global fire record indicates a general decline in biomass burning over the last 150 years and is attributable to global expansion of agriculture and advancements in fire management (Marlon, 2020; Marlon et al., 2013). The extreme 2020 and 2024 fire seasons serve as an example of how increasingly fire-conducive climate conditions have made the Rocky Mountain subalpine forests more flammable, more than at any point in the past two millennia, a trend expected to continue (Higuera et al., 2021). This overview of how humans have shaped Mountain landscapes through the Holocene clearly shows the power of high-resolution palaeoecological and archaeological studies when used together.

4.10 Applying palaeoecology

Palaeoecological studies on sedimentary deposits can reveal crucial information about ecosystem responses to climate change, fire occurrence, herbivore activity, and impacts of humans on landscapes over centennial to millennial time scales (Marchant et al., 2018; Marchant & Lane, 2014). There are several ways palaeoecological data can aid managers to conserve, maintain, and even restore biodiversity, ecosystems, and resources. Palaeoecological data are hugely important in understanding past ecosystem dynamics and are increasingly being used to inform contemporary management decisions, including setting restoration targets, dealing with invasive plants, and water management (Coffey et al., 2011). Palaeoecology can detect 'cold-stage refugia' or 'glacial refugia' as well as 'warm-stage refugia' or 'interglacial refugia' that can then be set as priority conservation targets (Willis et al., 2010). Palaeoecological studies can establish if a particular species of conservation concern is native, non-native, 'hidden', or 'alien' in an ecosystem or landscape, as applied in the Zagros Mountains (Djamali & Segarra-Moragues, 2021).

Palaeoecological methods open a window into the past by providing a possibility to study the long-term population dynamics, which cannot be recorded through experiments or observations due to their long lifespan and reproduction periods. Forests form an integral part of mountain systems (Chapter 3) and have constantly been subject to changes, particularly driven by humans from the Neolithic onwards. For example, the Hessian lower mountain ranges, which are nowadays part of the core area of *Fagus sylvatica* (European beech) forests (Stobbe & Gumnior, 2021), were previously populated by *Tilia* (Giesecke et al., 2017). A stronger anthropogenic influence through the Michelsberg culture (6300–5500 years ago) promoted the development of Hazel-rich mixed deciduous woodlands, in which *Tilia* dominated the canopy, especially in the Taunus mountains. In the pollen diagrams from across southern Germany, large-scale woodland clearance is evident, indicating substantial changes in land use from this time onward with the adoption of agriculture by Mesolithic people (Stobbe & Gumnior, 2021).

One way to chart these long-term perspectives is the planetary boundaries framework that was developed to define limits of critical Earth system processes, providing humanity with a designated and bounded 'safe operating space' (Bayles, 2025; Rockström et al., 2009). This has been used to characterise challenges and establish pathways that could effectively stabilise Earth system processes and guide adaptation when and where boundaries are crossed. Ecological and climatic processes combine to set boundaries to finite resilience; at some point, the combination of stressors would lead to inevitable system collapse and reorganisation. Hence, a pivotal determinant for effective planning and setting adaptation limits lies in identifying the

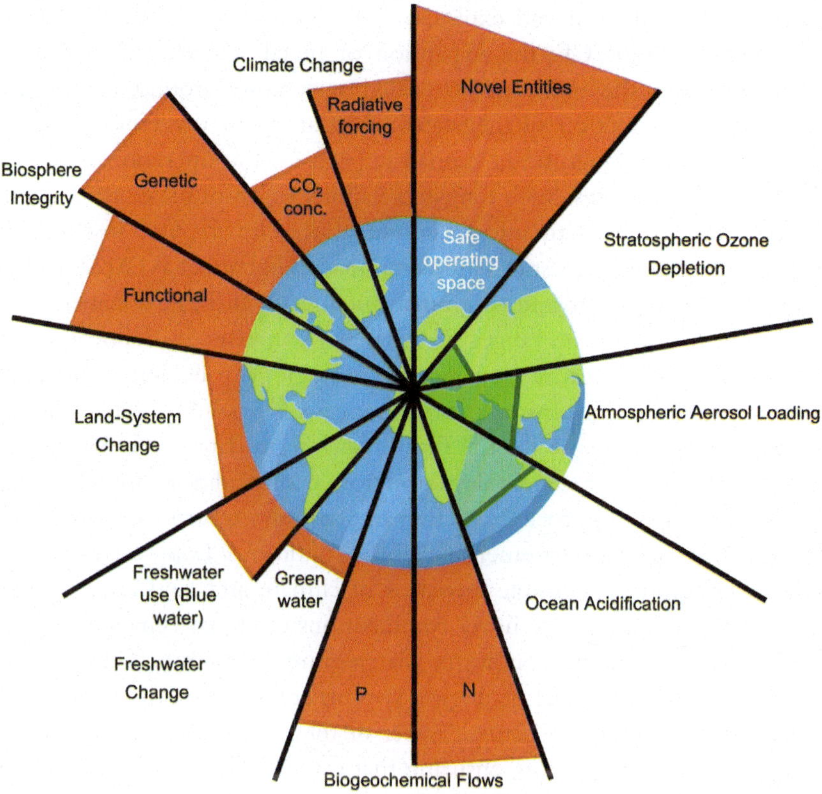

Figure 4.8 The 2023 update to the nine planetary boundaries, based on analysis by Richardson et al. (2023).

thresholds, critical if we are to effectively face the challenges and opportunities of the Anthropocene (Gillson et al., 2025). Continued land conversion, water abstraction, or biodiversity loss that will move the system beyond these thresholds may lead to destabilisation and ultimate collapse of key processes that will lead to the creation of novel states. Since its initial creation in 2009, six planetary boundaries have been crossed (Fig. 4.8) (Richardson et al., 2023). Although a 'full stabilisation' necessitates a return to 'Holocene-like' conditions (Steffen, 2021), this is impossible given the population growth, world momentum, and the emergence of new conditions (Ellis, 2024). Indeed, it is not about seeking this reset, but it is immensely useful to define how far we have departed from Holocene conditions. This long term perspective must extend beyond the Great Acceleration circa 1950 (Steffen et al., 2015), and include transformative periods such as; the Global

North industrialisation and explosions of European colonisation through the 1600s to 1900s CE, the expansion of metal working from 1000 to 6000 years ago, the transition to settled agriculture around 11,000 years ago, and the extinction of megafauna (beginning in the Pleistocene), all of which represent significant human impacts on the operationalisation of Earth system processes. Such long-term appreciation of pre-anthropogenic baselines can be used to inform conservation, climate, and sustainability targets, and to assess widespread human impact (Gibbard et al., 2022) and the deep histories of human and nonhuman entanglement (Braje & Lauer, 2020).

Exploring Holocene ranges of variability can assist in delineating safe operating spaces, as future management strategies can be substantially improved through insight from historical ranges of variability. While the past is an imperfect analogue for the future, long-term data proves essential in assessing the extent of current deviation from past conditions, thereby aiding the decision-making process, whether to attempt a return to former states of variability or adapt to emerging novel conditions. Long-term data can help in envisioning plausible scenarios of climate and land cover change, including situations where novel combinations of climate change and land use create no-analogue conditions that require adaptation (Gillson et al., 2025). As mountain social-ecological systems are dynamic and change over time, and variability is a natural aspect of these systems, it is also vital to define safe operating spaces, ensuring that critical ecological thresholds are not crossed or that social-ecological systems possess the resilience to adapt when reorganisation becomes inevitable, and Holocene-like conditions are unfeasible. Establishing such ranges that are safe operating spaces means either remaining within the Holocene range, where feasible, or within limits of adaptation, where unfeasible (Gillson et al., 2025). As we are moving from a warm to a warmer climate, the Holocene range is quickly being left behind, with the hope that we are going to stay within the limits of adaptation.

Methodologies are needed for monitoring progress and continued re-evaluation of safe operating spaces. Satellite remote sensing can assist in tracking degradation over recent decades and informing restoration that considers changes that have occurred since the 1980s. Such information can be especially valuable in tracking rates of recent changes, including changes in population size and distribution, and in monitoring the impact of, and recovery from, extreme events (e.g., droughts, fires, and floods). However, a scaled approach is really needed, which integrates the long palaeo timeframe and short satellite temporal scales, local knowledge with global data, and national policy goals with global boundaries. This joined-up multi-scalar approach can be used to realise the synergy among long-term data, landscape management, and the planetary boundaries community, and assist in planning for new, just, and sustainable transition pathways (Gillson et al., 2025).

The future is characterised by high uncertainty of how mountain social-ecological systems will respond to continued climate, social, and economic change and its ramifications. A scenarios-based approach that envisions different combinations of economic, climatic, and social-ecological trajectories can be used to explore what the future may bring. However, one weakness of such an approach lies in its rootedness in the present, leading to path dependency and resistance to transformative change beyond our current realities. Exploratory future visioning can be used to explore a diversity of possible futures with varying levels of plausibility. Such scenarios can be viewed through a resist/accept/direct framework that allows stakeholders to integrate objective (what is the trajectory and can we change it) and subjective (do we find this trajectory desirable) management considerations (Kariuki et al., 2018, 2021). These futures can draw on the full range of past possibilities that tap into deep social memory and palaeoecological data.

4.11 Conclusion

Clearly, there are strong legacies of past geological events on shaping mountains and the communities that live in them, along with ecosystems and the functions they perform. Mountain regions are shaped by long-term drivers which impart lasting legacies on contemporary environments. Understanding the temporal and spatial evolution of land use on mountains, as well as the resulting implications for ecological systems and human societies, contributes to our understanding of global change, human-land interactions, and sustainable land use. The interplay of climatic and anthropogenic factors during the Quaternary has left an indelible mark across the world's mountain landscapes, making this period one of profound transformation for the Earth's high-altitude regions. The complexity of ecosystems and human evolution in mountain systems needs to be better integrated into contemporary planning. Mountains can form an astonishing repository of scientific data about the past, with sedimentary layers that contain fossils from millions of years ago, to the much more recent past. Natural data archives contain complementary strands of data that can be woven together to reconstruct past climates, ecosystem and vegetation dynamics, and human activities. No one scientist has the capacity to be an expert in all these areas, and so interdisciplinary research teams are fundamental to help navigate the future. Mountains provide a remarkable natural laboratory, but bringing these past perspectives into the contemporary realm is fundamental to answering important questions of our time: how much forest has been lost? What were the targets of deforestation? When did this loss take place, and who were the beneficiaries? We have to adopt a long temporal frame of reference that combines data and model-based perspectives to deliver a rich interpretation of the interplay between environmental and social factors and a new understanding of mountain social ecological functioning at deep, medium, and shallow time, feeding into the present and future. An interdisciplinary synthesis of information, meta-datasets, and modelling will help define the different drivers and directions of changes in settlement, land-cover, and land-use, and the associated environmental histories and interactions with various cultures, technologies, and subsistence strategies through time and across space for the world's Mountains. This will produce nested temporal understandings that track the evolution of socio-ecological systems and quantify the character of the ecosystem services provided by mountain ecosystems at the household, village, catchment, national, and international levels. By synthesising and visualising existing data from the past, and adding new state-of-the-art data from targeted investigations, we can assess socio-ecological evolution and understand the resilience of socio-ecological systems to past, present, and future climate change. Such an interdisciplinary approach will fundamentally advance understanding of past human-environment interactions and highlight the relevance of these data for present-day conservation and management issues.

References

Acosta, G., Beramendi, L. E., Morales, P., Cienfuegos, E., Otero, F., González, G., Rivera, I., Eudave, I., Hernández, E., & Sánchez, S. (2018). Climate change and peopling of the neotropics during the Pleistocene-Holocene transition. *Boletin de la Sociedad Geologica Mexicana, 70*(1), 1–19. https://doi.org/10.18268/BSGM2018v70n1a1.

Ahrends, A., Bulling, M. T., Platts, P. J., Swetnam, R., Ryan, C., Doggart, N., Hollingsworth, P. M., Marchant, R., Balmford, A., Harris, D. J., Gross-Camp, N., Sumbi, P., Munishi, P., Madoffe, S., Mhoro, B., Leonard, C., Bracebridge, C., Doody, K., Wilkins, V., & …, Burgess, N. D. (2021). Detecting and predicting forest degradation: A comparison of ground surveys and remote sensing in Tanzanian forests. *Plants People Planet, 3*(3), 268–281. https://doi.org/10.1002/ppp3.10189.

Ahrends, A., Rahbek, C., Bulling, M. T., Burgess, N. D., Platts, P. J., Lovett, J. C., Kindemba, V. W., Owen, N., Sallu, A. N., Marshall, A. R., Mhoro, B. E., Fanning, E., & Marchant, R. (2011).

Conservation and the botanist effect. *Biological Conservation, 144*(1), 131–140. https://doi.org/10.1016/j.biocon.2010.08.008.

Allard, J. L., Hughes, P. D., & Woodward, J. C. (2021). A radiometric dating revolution and the Quaternary glacial history of the Mediterranean mountains. *Earth-Science Reviews, 223.* https://doi.org/10.1016/j.earscirev.2021.103844.

Ardelean, C. F., Becerra-Valdivia, L., Pedersen, M. W., Schwenninger, J. L., Oviatt, C. G., Macías-Quintero, J. I., Arroyo-Cabrales, J., Sikora, M., Ocampo-Díaz, Y. Z. E., Rubio-Cisneros, I. I., Watling, J. G., de Medeiros, V. B., De Oliveira, P. E., Barba-Pingarón, L., Ortiz-Butrón, A., Blancas-Vázquez, J., Rivera-González, I., Solís-Rosales, C., Rodríguez-Ceja, M., & …, Willerslev, E. (2020). Evidence of human occupation in Mexico around the Last Glacial Maximum. *Nature, 584*(7819), 87–92. https://doi.org/10.1038/s41586-020-2509-0.

Bach, E., Radic, V., & Schoof, C. (2018). How sensitive are mountain glaciers to climate change? Insights from a block model. *Journal of Glaciology, 64*(244), 247–258. https://doi.org/10.1017/jog.2018.15.

Baker, P. A., & Fritz, S. C. (2015). Nature and causes of Quaternary climate variation of tropical South America. *Quaternary Science Reviews, 124*, 31–47. https://doi.org/10.1016/j.quascirev.2015.06.011.

Baldermann, A., Wasser, O., Abdullayev, E., Bernasconi, S., Löhr, S., Wemmer, K., Piller, W. E., Rudmin, M., & Richoz, S. (2021). Palaeo-environmental evolution of Central Asia during the Cenozoic: New insights from the continental sedimentary archive of the Valley of Lakes (Mongolia). *Climate of the Past, 17*(5), 1955–1972. https://doi.org/10.5194/cp-17-1955-2021.

Bao, Y., Liu, Z., Thompson, L. G., Mosley-Thompson, E., Wan, L., & Lu, J. (2025). Climate simulations and ice core data highlight the Holocene conundrum over tropical mountains. *Communications Earth and Environment, 6*(1). https://doi.org/10.1038/s43247-025-02188-2.

Barros-Souza, Y., & Borges, L. M. (2023). Spatial- and lineage-dependent processes underpin floristic assembly in the megadiverse Eastern South American mountains. *Journal of Biogeography, 50*(2), 302–315. https://doi.org/10.1111/jbi.14527.

Battarbee, R. W., Jones, V. J., Flower, R. J., Cameron, N. G., Bennion, H., Carvalho, L., & Juggins, S. (2001). Diatoms. In J. P. Smol, H. J. B. Birks, & W. M. Last (Eds.), *Tracking environmental change using lake sediments.* Kluwer Academic Publishers.

Bayles, B. R. (2025). Paleoecology perspectives for planetary health. *Global Change Biology, 31*(1). https://doi.org/10.1111/gcb.70043.

Bazai, N. A., Cui, P., Carling, P. A., Wang, H., Hassan, J., Liu, D., Zhang, G., & Jin, W. (2021). Increasing glacial lake outburst flood hazard in response to surge glaciers in the Karakoram. *Earth-Science Reviews, 212.* https://doi.org/10.1016/j.earscirev.2020.103432.

Beneš, J., Ptáková, M., Kovačiková, L., Majerovičová, T., Bernardová, A., Budilová, K., Ayipey, P., Šitnerová, I., Bumerl, J., Komárková, V., Kovárník, J., Pokorná, A., Salova, Y., Vobejda, L., Šálková, T., Skružná, J., & Novák, J. (2023). Of humans and science. Laboratory of Archaeobotany and Palaeoecology and the Second Decade of Environmental Archaeology at the University of South Bohemia. *Interdisciplinaria Archaeologica, 14*(1), 121–136. https://doi.org/10.24916/iansa.2023.1.9.

Berger, A., & Loutre, M. F. (1991). Insolation values for the climate of the last 10 million years. *Quaternary Science Reviews, 10*(4), 297–317. https://doi.org/10.1016/0277-3791(91)90033-q.

Berglund, B. E., Gaillard, M. J., Björkman, L., & Persson, T. (2008). Long-term changes in floristic diversity in southern Sweden: Palynological richness, vegetation dynamics and land-use. *Vegetation History and Archaeobotany, 17*, 573–583. https://doi.org/10.1007/s00334-007-0094-x.

Bevan, A., Colledge, S., Fuller, D., Fyfe, R., Shennan, S., & Stevens, C. (2017). Holocene fluctuations in human population demonstrate repeated links to food production and climate. *Proceedings of the National Academy of Sciences of the United States of America, 114*(49), E10524. https://doi.org/10.1073/pnas.1709190114.

Birks, H. J. B. (2012). Ecological palaeoecology and conservation biology: Controversies, challenges, and compromises. *International Journal of Biodiversity Science, Ecosystem Services and Management, 8*(4), 292–304. https://doi.org/10.1080/21513732.2012.701667.

Bittner, L., Gil-Romera, G., Grady, D., Lamb, H. F., Lorenz, E., Weiner, M., Meyer, H., Bromm, T., Glaser, B., & Zech, M. (2022). The Holocene lake-evaporation history of the afro-alpine Lake Garba Guracha in the Bale Mountains, Ethiopia, based on δ 18 O records of sugar biomarker and diatoms. *Quaternary Research, 105*, 23–36. https://doi.org/10.1017/qua.2021.26.

Björklund, J., Seftigen, K., Stoffel, M., Fonti, M. V., Kottlow, S., Frank, D. C., Esper, J., Fonti, P., Goosse, H., Grudd, H., Gunnarson, B., Nievergelt, D., Pellizzari, E., Carrer, M., & von Arx, G. (2023). Fennoscandian tree-ring anatomy shows a warmer modern than medieval climate. *Nature, 620*(7972), 97–103. https://doi.org/10.1038/s41586-023-06176-4.

Bjorkman, A. D., Myers-Smith, I. H., Elmendorf, S. C., Normand, S., Rüger, N., Beck, P. S. A., Blach-Overgaard, A., Blok, D., Cornelissen, J. H. C., Forbes, B. C., Georges, D., Goetz, S. J., Guay, K. C., Henry, G. H. R., HilleRisLambers, J., Hollister, R. D., Karger, D. N., Kattge, J., Manning, P., & ..., Weiher, E. (2018). Plant functional trait change across a warming tundra biome. *Nature, 562*(7725), 57–62. https://doi.org/10.1038/s41586-018-0563-7.

Blischak, P. D., Barker, M. S., Gutenkunst, R. N., & Falush, D. (2020). Inferring the demographic history of inbred species from genome-wide SNP frequency data. *Molecular Biology and Evolution, 37*(7), 2124–2136. https://doi.org/10.1093/molbev/msaa042.

Bond, G., Kromer, B., Beer, J., Muscheler, R., Evans, M. N., Showers, W., Hoffmann, S., Lotti-Bond, R., Hajdas, I., & Bonani, G. (2001). Persistent solar influence on north atlantic climate during the Holocene. *Science, 294*(5549), 2130–2136. https://doi.org/10.1126/science.1065680.

Boom, A., Marchant, R., Hooghiemstra, H., & Sinninghe Damsté, J. S. (2002). CO_2- and temperature-controlled altitudinal shifts of C4- and C3-dominated grasslands allow reconstruction of palaeoatmospheric pCO2. *Palaeogeography, Palaeoclimatology, Palaeoecology, 177*(1-2), 151–168. https://doi.org/10.1016/s0031-0182(01)00357-1.

Bosson, J. B., Huss, M., Cauvy-Fraunié, S., Clément, J. C., Costes, G., Fischer, M., Poulenard, J., & Arthaud, F. (2023). Future emergence of new ecosystems caused by glacial retreat. *Nature, 620*(7974), 562–569. https://doi.org/10.1038/s41586-023-06302-2.

Braje, T. J., & Lauer, M. (2020). A meaningful anthropocene? Golden spikes, transitions, boundary objects, and anthropogenic seascapes. *Sustainability, 12*(16), 6459.

Brechin, S. R., & Lee, S. (2024). *Routledge handbook of climate change and society* (pp. 1–511). Taylor and Francis. https://doi.org/10.4324/9781003291206.

Brooks, J. P., Larocca, L. J., & Axford, Y. L. (2022). Little Ice Age climate in southernmost Greenland inferred from quantitative geospatial analyses of alpine glacier reconstructions. *Quaternary Science Reviews, 293*, 107701. https://doi.org/10.1016/j.quascirev.2022.107701.

Cai, H., Lyu, L., Shrestha, N., Tang, Z., Su, X., Xu, X., Dimitrov, D., & Wang, Z. (2021). Geographical patterns in phylogenetic diversity of Chinese woody plants and its application for conservation planning. *Diversity and Distributions, 27*(1), 179–194. https://doi.org/10.1111/ddi.13180.

Cantera, I., Carteron, A., Guerrieri, A., Marta, S., Bonin, A., Ambrosini, R., Anthelme, F., Azzoni, R. S., Almond, P., Alviz Gazitúa, P., Cauvy-Fraunié, S., Ceballos Lievano, J. L., Chand, P., Chand Sharma, M., Clague, J., Cochachín Rapre, J. A., Compostella, C., Cruz Encarnación, R., Dangles, O., & ..., Ficetola, G. F. (2024). The importance of species addition 'versus' replacement varies over succession in plant communities after glacier retreat. *Nature Plants, 10*(2), 256–267. https://doi.org/10.1038/s41477-023-01609-4.

Cauvy-Fraunié, S., & Dangles, O. (2019). A global synthesis of biodiversity responses to glacier retreat. *Nature Ecology and Evolution, 3*(12), 1675–1685. https://doi.org/10.1038/s41559-019-1042-8.

Chakrabarty, D. (2021). The climate of history in a planetary age. *The climate of history in a planetary age*. University of Chicago Press.

Chang, C. C., & Turner, B. L. (2019). Ecological succession in a changing world. *Journal of Ecology, 107*, 503–509.

Chanie, D., Azaze, D., Yirgu, T., & Zewudie, A. (2020). Soil loss and its impact on crop productivity under varying land uses in Demba-Gofa District, Southern Ethiopia. *Ethiopian Journal of Business and Social Sciences, 3*(1), 57–89.

Chen, L., Huang, Z., Niu, L., Dong, W., Xiao, S., Chen, S., Zhao, J., Wu, D., & Zhou, A. (2021). GDGTs-based quantitative reconstruction of water level changes and precipitation at Daye Lake, Qinling Mountains (central-east China), over the past 2000 years. *Quaternary Science Reviews, 267*, 107099. https://doi.org/10.1016/j.quascirev.2021.107099.

Clarke, C. L., Edwards, M. E., Gielly, L., Ehrich, D., Hughes, P. D. M., Morozova, L. M., Haflidason, H., Mangerud, J., Svendsen, J. I., & Alsos, I. G. (2019). Persistence of arctic-alpine flora during 24,000 years of environmental change in the Polar Urals. *Scientific Reports, 9*(1). https://doi.org/10.1038/s41598-019-55989-9.

Coffey, E. E. D., Froyd, C. A., & Willis, K. J. (2011). When is an invasive not an invasive? Macrofossil evidence of doubtful native plant species in the Galápagos Islands. *Ecology, 92*(4), 805–812. https://doi.org/10.1890/10-1290.1.

Cong, J., Wang, G., Han, D., & Gao, C. (2023). Organic matter sources in permafrost peatlands changed by high-intensity fire during the last 150 years in the northern Great Khingan Mountains, China. *Palaeogeography, Palaeoclimatology, Palaeoecology, 631*, 111821. https://doi.org/10.1016/j.palaeo.2023.111821.

Cortés, A. J., Garzón, L. N., Valencia, J. B., & Madriñán, S. (2018). On the causes of rapid diversification in the páramos: Isolation by ecology and genomic divergence in espeletia. *Frontiers in Plant Science, 871*. https://doi.org/10.3389/fpls.2018.01700.

Cui, H., Badingquiying, & Mou, H. (2022). Little Ice Age climate inferred from glacier-climate modeling: A case study of Gurla Mandhata, southwestern Tibetan Plateau. *Palaeogeography, Palaeoclimatology, Palaeoecology, 598*, 111034. https://doi.org/10.1016/j.palaeo.2022.111034.

Cui, L., Li, J., An, W., Qin, N., Song, H., & Liu, Y. (2024). The recent high occurrence of spring atmospheric droughts over central Hengduan Mountains is unprecedented in 669-year tree-ring records. *Palaeogeography, Palaeoclimatology, Palaeoecology, 649*, 112318. https://doi.org/10.1016/j.palaeo.2024.112318.

Cui, Y., Dong, G., Li, H., An, T., Liu, X., Wang, J., Wang, H., Ren, X., Li, X., & Chen, F. (2015). Early ceramic trade in Gansu and Qinghai regions, northwest China: A comparative elemental analysis on sherds of Majiayao culture, Yangshao culture and Qijia culture. *Journal of Archaeological Science: Reports, 3*, 65–72. https://doi.org/10.1016/j.jasrep.2015.05.018.

Cuni-Sanchez, A., Pfeifer, M., Marchant, R., Pompeu, P. V., & Burgess, N. D. (2018). Harvesting fodder trees in montane forests in Kenya: Species, techniques used and impacts. *New Forests, 49*(4), 511–528. https://doi.org/10.1007/s11056-018-9632-x.

Cuni-Sanchez, A., Sullivan, M. J. P., Platts, P. J., Lewis, S. L., Marchant, R., Imani, G., Hubau, W., Abiem, I., Adhikari, H., Albrecht, T., Altman, J., Amani, C., Aneseyee, A. B., Avitabile, V., Banin, L., Batumike, R., Bauters, M., Beeckman, H., Begne, S. K., & …, Zibera, E. (2021). High aboveground carbon stock of African tropical montane forests. *Nature, 596*(7873), 536–542. https://doi.org/10.1038/s41586-021-03728-4.

Dangles, O., Lana, S., Palomo, I., & Rabatel, A. (2025). *The voices of glaciers: Stories of grief and hope amidst shrinking glaciers in the tropics*. UNESCO Publishing.

Davis, B. A. S., Fasel, M., Kaplan, J. O., Russo, E., & Burke, A. (2024). The climate and vegetation of Europe, northern Africa, and the Middle East during the Last Glacial Maximum (21 000 yr BP) based on pollen data. *Climate of the Past, 20*(9), 1939–1988. https://doi.org/10.5194/cp-20-1939-2024.

Dillehay, T. D., Eling, H. H., & Rossen, J. (2005). Preceramic irrigation canals in the Peruvian Andes. *Proceedings of the National Academy of Sciences, 102*(47), 17241–17244. https://doi.org/10.1073/pnas.0508583102.

Djamali, M., & Segarra-Moragues, J. G. (2021). Palaeoecology and conservation of endangered hidden species; example of the liverwort Riella (Riellaceae). *Biodiversity and Conservation, 30*(10), 2731–2750. https://doi.org/10.1007/s10531-021-02218-3.

Dodson, J., Shi, G., Lu, F., & Yan, H. (2021). A 40,000 year record of vegetation, environment and climate change from Chongqing, Central China. *Palaeogeography, Palaeoclimatology, Palaeoecology, 573*, 110441. https://doi.org/10.1016/j.palaeo.2021.110441.

Dodson, J. R., Li, X., Zhou, X., Zhao, K., Sun, N., & Atahan, P. (2013). Origin and spread of wheat in China. *Quaternary Science Reviews, 72*, 108–111. https://doi.org/10.1016/j.quascirev.2013.04.021.

Ellis, E. C. (2024). The Anthropocene condition: Evolving through social–ecological transformations. *Philosophical Transactions of the Royal Society B: Biological Sciences, 379*(1893). https://doi.org/10.1098/rstb.2022.0255.

Erschbamer, B., & Caccianiga, M. (2016). Glacier forelands: Lessons of plant population and community development. *Progress in Botany, 78*, 256.

Fairhead, J., & Leach, M. (1996). *Misreading the African landscape: Society and ecology in a forest-savanna mosaic*: 90. Cambridge University Press.

Farinotti, D., Immerzeel, W. W., de Kok, R. J., Quincey, D. J., & Dehecq, A. (2020). Manifestations and mechanisms of the Karakoram glacier anomaly. *Nature Geoscience, 13*(1), 8–16. https://doi.org/10.1038/s41561-019-0513-5.

Fell, S. C., Carrivick, J. L., & Brown, L. E. (2017). The multitrophic effects of climate change and glacier retreat in mountain rivers. *BioScience, 67*(10), 897–911. https://doi.org/10.1093/biosci/bix107.

Ferro-Vázquez, C., & Stump, D. (2023). *The archaeological study of agricultural sustainability: The cases of Konso (S Ethiopia) and Engaruka (N Tanzania) people and Agrarian Landscapes: An archaeology of postclassical local societies in the western Mediterranean* (pp. 196–216). Archaeopress. https://www.archaeopress.com/Archaeopress/Products/9781803274379.

Ficetola, G. F., Marta, S., Guerrieri, A., Gobbi, M., Ambrosini, R., Fontaneto, D., Zerboni, A., Poulenard, J., Caccianiga, M., & Thuiller, W. (2021). Dynamics of ecological communities following current retreat of glaciers. *Annual Review of Ecology, Evolution, and Systematics, 52*, 405–426. https://doi.org/10.1146/annurev-ecolsys-010521-040017.

Finch, J., Wooller, M., & Marchant, R. (2014). Tracing long-term tropical montane ecosystem change in the Eastern Arc Mountains of Tanzania. *Journal of Quaternary Science, 29*(3), 269–278. https://doi.org/10.1002/jqs.2699.

Flantua, S. G. A., O'Dea, A., Onstein, R. E., Giraldo, C., & Hooghiemstra, H. (2019). The flickering connectivity system of the north Andean páramos. *Journal of Biogeography, 46*(8), 1808–1825. https://doi.org/10.1111/jbi.13607.

Gao, T., Li, J., Tang, Y., Jiang, X., & Huo, J. (2024). Response of runoff and its components to climate change and its future trend in the source region of the Yangtze River. *Hydrological Processes, 38*(6). https://doi.org/10.1002/hyp.15180.

Gavin, D. G., Fitzpatrick, M. C., Gugger, P. F., Heath, K. D., Rodríguez-Sánchez, F., Dobrowski, S. Z., Hampe, A., Hu, F. S., Ashcroft, M. B., Bartlein, P. J., Blois, J. L., Carstens, B. C., Davis, E. B., de Lafontaine, G., Edwards, M. E., Fernandez, M., Henne, P. D., Herring, E. M., Holden, Z. A., & …, Williams, J. W. (2014). Climate refugia: Joint inference from fossil records, species distribution models and phylogeography. *New Phytologist, 204*(1), 37–54. https://doi.org/10.1111/nph.12929.

Geissler, J., Mayer, C., Jubanski, J., Münzer, U., & Siegert, F. (2021). Analyzing glacier retreat and mass balances using aerial and UAV photogrammetry in the Ötztal Alps, Austria. *The Cryosphere, 15*(8), 3699–3717. https://doi.org/10.5194/tc-15-3699-2021.

Gibbard, P., Walker, M., Bauer, A., Edgeworth, M., Edwards, L., Ellis, E., Finney, S., Gill, J. L., Maslin, M., Merritts, D., & Ruddiman, W. (2022). The Anthropocene as an event, not an epoch. *Journal of Quaternary Science, 37*(3), 395–399. https://doi.org/10.1002/jqs.3416.

Giesecke, T., Brewer, S., Finsinger, W., Leydet, M., & Bradshaw, R. H. W. (2017). Patterns and dynamics of European vegetation change over the last 15,000 years. *Journal of Biogeography, 44*(7), 1441–1456. https://doi.org/10.1111/jbi.12974.

Gillson, L., & Marchant, R. (2014). From myopia to clarity: Sharpening the focus of ecosystem management through the lens of palaeoecology. *Trends in Ecology and Evolution, 29*(6), 317–325. https://doi.org/10.1016/j.tree.2014.03.010.

Gillson, L., Seddon, A., Mottl, O., Zhang, K., Kirsten, K., Gell, P., Marchant, R. A., Schwörer, C., Razanatsoa, E., Lane, P. J., Courtney-Mustaphi, C. J., & Dearing, J. (2025). Exploring the interface between planetary boundaries and palaeoecology. *Global Change Biology, 31*(1). https://doi.org/10.1111/gcb.70017.

Giraudi, C., & Giaccio, B. (2017). Middle pleistocene glaciations in the Apennines, Italy: New chronological data and preservation of the glacial record. *Geological Society, London, Special Publications, 433*(1), 161–178. https://doi.org/10.1144/sp433.1.

Githumbi, E. N., Finch, J., Kinyanjui, R. N., Courtney-Mustaphi, C., Musili, P., Rucina, S., Lejju, J., Liljenberg, P., & Marchant, R. (2025). Late Quaternary montane forest dynamics from equatorial east Africa: A biome perspective. *Journal of Biogeography, 52*(9). https://doi.org/10.1111/jbi.15173.

Grainger, T. N., Letten, A. D., Gilbert, B., & Fukami, T. (2019). Applying modern coexistence theory to priority effects. *Proceedings of the National Academy of Sciences, 116*(13), 6205–6210. https://doi.org/10.1073/pnas.1803122116.

Grunewald, K., & Scheithauer, J. (2010). Europe's southernmost glaciers: Response and adaptation to climate change. *Journal of Glaciology, 56*(195), 129–142. https://doi.org/10.3189/002214310791190947.

Guimarães, J. T. F., Sahoo, P. K., e Souza-Filho, P. W. M., da Silva, M. S., Rodrigues, T. M., da Silva, E. F., Reis, L. S., de Figueiredo, M. M. J. C., Lopes, K. S., Moraes, A. M., Leite, A. S., da Silva Júnior, R. O., Salomão, G. N., & Dall'Agnol, R. (2023). Landscape and climate changes in southeastern amazonia from Quaternary records of upland lakes. *Atmosphere, 14*(4). https://doi.org/10.3390/atmos14040621.

Guo, L., Zhao, H., Zhai, X., Wang, K., Liu, L., Wang, K., & Huang, D. (2021). Study on life histroy traits of Stellera chamaejasme provide insights into its control on degraded typical steppe. *Journal of Environmental Management, 291*, 112716. https://doi.org/10.1016/j.jenvman.2021.112716.

Haffer, J. (1969). Speciation in Amazonian forest birds. *Science, 165*(3889), 131–137. https://doi.org/10.1126/science.165.3889.131.

Hågvar, S., Gobbi, M., Kaufmann, R., Ingimarsdóttir, M., Caccianiga, M., Valle, B., Pantini, P., Fanciulli, P. P., & Vater, A. (2020). Ecosystem birth near melting glaciers: A review on the pioneer role of ground-dwelling arthropods. *Insects, 11*(9), 644. https://doi.org/10.3390/insects11090644.

Han, D., Sun, Y., Yu, Z., Jiang, M., Cong, J., Gao, C., & Wang, G. (2023). Diatom evidence for late Holocene environmental change in a permafrost peatland in the northern Greater Khingan Mountains, Northeast China. *Palaeogeography, Palaeoclimatology, Palaeoecology, 625*, 111665. https://doi.org/10.1016/j.palaeo.2023.111665.

Harris, S. (Ed.). (2022). *The nature, causes, effects and mitigation of climate change on the environment*. BoD–Books on Demand.

Higuera, P. E., Shuman, B. N., & Wolf, K. D. (2021). Rocky Mountain subalpine forests now burning more than any time in recent millennia. *Proceedings of the National Academy of Sciences, 118*(25). https://doi.org/10.1073/pnas.2103135118.

Hock, R., Rasul, G., Adler, C., Cáceres, B., Gruber, S., Hirabayashi, Y., Jackson, M., Kääb, A., Kang, S., Kutuzov, S., Milner, A., Molau, U., Morin, S., Orlove, B., & Steltzer, H. (2019).

High mountain areas. In H.-. O Pörtner, D. C. Roberts, V. Masson-Delmotte, P. Zhai, M. Tignor, E. Poloczanska, K. Mintenbeck, A. Alegriá, M. Nicolai, A. Okem, J. Petzold, B. Rama, & N. M. Weyer (Eds.), *IPCC special report on the ocean and cryosphere in a changing climate* (pp. 131–202). Cambridge University Press. https://www.ipcc.ch/srocc/.

Hooghiemstra, H., & Flantua, S. G. A. (2019). Colombia in the Quaternary: An overview of environmental and climatic change. In *The geology of Colombia* (pp. 43–104). Servicio Geológico Colombiano.

Hughes, P. D., Gibbard, P. L., & Ehlers, J. (2013). Timing of glaciation during the last glacial cycle: Evaluating the concept of a global 'Last Glacial Maximum' (LGM). *Earth-Science Reviews, 125*, 171–198. https://doi.org/10.1016/j.earscirev.2013.07.003.

Humanes-Fuente, V., Ferrero, M. E., Muñoz, A. A., González-Reyes, Á., Requena-Rojas, E. J., Barichivich, J., Inga, J. G., & Layme-Huaman, E. T. (2020). Two centuries of hydroclimatic variability reconstructed from tree-ring records over the Amazonian Andes of Peru. *Journal of Geophysical Research: Atmospheres, 125*(18). https://doi.org/10.1029/2020jd032565.

Hylander, K., Ehrlen, J., Luoto, M., & Meineri, E. (2015). Microrefugia: Not for everyone. *Ambio, 44*(Suppl 1), S60–S68. https://doi.org/10.1007/s13280-014-0599-3.

Immerzeel, W. W., Lutz, A. F., Andrade, M., Bahl, A., Biemans, H., Bolch, T., Hyde, S., Brumby, S., Davies, B. J., Elmore, A. C., Emmer, A., Feng, M., Fernández, A., Haritashya, U., Kargel, J. S., Koppes, M., Kraaijenbrink, P. D. A., Kulkarni, A. V., Mayewski, P. A., & ..., Baillie, J. E. M. (2020). Importance and vulnerability of the world's water towers. *Nature, 577*(7790), 364–369. https://doi.org/10.1038/s41586-019-1822-y.

IPCC. (2022). *Climate Change 2022: Impacts, adaptation, and vulnerability: Contribution of working group II to the sixth assessment report of the intergovernmental panel on climate change.* IPCC.

Jantz, N., Homeier, J., Behling, H., & Pillar, V. (2014). Representativeness of tree diversity in the modern pollen rain of Andean montane forests. *Journal of Vegetation Science, 25*(2), 481–490. https://doi.org/10.1111/jvs.12105.

Jia, X., Dong, G., Li, H., Brunson, K., Chen, F., Ma, M., Wang, H., An, C., & Zhang, K. (2013). The development of agriculture and its impact on cultural expansion during the late Neolithic in the Western Loess Plateau, China. *The Holocene, 23*(1), 85–92. https://doi.org/10.1177/0959683612450203.

Julier, A. C. M., Manzano, S., Razanatsoa, E., Razafimanantsoa, A. H. I., Githumbi, E., Hawthorne, D., Oden, G., Schüler, L., Tossou, M., & Bunting, J. (2021). *Modern pollen studies from tropical Africa and their use in palaeoecology* (pp. 317–348). Informa UK Limited. https://doi.org/10.1201/9781003162766-21.

Jump, A. S., Carr, M., Ahrends, A., & Marchant, R. (2014). Genetic divergence during long-term isolation in highly diverse populations of tropical trees across the Eastern Arc mountains of Tanzania. *Biotropica, 46*(5), 565–574. https://doi.org/10.1111/btp.12139.

Kappenberg, A., Amelung, W., Conze, N., Sirocko, F., & Lehndorff, E. (2021). Fire–vegetation relationships during the last glacial cycle in a low mountain range (Eifel, Germany). *Palaeogeography, Palaeoclimatology, Palaeoecology, 562*, 110140. https://doi.org/10.1016/j.palaeo.2020.110140.

Karachurina, S., Rudaya, N., Frolova, L., Kuzmina, O., Cao, X., Chepinoga, V., Stoof-Leichsenring, K., Biskaborn, B., Herzschuh, U., Nigmatullin, N., Vnukovskaya, Y., Grekov, I., & Pestryakova, L. (2023). Terrestrial vegetation and lake aquatic community diversity under climate change during the mid–late Holocene in the Altai Mountains. *Palaeogeography, Palaeoclimatology, Palaeoecology, 623*, 111623. https://doi.org/10.1016/j.palaeo.2023.111623.

Kariuki, R., Willcock, S., & Marchant, R. (2018). Rangeland livelihood strategies under varying climate regimes: Model insights from southern Kenya. *Land*, 7(2), 47. https://doi.org/10.3390/land7020047.

Kariuki, R. W., Western, D., Willcock, S., & Marchant, R. (2021). Assessing interactions between agriculture, livestock grazing and wildlife conservation land uses: A historical example from east africa. *Land, 10*(1), 1–19. https://doi.org/10.3390/land10010046.

Kaufman, D. S., & Broadman, E. (2023). Revisiting the Holocene global temperature conundrum. *Nature, 614*(7948), 425–435. https://doi.org/10.1038/s41586-022-05536-w.

Kistler, L., Maezumi, S. Y., Souza, J. G., Przelomska, N. A. S., Costa, F. M., Smith, O., Loiselle, H., Ramos-Madrigal, J., Wales, N., Ribeiro, E. R., Morrison, R. R., Grimaldo, C., Prous, A. P., Arriaza, B., Gilbert, M. T. P., Freitas, F. O., & Allaby, R. G. (2018). Multiproxy evidence highlights a complex evolutionary legacy of maize in South America. *Science, 362*(6420), 1309–1313. https://doi.org/10.1126/science.aav0207.

Kłapyta, P., Krzemień, K., Gorczyca, E., Krąż, P., & Dubis, L. (2020). Geomorphologic effects of human impact across the Svydovets Massif in the Eastern Carpathians in Ukraine. *Studia Geomorphologica Carpatho-Balcanica, vol. LIII-LIV*, 85–111.

Koffel, T., Boudsocq, S., Loeuille, N., Daufresne, T., & Grover, J. (2018). Facilitation- vs. competition-driven succession: The key role of resource-ratio. *Ecology Letters, 21*(7), 1010–1021. https://doi.org/10.1111/ele.12966.

Köse, O., Sarıkaya, M. A., Çiner, A., Candaş, A., Yıldırım, C., & Wilcken, K. M. (2022). Reconstruction of Last Glacial Maximum glaciers and palaeoclimate in the central Taurus Range, Mt. Karanfil, of the Eastern Mediterranean. *Quaternary Science Reviews, 291*. https://doi.org/10.1016/j.quascirev.2022.107656.

Lamb, H. H. (2013). *Climate: Present, past and future (Routledge revivals): Volume 2: Climatic history and the future*. Routledge.

Langlie, B. S. (2021). Origins of food production in the high Andes. *Oxford research encyclopedia of anthropology*. Oxford University Press https://doi.org/10.1093/acrefore/9780190854584.013.442.

Li, X., Dong, S., Li, Z., Peng, X., Shen, Y., Jin, Q., Yu, X., & Zhang, D. (2024). Surface soil phytolith assemblages across an altitudinal transect in the Qilian Mountains of Northwestern China, and their implications for palaeoecologic analysis in arid alpine regions. *Palaeogeography, Palaeoclimatology, Palaeoecology, 650*, 112374. https://doi.org/10.1016/j.palaeo.2024.112374.

Li, Y., Zhang, D., Zhang, Y., Sun, A., Li, X., Huang, X., Zhang, Y., & Li, Y. (2024b). Disentangling the late-Holocene human–environment interactions in the Altai Mountains within the Arid Central Asia. *Palaeogeography, Palaeoclimatology, Palaeoecology, 654*, 112466.

Liang, L., Cuo, L., & Liu, Q. (2018). The energy and mass balance of a continental glacier: Dongkemadi glacier in central Tibetan plateau. *Scientific Reports, 8*(1). https://doi.org/10.1038/s41598-018-31228-5.

Liang, L., Sun, Y., Beets, C. J., Prins, M. A., Wu, F., & Vandenberghe, J. (2013). Impacts of grain size sorting and chemical weathering on the geochemistry of Jingyuan loess in the northwestern Chinese Loess Plateau. *Journal of Asian Earth Sciences, 69*, 177–184. https://doi.org/10.1016/j.jseaes.2012.12.015.

Liniger, H., & Weingartner, R. (2000). Mountain forests and their role in providing freshwater resources. In *Forests in sustainable mountain development: A state of knowledge report for 2000. Task Force on Forests in Sustainable Mountain Development* (pp. 370–380). CABI Publishing. https://doi.org/10.1079/9780851994468.0370.

Liu, F., & Feng, Z. (2012). A dramatic climatic transition at ~4000 cal. yr BP and its cultural responses in Chinese cultural domains. *Holocene, 22*(10), 1181–1197. https://doi.org/10.1177/0959683612441839.

Liu, L., Zhang, H., Zhang, W., & Chai, L. (2022). Global last glacial maximum climate inferred from reconstructing the Eryehai Valley, Mount Taibai, Qinling Mountains, eastern China. *Palaeogeography, Palaeoclimatology, Palaeoecology, 590*, 110858. https://doi.org/10.1016/j.palaeo.2022.110858.

Ljungqvist, F. C., Piermattei, A., Seim, A., Krusic, P. J., Büntgen, U., He, M., Kirdyanov, A. V., Luterbacher, J., Schneider, L., Seftigen, K., Stahle, D. W., Villalba, R., Yang, B., & Esper, J. (2020). Ranking of tree-ring based hydroclimate reconstructions of the past millennium. *Quaternary Science Reviews, 230*. https://doi.org/10.1016/j.quascirev.2019.106074.

Los, S. O., Street-Perrott, F. A., Loader, N. J., Froyd, C. A., Cuní-Sanchez, A., & Marchant, R. A. (2019). Sensitivity of a tropical montane cloud forest to climate change, present, past and future: Mt. Marsabit, N. Kenya. *Quaternary Science Reviews, 218*, 34–48. https://doi.org/10.1016/j.quascirev.2019.06.016.

Losapio, G., Jordán, F., Caccianiga, M., & Gobbi, M. (2015). Structure-dynamic relationship of plant–insect networks along a primary succession gradient on a glacier foreland. *Ecological Modelling, 314*, 73–79. https://doi.org/10.1016/j.ecolmodel.2015.07.014.

Losapio, G., Norton Hasday, E., Espadaler, X., Germann, C., Ortiz-Sánchez, F. J., Pont, A., Sommaggio, D., & Schöb, C. (2021). Facilitation and biodiversity jointly drive mutualistic networks. *Journal of Ecology, 109*(5), 2029–2037. https://doi.org/10.1111/1365-2745.13593.

Ma, Y., Song, H., Liu, Y., Zhang, Q., Ren, M., Cai, Q., Zhang, T., Pourtahmasi, K., Duan, X., & Li, P. (2025). Drought reconstructions over the past 552 years based on minimum earlywood density in central Tianshan Mountains. *Palaeogeography, Palaeoclimatology, Palaeoecology, 667*, 112853. https://doi.org/10.1016/j.palaeo.2025.112853.

Marchant, R., Behling, H., Berrio, J. C., Hooghiemstra, H., van Geel, B., van der Hammen, T., Herrera, L., Melief, B., van Reenen, G., & Wille, M. (2004). Vegetation disturbance and human population in Columbia - A regional reconstruction. *Antiquity, 78*(302), 828–838. https://doi.org/10.1017/S0003598X0011347X.

Marchant, R., Cleef, A., Harrison, S. P., Hooghiemstra, H., Markgraf, V., Van Boxel, J., Ager, T., Almeida, L., Anderson, R., Baied, C., Behling, H., Berrio, J. C., Burbridge, R., BjörckS. , R. B., Bush, M., Duivenvoorden, J., Flenley, J., De Oliveira, P., Van Gee, B., & ..., Wille, M. (2009). Pollen-based biome reconstructions for Latin America at 0, 6000 and 18,000 radiocarbon years ago. *Climate of the Past, 5*(4), 725–767. https://doi.org/10.5194/cp-5-725-2009.

Marchant, R., & Hooghiemstra, H. (2004). Rapid environmental change in African and South American tropics around 4000 years before present: A review. *Earth-Science Reviews, 66*(3-4), 217–260. https://doi.org/10.1016/j.earscirev.2004.01.003.

Marchant, R., & Lane, P. (2014). Past perspectives for the future: Foundations for sustainable development in East Africa. *Journal of Archaeological Science, 51*, 12–21. https://doi.org/10.1016/j.jas.2013.07.005.

Marchant, R., Richer, S., Boles, O., Capitani, C., Courtney-Mustaphi, C. J., Lane, P., Prendergast, M. E., Stump, D., De Cort, G., Kaplan, J. O., Phelps, L., Kay, A., Olago, D., Petek, N., Platts, P. J., Punwong, P., Widgren, M., Wynne-Jones, S., Ferro-Vázquez, C., & ..., Wright, D. (2018). Drivers and trajectories of land cover change in East Africa: Human and environmental interactions from 6000 years ago to present. *Earth-Science Reviews, 178*, 322–378. https://doi.org/10.1016/j.earscirev.2017.12.010.

Marlon, J. R. (2020). What the past can say about the present and future of fire. *Quaternary Research, 96*, 66–87. https://doi.org/10.1017/qua.2020.48.

Marlon, J. R., Bartlein, P. J., Daniau, A. L., Harrison, S. P., Maezumi, S. Y., Power, M. J., Tinner, W., & Vanniére, B. (2013). Global biomass burning: A synthesis and review of Holocene paleofire records and their controls. *Quaternary Science Reviews, 65*, 5–25. https://doi.org/10.1016/j.quascirev.2012.11.029.

Mélières, M. A., & Maréchal, C. (2015). *Climate change: Past, present, and future*. John Wiley and Sons.

Millhauser, J. K., & Earle, T. K. (2022). Biodiversity and the human past: Lessons for conservation biology. *Biological Conservation, 272*. https://doi.org/10.1016/j.biocon.2022.109599.

Mölg, N., Huggel, C., Herold, T., Storck, F., Allen, S., Haeberli, W., Schaub, Y., & Odermatt, D. (2021). Inventory and evolution of glacial lakes since the Little Ice Age: Lessons from the case of Switzerland. *Earth Surface Processes and Landforms, 46*(13), 2551–2564. https://doi.org/10.1002/esp.5193.

Morales, M. S., Christie, D. A., Villalba, R., Argollo, J., Pacajes, J., Silva, J. S., Alvarez, C. A., Llancabure, J. C., & Soliz Gamboa, C. C. (2012). Precipitation changes in the South American Altiplano since 1300 AD reconstructed by tree-rings. *Climate of the Past, 8*(2), 653–666. https://doi.org/10.5194/cp-8-653-2012.

Muellner-Riehl, A. N., Schnitzler, J., Kissling, W. D., Mosbrugger, V., Rijsdijk, K. F., Seijmonsbergen, A. C., Versteegh, H., & Favre, A. (2019). Origins of global mountain plant biodiversity: Testing the 'mountain-geobiodiversity hypothesis'. *Journal of Biogeography, 46*(12), 2826–2838. https://doi.org/10.1111/jbi.13715.

Mumbi, C. T., Marchant, R., Hooghiemstra, H., & Wooller, M. J. (2008). Late Quaternary vegetation reconstruction from the Eastern Arc Mountains, Tanzania. *Quaternary Research, 69*(2), 326–341. https://doi.org/10.1016/j.yqres.2007.10.012.

Osman, M. B., Tierney, J. E., Zhu, J., Tardif, R., Hakim, G. J., King, J., & Poulsen, C. J. (2021). Globally resolved surface temperatures since the Last Glacial Maximum. *Nature, 599*(7884), 239–244. https://doi.org/10.1038/s41586-021-03984-4.

Pang, W., Li, Q., Liu, Y., Song, H., Sun, C., Wang, J., Yan, Y., Cai, Q., & Ren, M. (2023). Precipitation variations in China's Altay mountains detected from tree rings dating back to AD 1615. *Forests, 14*(7), 1496. https://doi.org/10.3390/f14071496.

Parks, K. E., & Mulligan, M. (2010). On the relationship between a resource based measure of geodiversity and broad scale biodiversity patterns. *Biodiversity and Conservation, 19*(9), 2751–2766. https://doi.org/10.1007/s10531-010-9876-z.

Pepin, N. C., Arnone, E., Gobiet, A., Haslinger, K., Kotlarski, S., Notarnicola, C., Palazzi, E., Seibert, P., Serafin, S., Schöner, W., Terzago, S., Thornton, J. M., Vuille, M., & Adler, C. (2022). Climate changes and their elevational patterns in the mountains of the world. *Reviews of Geophysics, 60*(1). https://doi.org/10.1029/2020rg000730.

Petek-Sargeant, N., & Lane, P. J. (2021). Weathering climate change in archaeology: Conceptual challenges and an east African case study. *Cambridge Archaeological Journal, 31*(3), 437–454. https://doi.org/10.1017/S0959774321000044.

Peterson, J. M., Shackleton, S., Higgins, J., Severinghaus, J., Yan, Y., Buizert, C., Kalk, M., Beaudette, R., Hishamunda, V., Carter, A., Kurbatov, A., Epifanio, J., Morgan, J., Bender, M., & Brook, E. (2024). Ice cores from the Allan Hills, Antarctica show relatively stable atmospheric CO2 and CH4 levels over the last 3 million years. *Research Square*. https://doi.org/10.21203/rs.3.rs-5610566/v1.

Pezo-Lanfranco, L., & Colonese, A. C. (2024). The role of farming and fishing in the rise of social complexity in the Central Andes: A stable isotope perspective. *Scientific Reports, 14*(1). https://doi.org/10.1038/s41598-024-55436-4.

Pianka, E. R. (1966). Latitudinal gradients in species diversity: A review of concepts. *The American Naturalist, 100*(910), 33–46. https://doi.org/10.1086/282398.

Piperno, D. R. (2011). The origins of plant cultivation and domestication in the New World Tropics patterns, process, and new developments. *Current Anthropology, 52*(4), S453. https://doi.org/10.1086/659998.

Piperno, D. R., & Smith, B. D. (2012). The origins of food production in Mesoamerica. In D. L. Nichols, & C. A. Pool (Eds.), *The Oxford handbook of mesoamerican archaeology* (pp. 151–164). Oxford University Press.

Porter, S. C., & An, Z. (1995). Correlation between climate events in the North Atlantic and China during the last glaciation. *Nature, 375*, 305–308. https://doi.org/10.1038/375305a0.

Pörtner, H. O., Roberts, D. C., Masson-Delmotte, V., Zhai, P., Tignor, M., Poloczanska, E., & …, Weyer, N. (2019). The ocean and cryosphere in a changing climate. In *Intergovernmental Panel on Climate Change (IPCC): 1* (p. 255). Cambridge University Press. https://doi.org/10.1017/9781009157964.

Raduła, M., Świerszcz, S., Nobis, M., Nowak, S., Nobis, A., & Nowak, A. (2021). Palaeoclimate has a major effect on the diversity of endemic species in the hotspot of mountain biodiversity in Tajikistan. *Scientific Reports, 11*(1), 18684.

Rahbek, C., Borregaard, M. K., Antonelli, A., Colwell, R. K., Holt, B. G., Nogues-Bravo, D., Rasmussen, C. M. Ø., Richardson, K., Rosing, M. T., Whittaker, R. J., & Fjeldså, J. (2019). Building mountain biodiversity: Geological and evolutionary processes. *Science, 365*(6458), 1114–1119. https://doi.org/10.1126/science.aax0151.

Rahman, A., Shah, R. A., Rathi, A., Yadava, M. G., & Kumar, S. (2024). Transport pathways of black carbon to a high mountain Himalayan lake during late Holocene: Inferences from nitrogen isotopes of black carbon. *Palaeogeography, Palaeoclimatology, Palaeoecology, 633*, 111865. https://doi.org/10.1016/j.palaeo.2023.111865.

Reitalu, T., Bjune, A. E., Blaus, A., Giesecke, T., Helm, A., Matthias, I., Peglar, S. M., Salonen, J. S., Seppä, H., Väli, V., & Birks, H. J. B. (2019). Patterns of modern pollen and plant richness across northern Europe. *Journal of Ecology, 107*(4), 1662–1677. https://doi.org/10.1111/1365-2745.13134.

Ren, G., Chan, J. C. L., Kubota, H., Zhang, Z., Li, J., Zhang, Y., Zhang, Y., Yang, Y., Ren, Y., Sun, X., Su, Y., Liu, Y., Hao, Z., Xue, X., & Qin, Y. (2021). Historical and recent change in extreme climate over East Asia. *Climatic Change, 168*(3-4). https://doi.org/10.1007/s10584-021-03227-5.

Richardson, K., Steffen, W., Lucht, W., Bendtsen, J., Cornell, S. E., Donges, J. F., Drüke, M., Fetzer, I., Bala, G., von Bloh, W., Feulner, G., Fiedler, S., Gerten, D., Gleeson, T., Hofmann, M., Huiskamp, W., Kummu, M., Mohan, C., Nogués-Bravo, D., & …, Rockström, J. (2023). Earth beyond six of nine planetary boundaries. *Science Advances, 9*(37). https://doi.org/10.1126/sciadv.adh2458.

Roberts, D. C., Forrest, A. L., Sahoo, G. B., Hook, S. J., & Schladow, S. G. (2018). Snowmelt timing as a determinant of lake inflow mixing. *Water Resources Research, 54*(2), 1237–1251. https://doi.org/10.1002/2017wr021977.

Roberts, P., Hunt, C., Arroyo-Kalin, M., Evans, D., & Boivin, N. (2017). The deep human prehistory of global tropical forests and its relevance for modern conservation. *Nature Plants, 3*. https://doi.org/10.1038/nplants.2017.93.

Rockström, J., Steffen, W., Noone, K., Persson, A., Chapin, F. S., Lambin, E., Lenton, T. M., Scheffer, M., Folke, C., Schellnhuber, H. J., Nykvist, B., de Wit, C. A., Hughes, T., van der Leeuw, S., Rodhe, H., Sörlin, S., Snyder, P. K., Costanza, R., Svedin, U., & …, Foley, J. (2009). Planetary boundaries: Exploring the safe operating space for humanity. *Ecology and Society, 14*(2). https://doi.org/10.5751/ES-03180-140232.

Rosero, P., Crespo-Pérez, V., Espinosa, R., Andino, P., Barragán, Á, Moret, P., Gobbi, M., Ficetola, G. F., Jaramillo, R., Muriel, P., Anthelme, F., Jacobsen, D., Dangles, O., Condom, T., Gielly, L., Poulenard, J., Rabatel, A., Basantes, R., Correa, B., & Cauvy-Fraunié, S. (2021). Multi-taxa colonisation along the foreland of a vanishing equatorial glacier. *Ecography, 44*(7), 1010–1021. https://doi.org/10.1111/ecog.05478.

Scaldaferro, M. A., Srsic, A., Romanutti, A. A., & Acosta, M. C. (2023). Mountains as refugia in the Dry Chaco: Phylogeography and species distribution modelling of the southernmost chilli pepper, *Capsicum chacoense* Hunz. (Solanaceae). *Biological Journal of the Linnean Society, 140*(1), 130–148. https://doi.org/10.1093/biolinnean/blad045.

Schickhoff, U., Bobrowski, M., Mal, S., Schwab, N., & Singh, R. B. (2021). The world's mountains in the Anthropocene. In *Mountain landscapes in transition: Effects of land use and climate change* (pp. 1–144). Springer International Publishing.

Schulte, L., Bernhardt, N., Stoof-Leichsenring, K., Zimmermann, H. H., Pestryakova, L. A., Epp, L. S., & Herzschuh, U. (2021). Hybridization capture of larch (Larix Mill.) chloroplast genomes from sedimentary ancient DNA reveals past changes of Siberian forest. *Molecular Ecology Resources, 21*(3), 801–815. https://doi.org/10.1111/1755-0998.13311.

Sha, Y., Shi, Z., Zhou, P., Lei, J., & Mao, X. (2023). Differentiation of Asian summer precipitation induced by the mountain building of the Tibetan Plateau and Central Asian Orogenic Belt. *Palaeogeography, Palaeoclimatology, Palaeoecology, 619*, 111547. https://doi.org/10.1016/j.palaeo.2023.111547.

Shipton, C., Roberts, P., Archer, W., Armitage, S. J., Bita, C., Blinkhorn, J., Courtney-Mustaphi, C., Crowther, A., Curtis, R., Errico, F. D., Douka, K., Faulkner, P., Groucutt, H. S., Helm, R., Herries, A. I. R., Jembe, S., Kourampas, N., Lee-Thorp, J., Marchant, R., & …, Boivin, N. (2018). 78,000-year-old record of middle and later stone age innovation in an East African tropical forest. *Nature Communications, 9*(1). https://doi.org/10.1038/s41467-018-04057-3.

Solís-Torres, Ó. R., Acosta-Ochoa, G., Arroyo-Cabrales, J., Flores-Granados, F., & Roberts, P. (2021). Pleistocene-Holocene human palaeoecology in southern Mexico: Stable isotopic evidence from the Santa Marta Cave, Chiapas. *Journal of Archaeological Science: Reports, 39*, 103131. https://doi.org/10.1016/j.jasrep.2021.103131.

Sommer, C., Malz, P., Seehaus, T. C., Lippl, S., Zemp, M., & Braun, M. H. (2020). Rapid glacier retreat and downwasting throughout the European Alps in the early 21st century. *Nature Communications, 11*(1). https://doi.org/10.1038/s41467-020-16818-0.

Song, M., Dodson, J., Lu, F., Shi, G., & Yan, H. (2023). A continuous paleorecord of vegetation and environmental change from Erxianyan Wetland over the past 60,000 years in central China. *Palaeogeography, Palaeoclimatology, Palaeoecology, 613*, 111399. https://doi.org/10.1016/j.palaeo.2023.111399.

Spicer, R. A., Farnsworth, A., & Su, T. (2020). Cenozoic topography, monsoons and biodiversity conservation within the Tibetan Region: An evolving story. *Plant Diversity, 42*(4), 229–254. https://doi.org/10.1016/j.pld.2020.06.011.

Stark, J. R., & Fridley, J. D. (2022). Microclimate-based species distribution models in complex forested terrain indicate widespread cryptic refugia under climate change. *Global Ecology and Biogeography, 31*(3), 562–575. https://doi.org/10.1111/geb.13447.

Staubwasser, M., & Weiss, H. (2006). Holocene climate and cultural evolution in late prehistoric—Early historic west Asia. *Quaternary Research, 66*(3), 372–387. https://doi.org/10.1016/j.yqres.2006.09.001.

Steffen, T., Huss, M., Estermann, R., Hodel, E., & Farinotti, D. (2022). Volume, evolution, and sedimentation of future glacier lakes in Switzerland over the 21st century. *Earth Surface Dynamics, 10*(4), 723–741.

Steffen, W. (2021). The Earth system, the great acceleration and the Anthropocene. In *Sustainability and the new economics: Synthesising ecological economics and modern monetary theory* (pp. 15–32). Springer International Publishing.

Steffen, W., Richardson, K., Rockström, J., Cornell, S. E., Fetzer, I., Bennett, E. M., Biggs, R., Carpenter, S. R., De Vries, W., De Wit, C. A., Folke, C., Gerten, D., Heinke, J., Mace, G. M., Persson, L. M., Ramanathan, V., Reyers, B., & Sörlin, S. (2015). Planetary boundaries: Guiding human development on a changing planet. *Science, 347*(6223). https://doi.org/10.1126/science.1259855.

Stobbe, A., & Gumnior, M. (2021). Palaeoecology as a tool for the future management of forest ecosystems in Hesse (Central Germany): Beech (Fagus sylvatica L.) versus Lime (Tilia cordata Mill.). *Forests, 12*(7), 924. https://doi.org/10.3390/f12070924.

Sun, C., Li, Q., Liu, Y., Song, H., Fang, C., Cai, Q., Ren, M., Ye, Y., Liu, R., & Sun, J. (2022). Tree rings reveal changes in the temperature pattern in eastern China before and during the Anthropocene. *Environmental Research Letters, 17*(12), 124034. https://doi.org/10.1088/1748-9326/aca68e.

Sun, Y., Xu, X., Zhang, L., Liu, J., Zhang, X., Li, J., & Pan, B. (2020). Numerical reconstruction of three Holocene glacial events in Qiangyong Valley, Southern Tibetan Plateau and their implication for Holocene climate changes. *Water, 12*(11), 3205. https://doi.org/10.3390/w12113205.

Thompson, L. G., Yao, T., Davis, M. E., Henderson, K. A., Mosley-Thompson, E., Lin, P.-N., Beer, J., Synal, H.-A., Cole-Dai, J., & Bolzan, J. F. (1997). Tropical climate instability: The last glacial cycle from a Qinghai-Tibetan Ice Core. *Science, 276*(5320), 1821–1825. https://doi.org/10.1126/science.276.5320.1821.

Tom, M., Wu, T., Baltsavias, E., & Schindler, K. (2022). Recent Ice Trends in Swiss Mountain Lakes: 20-year Analysis of MODIS imagery. *PFG - Journal of Photogrammetry, Remote Sensing and Geoinformation Science, 90*(4), 413–431. https://doi.org/10.1007/s41064-022-00215-x.

Tovar, C., Carril, A. F., Gutiérrez, A. G., Ahrends, A., Fita, L., Zaninelli, P., Flombaum, P., Abarzúa, A. M., Alarcón, D., Aschero, V., Báez, S., Barros, A., Carilla, J., Ferrero, M. E., Flantua, S. G. A., Gonzáles, P., Menéndez, C. G., Pérez-Escobar, O. A., Pauchard, A., & ..., Hollingsworth, P. M. (2022). Understanding climate change impacts on biome and plant distributions in the Andes: Challenges and opportunities. *Journal of Biogeography, 49*(8), 1420–1442. https://doi.org/10.1111/jbi.14389.

Vidaller, I., & Alonso-González, E. (2021). Changes in the frequency of global high mountain rain-on-snow events due to climate warming. *Environmental Research Letters, 16*(9), 094021.

von Hellermann, P. (2020). Partial stories: Repeat photography, narratives and environmental change in Tanzania. *Visual Anthropology, 33*(4), 363–391. https://doi.org/10.1080/08949468.2020.1791575.

Vorontsova, M. S., Besnard, G., Forest, F., Malakasi, P., Moat, J., Clayton, W. D., Ficinski, S. Z., Savva, G. M., Nanjarisoa, O. P., Razanatsoa, J., Randriatsara, F. O., Kimeu, J. M., Luke, W. R. Q., Kayombo, C., & Linder, H. P. (2016). Madagascar's grasses and grasslands: Anthropogenic or natural? *Proceedings of the Royal Society B: Biological Sciences, 283*(1823). https://doi.org/10.1098/rspb.2015.2262.

Wambulwa, M. C., Milne, R., Wu, Z. Y., Spicer, R. A., Provan, J., Luo, Y. H., Zhu, G. F., Wang, W. T., Wang, H., Gao, L. M., Li, D. Z., & Liu, J. (2021). Spatiotemporal maintenance of flora in the Himalaya biodiversity hotspot: Current knowledge and future perspectives. *Ecology and Evolution, 11*(16), 10794–10812. https://doi.org/10.1002/ece3.7906.

Wang, J., Zhou, W., Dong, G., Xian, F., Fu, Y., Zhang, L., & Ding, P. (2024). Repeated glacial fluctuations during the Last Glacial Maximum in the southeastern Tibetan Plateau: 10Be surface exposure dating of moraines in the Lahaku Valley, Haizishan Plateau, China. *Palaeogeography, Palaeoclimatology, Palaeoecology, 636*, 111959. https://doi.org/10.1016/j.palaeo.2023.111959.

Wang, L., Sun, L., Shrestha, M., Li, X., Liu, W., Zhou, J., Yang, K., Lu, H., & Chen, D. (2016). Improving snow process modeling with satellite-based estimation of near-surface-air-temperature lapse rate. *Journal of Geophysical Research, 121*(20), 12–30. https://doi.org/10.1002/2016JD025506.

Wende, Z., Hou, G., Gao, J., Chen, X., Jin, S., & Lancuo, Z. (2021). Reconstruction of cultivated land in the northeast margin of Qinghai–Tibetan Plateau and anthropogenic impacts on palaeo-environment during the mid-Holocene. *Frontiers in Earth Science, 9*. https://doi.org/10.3389/feart.2021.681995.

Wilkin, S., Ventresca Miller, A., Taylor, W. T. T., Miller, B. K., Hagan, R. W., Bleasdale, M., Scott, A., Gankhuyg, S., Ramsøe, A., Uliziibayar, S., Trachsel, C., Nanni, P., Grossmann, J., Orlando, L., Horton, M., Stockhammer, P. W., Myagmar, E., Boivin, N., Warinner, C., & Hendy, J. (2020). Dairy pastoralism sustained eastern Eurasian steppe populations for 5000 years. *Nature Ecology and Evolution, 4*(3), 346–355. https://doi.org/10.1038/s41559-020-1120-y.

Willis, K. J., Bailey, R. M., Bhagwat, S. A., & Birks, H. J. B. (2010). Biodiversity baselines, thresholds and resilience: Testing predictions and assumptions using palaeoecological data. *Trends in Ecology & Evolution, 25*(10), 583–591. https://doi.org/10.1016/j.tree.2010.07.006.

Wu, Z. Y., Sun, H., Zhou, Z. K., Peng, H., & Li, D. Z. (2005). Origin and differentiation of endemism in the flora of China. *Plant Diversity, 27*(6), 577–604.

Yang, Y., Yang, W., Li, Y., Liu, F., Xiao, J., Liu, B., Li, M., & Liu, G. (2023). Reconstruction of palaeoglaciers and palaeoclimate in Zheduo Shan, Eastern Tibetan Plateau, during the

Last Glacial Maximum. *Quaternary International, 673*, 18–28. https://doi.org/10.1016/j.quaint.2023.10.010.
Yuan, Y., Deng, Y., Gao, L., Fu, X., Xu, Y., & Gou, X. (2024). Unprecedented recent warming as recorded by tree-ring in the western Qinling Mountains, China. *Palaeogeography, Palaeoclimatology, Palaeoecology, 644*, 112202. https://doi.org/10.1016/j.palaeo.2024.112202.
Zarrillo, S., & Kooyman, B. (2006). Evidence for berry and maize processing on the Canadian plains from starch grain analysis. *American Antiquity, 71*(3), 473–499.
Zekollari, H., Huss, M., & Farinotti, D. (2019). Modelling the future evolution of glaciers in the European Alps under the EURO-CORDEX RCM ensemble. *Cryosphere, 13*(4), 1125–1146. https://doi.org/10.5194/tc-13-1125-2019.
Zhang, H., Xu, X., Sun, Y., Li, J., & Xu, B. (2024). Reconstructions of Little Ice Age glaciers and climate in the Tanggula Mountains, Central Tibet Plateau. *Palaeogeography, Palaeoclimatology, Palaeoecology, 637*, 112008. https://doi.org/10.1016/j.palaeo.2023.112008.
Zhang, S. Q., Zhang, Y., Li, J. S., & Gao, X. (2016). The broad-spectrum adaptations of hominins in the later period of Late Pleistocene of China—Perspectives from the zooarchaeological studies. *Science China Earth Sciences, 59*(8), 1529–1539. https://doi.org/10.1007/s11430-016-5287-7.
Zhang, Y., Chen, X., Yang, P., Gao, C., Tong, C., & Meyers, P. A. (2025a). Spatial differences in Holocene peat and carbon accumulation rates in three peatlands in the permafrost region of Altai Mountains, Northwestern China. *Palaeogeography, Palaeoclimatology, Palaeoecology, 667*, 112879. https://doi.org/10.1016/j.palaeo.2025.112879.
Zhang, Y., Cui, Q., Liang, C., Wang, Z., Lu, F., Lin, Y., Chen, L., & Zhou, A. (2025b). Quantitative pollen reconstruction of temperature and precipitation in the subtropical-temperate climate transition zone of East-Central China since the Last Glacial Maximum. *Quaternary Science Reviews, 367*, 109528. https://doi.org/10.1016/j.quascirev.2025.109528.
Zimmer, A., Beach, T., Klein, J. A., & Bullard, J. R. (2022). The need for stewardship of lands exposed by deglaciation from climate change. *Wiley Interdisciplinary Reviews: Climate Change, 13*(2). https://doi.org/10.1002/wcc.753.

CHAPTER 5

Mountain land use and societies – Living at altitude

5.1 Introduction

Humans settled in the European Alps, like many mountain areas around the world, as ice sheets retreated and warmer, wetter climates evolved (i.e., 12,000 years ago). There was a marked increase in population starting from the Neolithic and Bronze Age (i.e., 8000–3000 years ago) (Anselmetto et al., 2024) as settled food production developed. Since then, Alpine landscapes have been extensively modified by the exploitation of forests for timber and fuelwood, grazing, mining, and the maintenance of agricultural land. Mountain regions are home to diverse communities whose cultures, livelihoods, and identities are intricately tied to the rugged landscapes they inhabit and the food they produce. These 'mountain people' have developed unique ways of life and food production that reflect the challenges and opportunities presented by their environment. Rural communities occupying montane landscapes are often characterised by small-holder agrarian and pastoral economies, and have always been vulnerable to multiple environmental and socio-economic stressors (Frogley et al., 2025).

Globally, mountain population estimates vary greatly, from 0.344 billion to 2.29 billion in 2015. A recent analysis using one of the population datasets revealed that mountain populations have doubled from 1975 to 2015. The urban proportion of the total mountain population in 2015 ranged from 6% to 39%, depending on the combination of population and urban extent datasets used (Thornton et al., 2022). There are strong cultural and socio-economic characteristics of mountain people, resulting from their adaptation to high-altitude environments, the challenges they face, the food they produce, the social networks created, and the importance of preserving their unique heritage. There are a series of key livelihoods that support mountain communities, not so different from anywhere in the world, either growing vegetables and grains or keeping animals for milk and meat. Mountain agroecosystems deliver essential ecosystem services to society but are prone to climate change as well as socio-economic pressures, making

Mountain Diversity: Dynamics of High-Altitude Environments.
DOI: https://doi.org/10.1016/B978-0-443-22251-1.00005-8

multi-functional land systems increasingly central to sustainable mountain use (Platts et al., 2023). Over the past century, the human population in the Kilimanjaro region has increased tenfold, which has led to the retraction of large areas of natural habitat in the lowland and montane zones of the mountain, and to the expansion of anthropogenic ecosystems (Peters et al., 2019). These increasing human populations, and the resultant expansion and intensification of agriculture, have led to high population levels, for example, up to 1200 people km^2 in the Dilla Highlands of Ethiopia, with an impact on local services. Changing climate and human demographics in the world's mountains will have increasingly profound environmental and societal consequences across all elevations (Thornton et al., 2022). These changes will be largely played out through the lens of land use and land cover change that will reflect changing choices, societal demands, and values that people get from the mountains. These major uses of mountain land for agriculture, pasture, and non-food production will be explored here, with the challenges faced and sustainable mountain futures a focus of subsequent chapters.

5.2 Cultural and socio-economic characteristics of mountain people

Mountain people often live in remote, isolated communities where traditional customs and practices can be strongly preserved. In common with mountain ecosystems, geology, topography, and climate explored in previous chapters, mountain regions are characterised by a high degree of cultural diversity. This is partly due to the geographic isolation of mountain communities, which fosters the development of unique and distinct languages, traditions, and social structures. For example, the Himalayas are home to multiple ethnic groups, such as the Sherpas, Gurungs, and Bhutias, each with its own language, religious practices, and cultural heritage (Bhutia, 2021). The Quechua and Aymara people, among the largest indigenous groups in the Andes, have preserved their languages, traditional clothing, and agricultural practices. The Quechua language, a legacy of the Inca Empire, is still widely spoken across Peru, Bolivia, and Ecuador. Traditional festivals, such as Inti Raymi, 'the Festival of the Sun', continue to be celebrated, reflecting the deep spiritual connection between these communities and their mountain environment. The Cherokee people of the Southern Appalachians have maintained their cultural heritage despite significant challenges, including

Plate 5.1 Sappada, Plodn, stands out clearly due to its linguistic and cultural characteristics that are unique to this small area of the Italian Dolomites. *(Photograph Rob Marchant).*

forced removal during the colonisation of their lands during the 'Trail of Tears' (Greene, 2022). The Ladin people, who inhabit the Dolomites in northern Italy, are a mountain community which has maintained its unique language and culture despite external pressures. The Ladin language, a Romance language, is still spoken by around 30,000 people, and the community has preserved traditional crafts such as woodcarving and lace-making, as well as their distinctive architecture (Connor, 2023). Similarly, in a small area around Sappada, Plodn stands out clearly from the surrounding areas of Carnia and Cadore due to its linguistic and cultural peculiarities. This area forms a 'linguistic island' of German origin and a survival of Middle High German that has continued in use since the Middle Ages, along with traditions, folklore, food, and arts and crafts (Plate 5.1).

The Balti people of the Karakoram range in Pakistan practice subsistence farming, cultivating crops like barley, wheat, and peas on terraced fields, and rely on livestock for milk, meat, and wool. The Baltis have also historically been involved in trade and transport, serving as porters for mountaineering expeditions in the Karakoram. The Sherpas of the Himalayas, particularly in Nepal, are one of the most well-known mountain communities. Renowned for their skills in mountaineering, the Sherpas have maintained their cultural heritage through language, religion (Tibetan Buddhism), and traditional practices. The Sherpa's spiritual connection to the mountains, particularly Mount Everest (known locally as Chomolungma), is central to their cultural identity, and their festivals and rituals reflect this deep reverence for their environment (Bhutia, 2021). The Ngai Tahu iwi (tribe) of New Zealand's South Island has a deep cultural connection to the Southern Alps (Kā Tiritiri o te Moana), which are central to their identity and spirituality. The Ngai Tahu maintain traditional practices such as mahinga kai (food gathering), and their relationship with the land is enshrined in the Ngai Tahu Claims Settlement Act of 1998, which recognises their rights to natural resources and supports their cultural preservation efforts, particularly under current environmental and social challenges (Tumahai, 2024).

5.3 Subsistence agriculture and livelihoods

Most agricultural production takes place in the highland areas from 1500 to 3200 m above sea level, where the highland productivity coincides with the densest human populations. Mountain regions pose significant challenges for agricultural production due to their steep slopes, limited arable land, diversity of climates, particularly seasonal extremes (Chapter 2), and susceptibility to erosion. The challenging terrain and limited arable land often necessitate innovative agricultural practices, such as agroforestry and terracing. Agroforestry systems in the world are deeply rooted in Indigenous knowledge and practices that have evolved over thousands of years to support very high population densities up to 1200 people km^2 in the Ethiopian Highlands. Terracing addresses topographical challenges by transforming steep slopes into a series of flat, horizontal surfaces, making them more suitable for crop cultivation. Terracing is particularly important for ensuring food security in mountainous areas where flat land is scarce, but with changing relationships between societies and land, many places have seen terraces abandoned and the infrastructure fall into disrepair. Traditional agricultural systems, both agroforestry and terracing, are receiving increased attention at an international level due to their multifunctional role in growing food, supporting societies, and wider functions of mountain systems. As such, there will be quite a focus on these systems in this chapter.

5.3.1 Agroforestry

Agroforestry systems offer ideal solutions to life on a mountain as they can be developed in areas where production would either rapidly degrade the land, leave the soils prone to erosion, or otherwise not be possible. Agroforestry is a land use system that combines multiple crop species with trees and woody shrubs. These systems are characterised by the intentional integration of trees, crops, and sometimes livestock in the same space to create diverse, sustainable, productive landscapes. Some of the indigenous community agroforestry systems have extended some 8000 years in Indonesia, where swidden agriculture has retained valuable trees like rubber and teak (Fig. 5.1). Over time, these led to 'anthropogenic forests' rich in useful species from a commercial perspective or from their role in fertilising and stabilising the soil. Agroforestry systems emphasise local, small-scale farming with ecologically friendly, resource-efficient agricultural practices that can increase environmental sustainability while being very productive (Roy et al., 2025). Agroforestry systems support some 1.2 billion people with about 17% of agricultural land under agroforestry production (Gebrekirstos, 2025).

Agroforestry has been widely practised across the world's mountain systems; for example, by farmers on Mount Merapi, Indonesia, where there is a volcanic risk and reward from working the fertile volcanic soils (Rozaki et al., 2021). Each area in Mount Merapi has different types of agroforestry such as agrisilviculture, silvopasture, and agrosilvopasture; these systems have value for both production and conservation, although unfortunately, the latter is not often economically rewarded (Buckley et al., 2023; Rozaki et al., 2021). In New Zealand, the Waitangi Tribunal has ensured that Māori communities have a voice in the management of their traditional lands and has led to the return of significant lands to Māori ownership, including areas in the Southern Alps, helping to preserve cultural practices and promote sustainable management (Buckley et al., 2023). Madagascar is one of the countries where agroforestry plays a central socio-economic, cultural, and environmental role. Madagascar is a major producer of some of the most demanded agroforestry crops, such as vanilla, clove, coffee, cocoa, and lychee, which constitute a critical input to the country's economy (Andriatsitohaina et al., 2024). Agroforestry is widely regarded as a model for balancing tradition and modernity in a rapidly changing world (Plate 5.2). One defining characteristic of agroforestry systems is that they are multifunctional, having multiple layers of trees and crops with careful selection and nurturing of the canopy trees. The Chagga's traditional

Figure 5.1 Agroforestry is much more than just providing shade! There are direct benefits for: climate action through stores of carbon and water, biodiversity as agroforestry provides habitat for pollinators and wildlife, livelihoods as agroforestry boosts farmer income and food security, provides multiple ecosystem services, and the bioeconomy supports sustainable value chains such as in timber, fruit, cocoa, and coffee. These five factors allow communities to be more resilient and adapt to climate change. *(All Photographs by Rob Marchant)*.

agroforestry practices on the slopes of Kilimanjaro, known as the 'kihamba' system, involve the cultivation of multiple crops on the same plot of land, mimicking the natural forest structure and providing food security for the community (Fig. 5.2). The Appalachian Mountains have a long history of subsistence farming, where families have relied on small plots of land to grow crops such as corn, beans, and tobacco. In addition, the harvesting of

Plate 5.2 Agroforestry systems in the Ethiopian Highlands can have upwards of 70 crops in a small plot, along with high biodiversity. *(Photograph Rob Marchant)*.

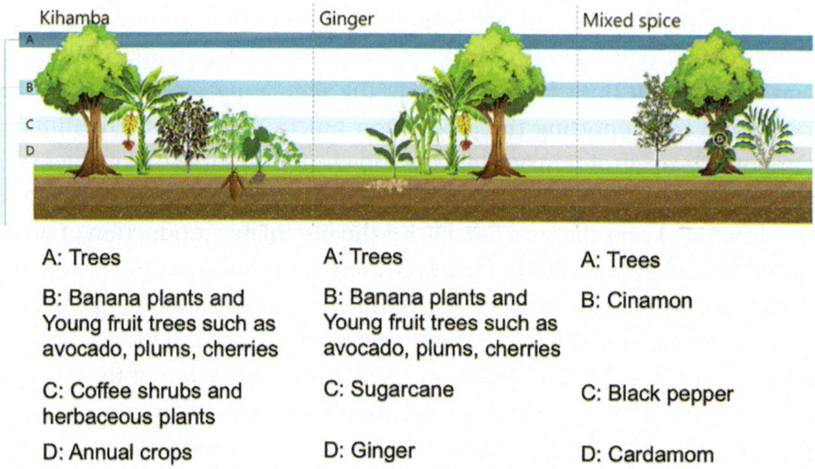

Figure 5.2 Overview of different homegarden agroforestry systems in mountain regions of Tanzania: Kihamba (*left*), Ginger (*centre*), and Mixed spices agroforestry (*right*). After Kimaro et al. (2018).

wild ginseng, a plant native to the Appalachian Mountains, has provided a valuable source of income for generations of mountain people. This practice is regulated to ensure sustainability, reflecting the community's deep knowledge of and respect for their natural environment (Law, 2022).

Agroforestry can be a sustainable land use approach with the potential to address multiple issues (Kadigi et al., 2021). Agroforestry, if practised at the right intensity, can support biodiversity, mitigate climate change impacts, support socio-economic systems, and provide ecosystem resilience. Addressing interconnected social and environmental issues, including poverty, food security, climate change, and biodiversity loss, requires such integrated solutions that are the right system for the right location. Diversification of income sources is an important strategy for reducing income inequality and supporting resilience in mountain areas. Policies and initiatives that aim to promote diversification of livelihoods are more likely to reduce income inequality in these areas (Kadigi et al., 2021). A typical agroforestry system can have up to 70 crops growing in a single plot with banana, plantains, maize, chillies, herbs, beans, and leafy greens growing next to, and on top of, each other. Diversification epitomised!

Agroforestry systems are multifunctional and can support livelihoods by providing vital ecosystem services of food, fodder, and firewood to smallholder farmers (Buckley et al., 2023). They can also store substantial amounts of carbon in the soil and are important drivers of carbon sequestration (Denu et al., 2016). Agroforestry systems have the potential to mitigate land degradation both by maintaining the regulating service of mountain forests and by supporting the production functions played by the different components of agroforestry; reducing soil erosion to some extent, increasing soil fertility, reducing salinity, buffering pH, and ultimately maintaining soil health, which keeps the land suitable for the sustainable production of diverse products (Kadigi et al., 2021). However, these ecosystems are also facing some challenges. Loss of diversity and native tree species in the overstory layer has been a growing concern in agroforestry worldwide, yet the drivers behind it remain inadequately understood, and hence, the true value of these systems is not realised. For example, in many mountain areas of Ethiopia, it makes business sense to convert diverse long-lived agroforestry systems to sugar cane production, moving from a system that supports hundreds of species per plot to one. Who can blame the farmer with bills to pay and children to educate? They have to maximise the profit from the plot, and often this is focused on the short-term benefits of growing the most profitable crop. With a small incentive for paying for the multiple benefits, we need to shift the economy of food production and support extensive, high-quality, low-input, slow production systems where the wider value of agroforestry systems can be realised.

Depending upon the selected tree species and its management, the incorporation of trees into crop fields and agricultural landscapes has significant social, economic, and ecological benefits (Denu et al., 2016; Gebrekirstos et al., 2011). The higher proportions of non-native tree species in mixed spice agroforestry, where ginger is one of the main cash crops, are dictated by economic needs for timber, fuel, and sun-requiring cash crops. Policies to increase resilience and restore the native tree species cover can only be successful if they are based on the knowledge of native species, their traits, and wider ecosystem service potential, such as the ability of trees to fix nitrogen or let just the right amount of sunlight to filter through to the production layer. Furthermore, there should be a balance of conservation and livelihood, acknowledgement of the complex mix of pressures on farmers' livelihoods, and proposed measures tailored to the areas' salient features and specific challenges (Kimaro et al., 2018). However, doing so is knowledge-intensive as it necessitates careful consideration of potential negative interactions of the different components and potential impacts on biodiversity. Further, it necessitates a realisation that the farmer is making a positive contribution to aspects such as water regulations, carbon sequestration, and biodiversity conservation benefits that have an impact far outside the farm boundaries, and, as such, the farmer should be paid for this export.

Matching the right trees and management practices to the right production system, in the right place, so that the wider ecosystem benefits can be maximised whilst avoiding interventions that threaten biodiversity, is crucial. For example, in the Ethiopian highlands, *Eucalyptus* species currently dominate many farm woodlots. While *Eucalyptus* may have made significant contributions to smallholder livelihoods, due to their rapid growth and ability to provide large amounts of timber and fuel, their high direct water consumption and oily foliage (that takes a long time to decompose) have negative impacts on soil health. Indeed, the rapid spread and update of the *Eucalyptus* for quick wins has resulted in severe negative consequences for regulating and underpinning ecosystem services, particularly enhanced water stress and declining soil fertility (Gebrekirstos, 2025). There is a need to promote portfolios of diverse tree species and management options that meet income, fodder, fruit, fuelwood, and soil fertility needs, without compromising biodiversity conservation goals. Moreover, as trees are vulnerable

Plate 5.3 Selecting and grading coffee cherries in the Dilla Highlands, Ethiopia. *(Photograph Rob Marchant).*

to climate variability and change (Gebrekirstos, 2025), identifying resilient species for climate-resilient development and biodiversity enhancement is essential to support the agroforestry system and the livelihood (Plate 5.3).

Biodiversity associated with traditional agricultural practices is particularly important for ensuring food and nutrition to local communities (Santoro et al., 2020). Agrobiodiversity can be locally high; for example, 20 varieties of date palm, 21 fruit tree species, 21 vegetable species, and two fodder crops are present in a Tunisian agroforestry system. Such agroforestry and agricultural heritage systems continue to play a crucial role in maintaining genetic resources and agrobiodiversity (Santoro et al., 2020). Farmers who continue to cultivate by applying traditional practices are the main actors who practice real conservation of genetic resources and diversity by maintaining traditional cultivars and a diversified landscape structure. However, these farmers are often not appropriately rewarded for this wider contribution and sell their agricultural produce on the open market, where they compete with more intensive monocultural systems. In Europe, food from these systems can reach a premium under the 'slow food' of local origin banners, selling into more niche markets where the consumer has the

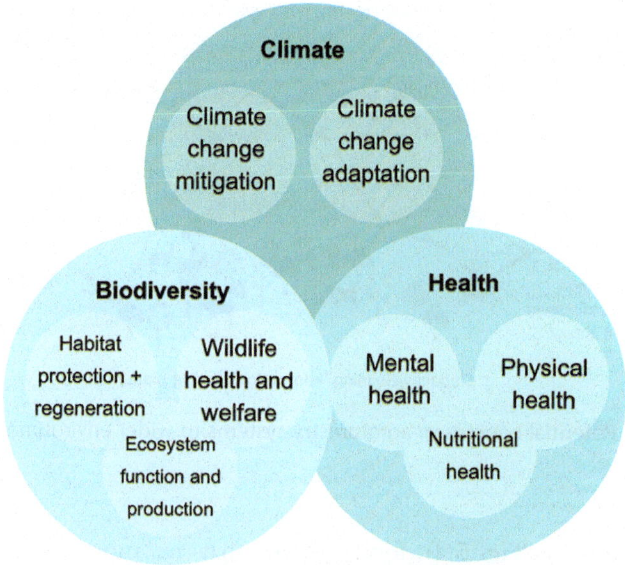

Figure 5.3 Potential benefits of agroforestry systems to health.

luxury of being interested in the provenance of the food they are consuming (Röhrig et al., 2021). Such a situation, where the price for food reflects the wider benefits and costs of the system that is producing the food, is needed.

Agroforestry systems contribute to public health in many ways (Fig. 5.3); through improved food and nutritional access, enhanced food sovereignty and community well-being, increased environmental quality and health, and impacts on communicable and non-communicable disease (Roy et al., 2025). Despite the various ways that agroforestry systems may contribute to public health and planetary health, it is not well described in the health literature and has not gained traction as a public health practice. Increased agroforestry research and support from the public health community may benefit short and long-term human and ecosystem health in our changing environment. Agroforestry can strengthen climate change resilience, while contributing to food access, income, health, and environmental stability through environmental benefits which are rarely quantified (Duffy et al., 2021). Climate resilience is increasingly necessary for creating reliable food systems that can withstand environmental impacts, and thus can be a key component to the resilience of public health interventions (Cuni-Sanchez et al., 2025; Roy et al., 2025). Through reimagining the benefits of healthy food systems to include the health of the surrounding environment

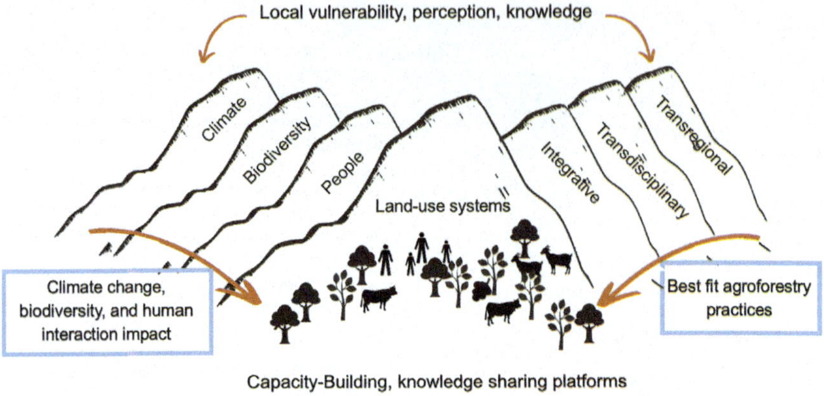

Figure 5.4 Potential benefits of agroforestry systems to wider environmental stability and resilience.

and community (Fig. 5.4), food systems that are more environmentally resilient against climate pressures give communities the power to adapt agricultural systems to fit their context-specific needs and priorities (Roy et al., 2025). Indigenous deep-rooted agroforestry systems can re-revolutionise climate-smart mountain landscapes. The multi-functionality of these systems and their impact on climate change can contribute to the global good and provide climate-smart solutions for managing degraded agricultural landscapes and biodiversity conservation. By enhancing soil fertility, these systems can therefore make the smallholder tropical agricultural landscapes more resilient. The increase of global climate change impacts on mountain communities is a glaring indication that more pragmatic and direct approaches to tree and forest management are urgently needed (Cuni-Sanchez et al., 2025). Agroforestry offers such a solution, as well as preferentially supporting income generation from women; critical as the women and youth involved in farming activities were generally poor, illiterate, and faced issues of land insecurity (Galabuzi et al., 2021).

Agriculture faces the unprecedented task of feeding a world population of 9 billion people by 2050, while simultaneously avoiding harmful environmental and social effects. There is great potential for agroforestry to improve agronomic productivity, carbon sequestration, nutrient cycling, soil biodiversity, water retention, and pollination; to reduce soil erosion and fire; and to provide recreational and cultural benefits. However, there are several challenges due to the high costs of implementation, lack of financial incentives, inadequate product marketing and pricing, lack of awareness and education, and lack of field demonstrations. These challenges need to be addressed to encourage farmers to adopt agroforestry systems (Sollen-Norrlin et al., 2020). One effort to meet this challenge has been organic farming, with outcomes that are generally positive. However, a number of challenges remain. Organic yields lag behind those in conventional agriculture, and greenhouse gas emissions and nutrient leaching remain somewhat problematic (Wilson & Lovell, 2016). By implementing systems that mimic nature's functions, agroforestry has the potential to remain productive whilst supporting a range of ecosystem services; however, for this potential to be realised, the economics have to match. Multifunctional agroforestry systems can reduce poverty, enhance food security, conserve biodiversity, and enhance adaptive capacity to climate change challenges through sustaining agro-ecosystem productivity and enhancing income generation by planting a mixture of the right trees in the right place to support landscapes and agroforestry farming systems. Multifunctional agroforestry interventions most suitable for selected landscapes must view the challenge at multiple nested scales (i.e., viewing the issue spatially from plot to landscape and vice versa, and considering biophysical, socio-economic, and climate conditions to ensure the potential for scaling is rooted in farmers' and landscapes' context and vision) at the local, regional, and national scales. Due to the importance of agroforestry systems, there really needs to be a new economic model that is more supportive of agroforestry, and that captures and recognises the wider societal benefits rather than just the crop production.

5.3.2 Coffee agroforestry

Coffee is grown in more than 80 tropical countries and provides an income to millions of families, making it one of the most traded commodities on the global market (Cassamo et al., 2023). Coffee agroforestry is a traditional and complex system in which coffee plants are cultivated under the dappled shade of trees, mimicking a natural forest environment. There is a wide range of production that extends from the intensive to the extensive, with a trade-off along this gradient from mass coffee production to mixed and diverse agroforestry systems that provide multiple benefits. Coffee is an understory shrub that originated in the Ethiopian Highlands, but is grown extensively around the world's mountains. The canopy cover is to reduce direct sunlight as well as enhance soil nutrients by selecting key trees such as *Hagenia, Inga, Erythrina,* and *Grevillea.* There can then be a mid-layer of fruit trees (avocados,

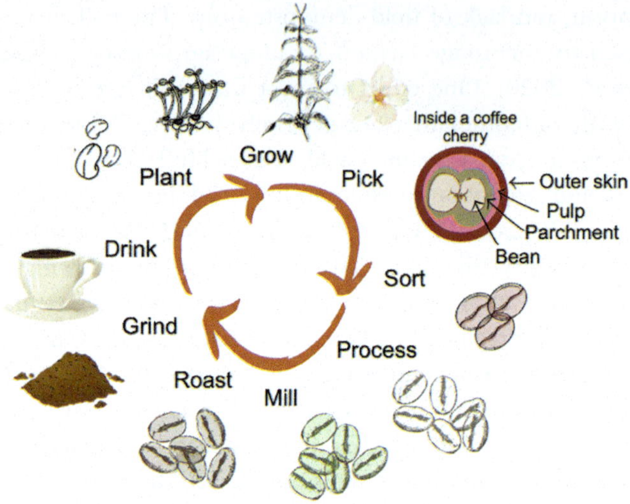

Figure 5.5 The Journey of the Coffee Bean: As the plant grows, environmental factors such as altitude, soil, and climate affect the flavour. Each plant will produce around 500 g of coffee 5 years after planting. The coffee cherries ripen from green to red fruits that are selectively harvested and sorted by hand to ensure only the ripest are processed. The outer pulpy fruit is removed from the berry mechanically or through a fermentation process, leaving the green seeds (beans). Following a drying process, these are roasted for between 7 and 14 minutes to transform green coffee into brown or black coffee beans. Once roasted, beans start to lose their freshness; when grinding beans release 60% of their aroma.

cherries, citrus), which are nitrogen-fixing legumes, below which there are coffee plants. The genus *Coffea* has 124 species widespread and adapted to tropical and subtropical areas, particularly in Africa, the centre of endemism (Davis et al., 2019), although *Coffea arabica* L. (Arabica coffee) and *Coffea canephora* Pierre ex A. Froehner (Robusta coffee) species dominate the coffee trade, representing about 99% of the coffee global economy (Anhar et al., 2021) (Fig. 5.5).

The coffee – enset – tree agroforestry system in the Gamo highland (Ethiopia) has recently been recognised by UNESCO as a site of World Heritage due to the contribution to improved food security, regional and national economies, and environmental resilience (Seid & Kebebew, 2022). These mixed systems can enhance biodiversity and provide multiple sources of income or food (Plate 5.4). Coffee forests can be considered the keys to sustainable forest management as they provide a variety of ecosystem services, including biodiversity conservation, an alternative source of income

Plate 5.4 Coffee farming has spread across many mountain systems, particularly in Latin America, East Africa (e.g., Tanzania) and Southeast Asian countries (e.g., Thailand). *(All Photographs Rob Marchant).*

to address food security, and mechanisms of carbon storage in both living biomass and soil. The expansion of coffee forest farming and the initiation of coffee carbon credits act as strategies to encourage farmers to grow more organic coffee plants, conserve shade trees, and ultimately contribute to climate change mitigation (Niguse et al., 2022). Coffee plants sequester substantial amounts of carbon in the coffee agroforestry systems. The carbon sequestration potential of shade-grown coffee plants is influenced by the compatibility of the shade trees (Denu et al., 2016). Furthermore, coffee agroforestry has emerged as a promising land-use system for reducing or offsetting deforestation. This is because coffee provides an alternative source of income that would otherwise be obtained from forest cutting.

Climatic changes and global warming will affect crop distribution, particularly *C. arabica*, which only grows in a relatively narrow climatic range (Cassamo et al., 2023). The impact is not only on the direct environmental niche, but also on the spread of coffee pathogens such as the coffee cherry borer and leaf rusts that are being advantaged by the changing climate, particularly the great moisture availability of many mountains (Chapter 6) (de Carvalho Alves & Sanches, 2022). However, this is not a simple case of coffee production being affected by changing climates. The impact of CO_2 concentration on stomatal conductance is also impacting *Coffea* genotypes. The concentration of CO_2 can greatly strengthen coffee vigour, growth, and yield, and is closely associated with significant increases in net C-assimilation close to or above 50% in the range of 550–700 ppm of air CO_2 (Rakocevic et al., 2021). Importantly, eCO_2 can promote/reinforce the triggering of protective responses to drought and supra-optimal temperatures, improving coffee plant resilience against environmental shifts (Cassamo et al., 2023).

5.4 Terracing systems of the world

Agricultural terracing, a millennia-old technique, has been employed across diverse landscapes worldwide to mitigate soil erosion, improve water management, and optimise land use for crop cultivation. This section examines the historical evolution of agricultural terracing, the engineering principles underlying terraced landscapes, and the contemporary relevance of terracing in sustainable agriculture. Terracing has been used by civilisations worldwide, from ancient Mesopotamia and the Mediterranean to the highlands of Africa, Asia, and the Americas. This section explores the historical roots, technical aspects, and contemporary applications of agricultural terracing, emphasising its role as a sustainable land management strategy in diverse mountain contexts.

One of the primary benefits of terracing is its ability to prevent soil erosion. In mountainous regions, rainfall can quickly wash away topsoil on steep slopes, leading to reduced soil fertility and increased risk of landslides. Terraces help to slow down the flow of water, allowing it to infiltrate the soil before it travels down the slope. For example, in the Ethiopian Highlands, terracing has been widely implemented as part of soil and water conservation efforts, where the steep terrain and heavy seasonal rains make the region particularly vulnerable to soil erosion (Plate 5.5). Terracing has helped to stabilise the soil, improve water retention, and increase agricultural

Plate 5.5 Terracing systems in the Adi Tesfo region, northern Ethiopia. *(Photograph Rob Marchant).*

productivity, contributing to the livelihoods of local farmers. Terracing is also an effective water management strategy, particularly in regions with irregular rainfall patterns. By creating flat surfaces on mountain slopes, terraces help to capture and retain rainwater, reducing runoff and allowing more water to seep into the soil. This is especially important in arid and semi-arid regions where water is a scarce resource. The Inca civilisation in the Andes is renowned for its sophisticated use of terracing for water management. The terraces, known as 'andenes', were designed to capture and channel water from mountain springs and rainfall, ensuring a steady supply of water for crops throughout the growing season (Frogley et al., 2025). These terraces, some of which are still in use today, demonstrate the Incas' advanced understanding of hydrology and sustainable agriculture. Some systems, such as the Engaruka system in northern Tanzania, have a fascinating history of water and soil management (Plate 5.6). Some people call the Engaruka terracing system the most important archaeological site in Africa, much more so than the pyramids in Egypt or the stela in Axum! The

Plate 5.6 Terraces from the Engaruka system have a fascinating history of water and sediment management. The sediments are increasingly eroded from the Ngorongoro Highlands as forested areas have been cleared and soils exposed. *(Photograph Rob Marchant).*

amazing thing about the Engaruka terraces is that they span some 20 km^2 and clearly demonstrate an agriculturally advanced society, living in what is now a dry marginal landscape some 1000 years ago and able to manage, through an elaborate system of terraces, the flow of water and soil from the Ngorongoro Highlands to the lowlands.

5.4.1 History of terracing

Terracing is a traditional agricultural practice that has been used for thousands of years by communities living in mountainous regions around the world. This method involves the creation of stepped levels, or terraces, on the slopes of mountains. Terraced landscapes are repositories of cultural heritage, embodying the knowledge, skills, and traditions of past generations. Preserving and revitalising terraced agriculture not only sustains local livelihoods but also fosters cultural identity, community cohesion, and tourism opportunities. That said, the maintenance of terracing faces many

Plate 5.7 Terracing is commonly used to support agricultural production in many mountain areas, such as Tenerife (*left*), Nepal (*middle*), and the Konso region of southern Ethiopia, which is renowned for sophisticated terracing systems, which have been recognised by UNESCO as a World Heritage Site and often portrayed as an example of sustainable mountain agriculture. *(Photographs Rob Marchant).*

challenges related to land tenure, labour availability, and investment costs, particularly in marginalised rural areas. Addressing these constraints requires supportive policies, institutional capacity-building, and targeted investments in agricultural infrastructure. Revitalising terraced agriculture necessitates knowledge transfer mechanisms, farmer training programs, and innovation platforms that integrate traditional wisdom with modern technologies. Participatory approaches, farmer-to-farmer networks, and interdisciplinary collaborations can facilitate the co-creation of context-specific terracing solutions.

Terracing has a deep historical significance across the world's mountains, particularly in regions where indigenous communities have adapted to mountainous landscapes to sustain their agricultural practices. For example, the people in the Konso region of southern Ethiopia are renowned for their sophisticated terracing systems, which have been recognised by UNESCO as a World Heritage Site. The Konso terraces are meticulously constructed with stone walls and are designed to manage water flow and prevent soil erosion on the steep hillsides (Plate 5.7). These terraces have enabled the Konso people to practice sustainable agriculture to cultivate crops such as teff, rice, barley, and wheat in a challenging environment for over 400 years. The Himalayan region, which spans several countries including India, Nepal, Bhutan, and China, also has a long tradition of terracing. Archaeological evidence suggests that terracing was practised as early as 7000 years ago

Plate 5.8 Terraces are often meticulously constructed with stone walls and are designed to manage water flow and prevent soil erosion on the steep hillsides, such as in Nan Province, northern Thailand. *(Photograph Rob Marchant).*

in the Middle East and China (Wang et al., 2021), as seen in the Xinjiang Highlands, where terracing dates back thousands of years (Yin et al., 2023). Terraces help manage the impact of the region's monsoon rains, reducing the risk of landslides and soil erosion.

In northern Thailand, terraced fields are a common sight in the hilly regions, particularly in areas like the Nan region. These terraces, often built by hand, allow farmers to grow maize, the staple crop, on slopes that would otherwise be too steep for cultivation (Plate 5.8). China has a long history of terracing, particularly in its southern and southwestern mountainous regions. The Hani people of Yunnan Province are especially known for their rice terraces, which have been cultivated for over 1300 years (Aimar, 2024). These terraces are not only important for food production but also for their cultural and environmental significance. The Honghe Hani Rice Terraces in Yunnan, China, are an impressive example of terraces over 16,000 hectares, which have also been recognised by UNESCO as a World Heritage Site. The Hani people developed these terraces to cultivate rice on the steep slopes of the Ailao Mountains, using a complex system of channels to divert water from the forests above to the fields below (Aimar, 2024). Terrace systems have shaped the communities' landscape and the resulting spatial diversity through the interlocking mediation of natural resource management with indigenous knowledge systems (Wang et al., 2021).

The Inca Empire, which flourished in the Andes from the 13th to the 16th century, is perhaps the most well-known example of a civilisation that extensively utilised terracing. Terracing was not unique to the Incas; earlier Andean civilisations, such as the Wari and the Tiwanaku, also developed terrace agriculture (Mader et al., 2025). The Wari civilisations, like many mountain communities, recognised the need to manage the steep terrain and created terraces to sustain their growing populations. The knowledge and techniques developed by these early cultures laid the foundation for the more extensive terracing systems later adopted by the Incas. The Wari culture, which existed from about 1500 to 1000 years ago, built extensive terraces in the Ayacucho region of Peru, which they used for growing staple crops such as maize and potatoes, and played a crucial role in supporting the Wari's urban centres and their expansion across the Andean highlands (Mader et al., 2025). For at least the last two millennia, agriculturalists have been living in the highland Andean landscapes of the Sacred Valley of the Incas, located near Cusco in Peru. The terraces at sites such as Moray and Ollantaytambo demonstrate their advanced understanding of agriculture, hydrology, and engineering. These terraces allowed the Incas to grow a variety of crops, including maize, potatoes, and quinoa, at different altitudes, thereby taking advantage of the region's diverse microclimates. A particular challenge across the region would have been the relatively rapid warming in conditions during the Medieval Climate Anomaly (1000–1200 years ago), forcing the agriculturalists to adapt (Lüning et al., 2019). During this period, populations moved up valleys in response to climatic warming, expanded maize agriculture from mid-elevations up to *c.* 3400 m, shifted potato cultivation to even higher elevations, and diversified crop varieties, accompanied by the construction of large-scale terraces and associated irrigation infrastructure, and the adoption of agroforestry techniques (Frogley et al., 2025). In the Peruvian Andes, many terraces fell into disrepair following the Spanish conquest due to the disruption of traditional agricultural practices and labour systems. However, recent efforts by NGOs and local communities to rehabilitate these terraces have shown that, with the right support, it is possible to restore and maintain these ancient agricultural systems, ensuring their continued use for future generations (Vanacker et al., 2022).

Understanding ancient terrace agricultural systems can inform the development of sustainable land use strategies to address contemporary environmental and social challenges, including climate change, water scarcity, and economically driven depopulation (Mader et al., 2025). The Ancestral Puebloans (Anasazi) of the Southwestern United States, particularly in the regions of present-day Arizona, New Mexico, and Colorado, developed terracing techniques to cultivate maize, beans, and squash on the steep canyon walls and mesas (Lentz et al., 2021). These terraces were often reinforced with stone walls to create stable planting surfaces and to capture water during the infrequent rains. This method of farming enabled the Ancestral Puebloans to sustain large populations in arid environments with limited arable land (Lentz et al., 2021).

Terracing is a vital agricultural practice in mountainous regions around the world, providing a sustainable solution to the challenges of steep slopes, soil erosion, and water management. Given the importance of terracing for sustainable agriculture in mountainous regions, efforts are being made to preserve and promote this practice through various initiatives and policies. In the Bolivian Andes, terracing has been reintroduced as part of soil conservation programs aimed at combating land degradation and desertification by stabilising soils, improving agricultural productivity, and reducing the risk of landslides in areas where traditional farming practices have led to severe soil erosion. In the Colca Valley of Peru, ancient terraces continue to be used by local farmers to manage water resources efficiently. These terraces, originally built by the Collagua and Cabana cultures, are designed to capture water from mountain springs and rainfall, distributing it evenly across the terraces (Leceta et al., 2024); thus allowing farmers to cultivate crops such as maize and quinoa, even in a region with limited rainfall.

5.4.2 Benefits of terracing

By converting steep slopes into flat, cultivable land, terracing allows mountain communities to expand their agricultural activities and increase crop yields. This is particularly important in regions with limited arable land, where every piece of cultivable land is valuable. Terracing also enables the cultivation of a wider variety of crops by creating microclimates on different terraces, which can be used to grow different crops suited to varying conditions. Terraces facilitate efficient water distribution and retention, particularly in regions with erratic rainfall or monsoonal climates. The construction of contour terraces, bunds, and check dams helps to capture rainwater, recharge groundwater, and sustain crop growth during dry periods (Leceta et al., 2024). Terraces act as barriers that intercept runoff, reducing soil erosion and sedimentation downstream. By breaking the slope gradient into manageable steps, terracing also helps to slow down water flow, allowing sediments to settle and nutrients to infiltrate the soil (Ferro-Vázquez & Stump, 2023). Terraced landscapes promote soil conservation and fertility through the deposition of organic matter, nutrient retention, and microbial

Plate 5.9 Extensive terraces extend across the Tigray region of northern Ethiopia. *(Photographs Rob Marchant).*

activity. The construction of terrace walls, backfilled with organic material or crop residues, creates microclimatic conditions conducive to plant growth and soil health (Zhu et al., 2021).

It is important to view terraces not as an individual strip of land but as part of a series of diverse cultivation characters connected across the landscape. For example, in the world-famous Konso system, the terraces at lower altitudes and closer to the river courses are highly fertile and prized land, whereas those on the upper slopes are characterised by thinner soils that do not retain so much water and are less productive (Ferro-Vázquez & Stump, 2023). In the Yungas region of Bolivia, in the transition between the Andes and the Amazon Basin, improper terrace construction has occasionally led to increased erosion and landslides (Meddens et al., 2025). This has prompted efforts to integrate traditional knowledge with modern engineering practices to ensure that terracing remains sustainable and beneficial to the environment. In West Africa, particularly in the highlands of Mali and Guinea, terracing has been used for centuries to manage the cultivation of crops on steep slopes. The Dogon people of Mali, for example, have developed intricate terracing systems to grow millet, sorghum, and other crops in an arid and rocky landscape. The Dogon's terracing techniques are not only ingenious but are built to capture and retain water in a region where rainfall is scarce, allowing for the successful cultivation of crops despite the harsh conditions (Tappan & McGahuey, 2007). In the Tigray region of northern Ethiopia, where soil erosion has been a significant problem, terracing has been widely adopted as part of soil and water conservation programs (Plate 5.9). These efforts have helped to stabilise soils, reduce erosion, and improve agricultural productivity, ensuring that the land remains fertile and productive for future generations (Gebremikael et al., 2024). In the Lesotho Highlands, terracing is used to manage water resources in a region with limited and highly variable rainfall. The terraces help to capture and store rainwater, which is then used to irrigate crops such as maize and beans. This practice is crucial for sustaining agriculture in a region where water scarcity is a constant challenge (Showers, 1989). In Bhutan, terracing is integrated into the traditional practice of agroforestry, where crops are grown alongside trees and other vegetation. This approach helps to maintain biodiversity, protect watersheds, and enhance the resilience of the farming system. The terraces in Bhutan not only support rice and maize cultivation but also provide habitats for a variety of native plant and animal species (Namgyel et al., 2023). The Banaue Rice Terraces in the Philippines, often referred to as the 'Eighth Wonder of the World', are a testament to the ingenuity of the Ifugao people. These terraces, carved into the mountains over 2000 years ago, allow for the cultivation of rice in a region with steep terrain and limited flat land. The terraces have enabled the Ifugao people to sustain their agricultural practices for centuries, providing food security and supporting their cultural heritage (Habito et al., 2023).

5.4.3 Terracing and biodiversity

Terracing contributes to the preservation of mountain biodiversity by maintaining a mosaic of different habitats at various altitudes. The terraces create microclimates that can support a variety of plant species, both cultivated and wild. This diversity is crucial for the resilience of agricultural systems as it reduces the risk of crop failure due to pests, diseases, or climatic variations (Zhang et al., 2024). For example, in the Sierra Nevada, where terracing has been used for small-scale farming, the practice has helped to maintain a diverse range of native plants and animals. In the Usambara Mountains of Tanzania, terracing has helped to preserve native vegetation and biodiversity. The region is known for its rich biodiversity, including many endemic plant species, and terracing has played a role in maintaining the health of these ecosystems by reducing soil erosion and supporting sustainable agriculture (Tripathi et al., 2022). In Bhutan, terracing is integrated into the traditional practice of agroforestry, where crops are grown alongside trees and other vegetation. This approach helps to maintain biodiversity, protect watersheds, and enhance the resilience of the farming system. The terraces in Bhutan not only support rice and maize cultivation but also provide habitats for a variety of native plant and animal species (Nepal, 2022). In the Shouf

Biosphere Reserve, Lebanon, a program of restoration of abandoned terraces combined with biodiversity monitoring activities observed 332 species, with significantly higher diversity found in abandoned and restored terraces compared to intensively farmed terraces (Martinoli et al., 2022).

The integrated rice-fish culture system is a globally important agricultural heritage system and is widely practised in Southeast Asian countries, where it reduces costs and increases the economic benefits and quality of both rice and fish (Li et al., 2021). In this way, many terraced landscapes have significant cultural and historical value, and efforts are being made to preserve these sites for future generations. In some cases, terraced landscapes have been designated as UNESCO World Heritage Sites, recognising their cultural importance and promoting their conservation. Additionally, terraced landscapes have become popular tourist destinations, providing economic benefits to local communities and raising awareness on the importance of sustainable agriculture. A classic example is the terraces of Machu Picchu in Peru, part of the UNESCO World Heritage Site, and a major tourist attraction, although not without its challenges (Schlauderaff et al., 2022). In the Ecuadorian Andes, the use of terracing has helped to preserve traditional crop varieties, such as Andean tubers and grains, which are adapted to the specific conditions of the region. These crops are not only important for food security but also for maintaining the genetic diversity of Andean agriculture (Pancorbo-Olivera et al., 2024). In the Peruvian Andes, organisations like the International Potato Centre and local NGOs have been working with farmers to rehabilitate ancient terraces and promote sustainable farming practices. These efforts have helped to increase agricultural productivity, improve food security, and preserve traditional knowledge and cultural heritage. In Rwanda, where land is scarce and the population is dense, terracing has been promoted by the government as a way to improve agricultural productivity, maintain biodiversity, and prevent soil erosion (Plate 5.10). A nationwide program to promote terracing as part of its efforts to improve soil health, water management, biodiversity, food security, and improve livelihoods includes developing new techniques and technologies for terrace construction and maintenance.

5.4.4 Environmental and cultural challenges associated with terracing

While terracing offers numerous benefits, it also presents several challenges, particularly in terms of construction, maintenance, and socio-economic

Plate 5.10 Rwanda terracing can stabilise these slopes, but is often characterised by *Eucalyptus* trees. *(Photographs Rob Marchant).*

factors. Constructing terraces on mountain slopes is a labour-intensive process that requires significant time, effort, and resources. Building terraces involves moving large amounts of soil and stone, levelling the land, and constructing retaining walls to prevent collapse. Once built, terraces require ongoing maintenance to ensure their continuing stability and functionality. This can be a significant burden for small-scale farmers, particularly in regions with limited access to modern tools and machinery. In some regions, the lack of secure land tenure can discourage farmers from investing in terrace construction and maintenance, as they may not reap the long-term benefits. Additionally, terracing often requires collective action and cooperation among community members, which can be challenging in areas with social or political conflicts. For example, in the Tigray region of Ethiopia, land tenure reforms have been implemented to encourage farmers to invest in soil and water conservation practices, including terracing. These reforms have provided farmers with greater security and incentives to improve their land, leading to increased adoption of terracing and other sustainable agricultural practices.

Concerningly, many mountain farmers in Ethiopia must contend with increasingly degraded landscapes and severe soil erosion, as their fields on steep slopes should not be under continuous cultivation, even under a terrace system. Scarcity of fuel wood compels farm households to burn manure and crop residues for household energy, thus resulting in further

Plate 5.11 Abandonment of traditional agricultural terraces as mountain economies change. *(Photograph Rob Marchant).*

decreased soil fertility, agricultural yields, and animal feed, creating more pressure on remaining forests. This downward spiral in rural productivity is compounded by recurrent droughts and floods due to adverse natural climate conditions, which are often aggravated by increasing climate change and climate variability (Capitani et al., 2019).

In the face of climate variability and extreme weather events, terraced agriculture offers some resilience by minimising soil erosion, conserving water resources, and buffering against land degradation. Terraces can be adapted to changing climatic conditions through slope stabilisation measures, agroforestry integration, and diversified cropping systems, although there is no infinite adaptability. In some cases, poorly designed terraces can exacerbate soil erosion and increase the risk of landslides, particularly in regions with unstable geological conditions. For example, in the Rwenzori Mountains of Uganda, improper terracing techniques have sometimes led to increased soil erosion and landslides (Barasa et al., 2022). The abandonment of traditional agricultural land is a global phenomenon, especially in mountainous areas. The dynamic magnitude of landscape changes has increased since the 19th century across many European mountains, particularly accelerating after the Second World War (Sobala, 2023). Natural reforestation is one of the dominant processes in marginal mountain areas of the Northern hemisphere (Anselmetto et al., 2024) (Plate 5.11). The industrial revolution of the 19th century led to an intense exodus from the marginal valleys of the Alps towards urban and industrial centres, resulting in widespread natural reforestation following agricultural abandonment (Anselmetto et al., 2024). Patterns of abandonment may vary locally, with an intensification of agriculture in districts with more favourable environments, greater productivity, and closer connections to transportation corridors and population centres, but an enduring abandonment of environmentally harsh and remote areas (Anselmetto et al., 2024; MacDonald et al., 2000). Although the abandonment of traditional agricultural land may have positive effects on ecosystem services, such as an increase in carbon storage due to forest regrowth, it also has significant negative effects, including soil erosion, a reduction in biodiversity, water provisioning, landscape heterogeneity, and cultural and aesthetic amenity (Imai et al., 2023; Sobala, 2023).

5.5 Mountain pastoralism

Pastoralism is the most widespread land use worldwide (Scoones, 2021). Practised in over 100 countries, mountain rangelands and pastoral populations play vital roles in the functioning of mountain systems, supporting livestock grazing, hunting, and harvesting of plants (Manzano et al., 2021). Mountain rangelands provide livelihoods for many communities across the world's mountains. Pastoral systems are also key for water supply to lowland areas and wider watershed health, carbon sequestration and storage, fire regulation, and, increasingly, recreational use and tourism. Mountain rangelands are vital for the conservation of mountain species and habitats, including endemic species and habitats exclusive to mountains (Azevedo et al., 2020). Many of these ecosystems require management to maintain species diversity and productivity and keep supplying the range of high-quality ecosystem services to mountain communities, and those far downstream.

Mountain pastoralism is a unique way of life and a production system creating strong cultural connections to the mountains. It encompasses diverse herding practices focused on cows, sheep, goats, lama, yaks, and vicuna (Plate 5.12), that link natural and social systems at multiple social and ecological scales. Mountain pastoralists often move their livestock seasonally to respond to growing seasons of pasture or extensive winter snow cover and take advantage of new pastures, a practice known as transhumance pastoralism (Azevedo et al., 2020). Examples include the optimal use of pastures in mountain and lowland systems with transhumance corridors extending over thousands of kilometres (Manzano et al., 2021). Indeed, pastoral systems can be complex and frequently rely on multiple types of social institutions and knowledge (e.g., traditional, ecological, and indigenous)

Plate 5.12 Pastoral systems can be quite diverse and include cattle, sheep, goats, lama, yaks, and vicuna. *(Photographs Rob Marchant).*

for managing mountain environments, including the sharing of labour and kin relationships. Pastoralism may also involve sophisticated technology and innovative management practices to improve feeding patterns and strategies, track animal locations and movement, implement rotational grazing systems, or prevent attacks from large carnivores (Azevedo et al., 2020).

A long history of summer grazing on Alpine pastures by cattle and sheep has completely transformed many mountain ecosystems (Plate 5.13). For example, in the European Alps, grazing has pushed the treeline to areas below the elevation threshold at which local environmental conditions naturally limit tree growth. The treeline belt consists of a complex mosaic of coniferous stands, patches of dwarf shrubs, and grassland that can stretch over an elevational range of a few hundred metres. This heterogeneous

Plate 5.13 Alpine biodiversity-rich grasslands are changing following intensified nutrient and changed management regimes. *(Photograph Rob Marchant).*

anthropogenically crafted habitat harbours a rich biodiversity, which is now threatened by a dichotomous mix of agricultural abandonment and management intensification of traditional pastures and meadows (Tasser et al., 2017). The management of biodiversity-rich grasslands is also changing following intensified fertilisation, particularly increased nitrogen input to enhance fodder productivity (Lavorel et al., 1999, 2022; Sa Rego et al., 2022), which is resulting in some of the diverse flower meadows declining in diversity due to the removal of the disturbance regime (grazers) and growth enhancement (nutrient addition) (Plate 5.13).

Changing markets and demand for sustainable products in other world regions, as well as new forms of communication, can have negative or positive impacts on pastoralists' lives. Pastoral systems in the mountains are changing rapidly. In some parts of the world, this change results from overgrazing, overharvesting, or intensification of livestock production related to growing populations and market demands; further drivers of change include competing land uses encroaching on pastoral territories and resources, as well as new borders restricting the movement of transhumant populations (Scoones, 2021). In other mountain regions, depopulation, abandonment, and under-grazing have led to new land uses and landscapes, favouring the spread of alien invasive species (Plate 5.14), reforestation, possible biodiversity loss, and altered fire regimes.

Coupled with climate change and other drivers and disturbances, these changes in mountain pastoral social-ecological systems are affecting their

Plate 5.14 *Opuntia* was introduced to Tenerife for the cochineal industry, as synthetic red dyes were substituted, this was no longer needed, although it spread rapidly across the island, particularly as terraced agriculture was abandoned as people moved into working in the tourism industry since the 1960s. *(Photograph Rob Marchant).*

resilience. This in turn leads to degradation, thereby reducing the production and supply of ecosystem services, causing loss of livelihoods, and most importantly, threatening the maintenance of these systems and the benefits that can be accrued across the landscape, and to downstream communities. Indeed, the global relevance and urgency of addressing the challenges for pastoral social–ecological systems is widely acknowledged. For example, pastoralism has been valued by the European Union as a vector of environmental management and preservation of the mountainous areas. In Switzerland, numerous policies and schemes to compensate pastoralists for managing stocking densities and types of animals have created a network of pasture managers with a focus on maintaining high-value alpine meadows (Plate 5.15).

Transhumance pastoralism, involving movements between summer and winter pastures, is a centuries-old practice in high mountain areas

Plate 5.15 In many countries, cattle move into and out of mountains in response to the seasonal availability of pasture. This often can create high diversity pastures that, in some countries, such as Switzerland, are actively promoted. *(Photographs Rob Marchant).*

(Haq et al., 2021). Across high mountain Asia, European ranges, East Africa, and across the Andes, transhumance pastoralism is declining due to both climatic factors, including changes in snow distribution and glaciers, and non-climatic factors, particularly increasing occupied land, which is fenced, thus preventing the free range of livestock over long distances. As populations are continuing to grow, land is continuing to be occupied, and fences are continuing to be erected, leading to land being dissected and connectivity lost (Plate 5.16). It seems inevitable that free-ranging pastoralism is projected to continue declining (Hock et al., 2019). The traditional practice of alpine

Plate 5.16 Fences are expanding across mountain landscapes, such as here on the Maloti-Drakensburg, leading to land being dissected and connectivity lost. *(Photograph Rob Marchant).*

transhumance, where livestock are moved between lowland winter pastures and high-altitude summer pastures, is still practised in the Alps but at much reduced levels, partly due to the challenges of moving animals across the landscape. This sustainable form of pastoralism not only supports local economies but also helps to maintain the biodiversity of alpine meadows by preventing overgrowth and maintaining open landscapes. Other climate change impacts, including erratic snowfall patterns and a decrease in rainfall, are perceived by herders in Afghanistan, Nepal, and Pakistan to have resulted in vegetation of lower quality and quantity, putting further pressure on pastoral livelihood (Shaoliang et al., 2012).

In New Zealand's Southern Alps, the practice of high-country sheep farming has been a way of life for generations of settlers. These farmers, known as high-country runholders, manage vast tracts of mountainous land, grazing sheep on native tussock grasslands. This sustainable form of pastoralism not only supports local economies but also plays a key role in maintaining the unique landscapes of the Southern Alps. Mediterranean mountain landscapes have a long history of human occupation and land cover changes going back to at least Mesolithic times (Alcolea et al., 2017). This has resulted in highly complex landscapes that range from eroded hillslopes to well-preserved bench-terraced fields in which hedgerows, isolated trees, and small woods enhance biodiversity. Such mosaic landscapes are the consequence of numerous political, economic, and demographic events spanning thousands of years, all of which have left imprints that can be partially unravelled (García-Ruiz et al., 2015). In general, cultivated areas expanded until the mid-19th century, when the population pressure peaked before many cultivated slopes were abandoned, with the fields becoming covered with shrubs and forest within a few decades (Tasser et al., 2017). The decline of transhumance systems also resulted in a progressive uplift of the treeline and a consequent decline in the area occupied by summer grasslands. This collapse of the landscape along the different altitudinal belts of the Mediterranean mountains paralleled a dramatic decline in the human population and the ageing of farmers, both of which reflected a socio-economic change and massive migration towards the main cities. In most Mediterranean mountain areas, both agriculture and stockbreeding are closely related and are integrated in a holistic land management system where the stock graze on cultivated fields once the crop has been harvested.

While some pastoral regions' economies have seen a revival through the influx of small-scale artisanal producers, in the French Pyrenees, with investment via tourism, the nature of European pastoralism has dramatically shifted. There are fewer owners, more consolidated flocks and herds, and more migrant herders, with much livestock production shifting away from the hills and mountains to the plains, as part of intensified, mixed farming systems. Conflicts between new wildlife uses in mountain areas and traditional pastoralism have escalated, especially with the reintroduction of bears and wolves (Plate 5.17). As a result of these changes, new social-cultural and class relations emerge between long-standing pastoral families and newcomers into the mountain environment (Scoones, 2021).

Insights into pastoral systems need to be effectively integrated into global science-policy interfaces and assessments, to better recognise pastoralism in global environmental and economic forums or sustainability agendas. Understanding of trajectories in pastoralist societies, and the characterisation of the factors that drive them through globally accepted indicators, should be used to find solutions to manage sustainable pasture futures by informing national policies (Manzano et al., 2021). There needs to be better-informed and more effective policies and interventions, which will ensure pastoralism resilience and sustainability in the future, possibly as much for their role in

Plate 5.17 Conflicts between new wildlife uses in mountain areas and traditional pastoralism have escalated, especially with the reintroduction of bears and wolves. *(Photograph Rob Marchant).*

creating biodiverse mountain landscapes as for the production of meat, milk, and cheese.

5.6 Conclusions

Mountain communities and mountain food production systems, be it agroforestry, terraced agriculture, or pastoral, face the unprecedented task of feeding a growing world population while simultaneously avoiding harmful environmental and social effects. The most successful attempts to build resilience in mountain regions in the face of increasing environmental uncertainty have involved a combination of climate-smart agricultural strategies, coupled with local, community-led initiatives with positive incentives that lean on traditional ecological knowledge (Frogley et al., 2025). Such a combined, joined-up approach has been applied in multiple settings for improving the livelihoods of many vulnerable highland rural communities in the Andes and, indeed, across the globe. The effort to meet this challenge, when combined with schemes to maximise additional value to mountain produce, such as organic farming, can have positive outcomes. However, a number of challenges remain. Organic yields lag behind those in conventional agriculture, and greenhouse gas emissions and nutrient leaching remain somewhat problematic. Agroforestry systems that mimic nature's function have the potential to remain productive while supporting a range of ecosystem services, and for this potential to be realised, the economics have to match. There has to be recognition that the farmer is making a positive contribution to aspects such as water regulations, carbon sequestration, biodiversity conservation, etc., that have benefits far outside of the farm boundaries, and as such should be suitably compensated for this export. We need to shift the economy of food production and support extensive, high-quality, low-input, slow production systems that, where possible, support mountain communities.

References

Aimar, F. (2024). The "Cultural Landscape of Honghe Hani Rice Terraces": New identities and continuation of a cultural landscape. In *The resilience of cultural landscapes* (pp. 225–245). Springer. https://doi.org/10.1007/978-3-031-55861-0_8.

Alcolea, M., Domingo, R., Piqué, R., & Montes, L. (2017). Landscape and firewood at Espantalobos Mesolithic site (Huesca, Spain). First results. *Quaternary International, 457*, 198–210. https://doi.org/10.1016/j.quaint.2016.10.007.

Andriatsitohaina, R. N. N., Laby, P., Llopis, J. C., & Martin, D. A. (2024). Agroforestry in Madagascar: Past, present, and future. *Agroforestry Systems, 98*(6), 1659–1680. https://doi.org/10.1007/s10457-024-00975-y.

Anhar, A., Abubakar, Y., Widayat, H. P., Muslih, A. M., & Romano, A. B. (2021). Altitude, shading, and management intensity effect on Arabica coffee yields in Aceh, Indonesia. *Open Agriculture, 6*(1), 254–262. https://doi.org/10.1515/opag-2021-0220.

Anselmetto, N., Weisberg, P. J., & Garbarino, M. (2024). Global change in the European Alps: A century of post-abandonment natural reforestation at the landscape scale. *Landscape and Urban Planning, 243*, 104973. https://doi.org/10.1016/j.landurbplan.2023.104973.

Azevedo, J. C., Clark, V. R., Millar, J., Mukwada, G., Postigo, J. C., & Wurzinger, M. (2020). Focus issue: Pastoralism and rangelands in mountains. *Mountain Research and Development, 41*(4), 1–2.

Barasa, B., Nakileza, B., Mugagga, F., Nseka, D., Opedes, H., Gudoyi, P. M., & Ssentongo, B. (2022). Natural hazards magnitude, vulnerability, and recovery strategies in the Rwenzori mountains, Southwestern Uganda. In *Remote sensing of African mountains: Geospatial tools toward sustainability* (pp. 83–116). Springer International Publishing. https://doi.org/10.1007/978-3-031-04855-5_5.

Bhutia, K. D. (2021). Death by poisoning: Cautionary narratives and inter-ethnic accusations in contemporary Sikkim. *Journal of Ethnology and Folkloristics, 15*(1), 65–84. https://doi.org/10.2478/jef-2021-0005.

Buckley, H. L., Hall, D., Jarvis, R. M., Smith, V., Walker, L. A., Silby, J., Hinchliffe, G., Stanley, M. C., Sweeney, A. P., & Case, B. S. (2023). Using long-term experimental restoration of agroecosystems in Aotearoa New Zealand to improve implementation of Nature-based Solutions for climate change mitigation. *Frontiers in Forests and Global Change, 5*. https://doi.org/10.3389/ffgc.2022.950041.

Capitani, C., Garedew, W., Mitiku, A., Berecha, G., Hailu, B. T., Heiskanen, J., Hurskainen, P., Platts, P. J., Siljander, M., Pinard, F., Johansson, T., & Marchant, R. (2019). Views from two mountains: Exploring climate change impacts on traditional farming communities of Eastern Africa highlands through participatory scenarios. *Sustainability Science, 14*(1), 191–203. https://doi.org/10.1007/s11625-018-0622-x.

Cassamo, C. T., Draper, D., Romeiras, M. M., Marques, I., Chiulele, R., Rodrigues, M., Stalmans, M., Partelli, F. L., Ribeiro-Barros, A., & Ramalho, J. C. (2023). Impact of climate changes in the suitable areas for *Coffea arabica* L. production in Mozambique: Agroforestry as an alternative management system to strengthen crop sustainability. *Agriculture, Ecosystems and Environment, 346*. https://doi.org/10.1016/j.agee.2022.108341.

Connor, A. (2023). *Ladin perspectives on language and identity in the Central Dolomites of Northern Italy (Doctoral dissertation*. University of Sheffield.

Cuni-Sanchez, A., Aneseyee, A. B., Baderha, G. K. R., Batumike, R., Bitariho, R., Imani, G., Jha, N., Kaganzi, K. R., Kaplin, B. A., Klein, J. A., Leite, A., Marchant, R. A., Martin, E. H., Mcharazo, F., Mwangi, B., Ngute, A. S. K., Nkengurutse, J., Nkurunziza, A., Olaka, L., & …, Zafra-Calvo, N. (2025). Perceived climate change impacts and adaptation responses in ten African mountain regions. *Nature Climate Change, 15*(2), 153–161. https://doi.org/10.1038/s41558-024-02221-w.

Davis, A. P., Chadburn, H., Moat, J., O'Sullivan, R., Hargreaves, S., & Lughadha, E. N. (2019). High extinction risk for wild coffee species and implications for coffee sector sustainability. *Science Advances, 5*(1). https://doi.org/10.1126/sciadv.aav3473.

de Carvalho Alves, M., & Sanches, L. (2022). Potential effects of spatio-temporal temperature variation for monitoring coffee leaf rust progress under CMIP6 climate change scenarios. *Earth Systems and Environment, 6*(2), 421–436. https://doi.org/10.1007/s41748-021-00286-7.

Denu, D., Platts, P. J., Kelbessa, E., Gole, T. W., & Marchant, R. (2016). The role of traditional coffee management in forest conservation and carbon storage in the Jimma Highlands, Ethiopia. *Forests Trees and Livelihoods, 25*(4), 226–238. https://doi.org/10.1080/14728028.2016.1192004.

Duffy, C., Toth, G. G., Hagan, R. P. O., McKeown, P. C., Rahman, S. A., Widyaningsih, Y., Sunderland, T. C. H., & Spillane, C. (2021). Agroforestry contributions to smallholder farmer food security in Indonesia. *Agroforestry Systems, 95*(6), 1109–1124. https://doi.org/10.1007/s10457-021-00632-8.

Ferro-Vázquez, C., & Stump, D. (2023). *The archaeological study of agricultural sustainability: The cases of Konso (S Ethiopia) and Engaruka (N Tanzania) people and Agrarian landscapes: An archaeology of postclassical local societies in the western Mediterranean* (pp. 196–216). Archaeopress,. https://www.archaeopress.com/Archaeopress/Products/9781803274379.

Frogley, M. R., Chepstow-Lusty, A., Thiele, G., & Chutas, C. A. (2025). Trees, terraces and llamas: Resilient watershed management and sustainable agriculture the Inca way. *Ambio, 54*(5), 793–807. https://doi.org/10.1007/s13280-024-02121-5.

Galabuzi, C., Agaba, H., Okia, C. A., Odoul, J., & Muthuri, C. (2021). Women and youths participation in agroforestry: What counts and what doesn't around Mount Elgon, Uganda? *Journal of Mountain Science, 18*(12), 3306–3320. https://doi.org/10.1007/s11629-021-6812-5.

García-Ruiz, J. M., Beguería, S., Nadal-Romero, E., González-Hidalgo, J. C., Lana-Renault, N., & Sanjuán, Y. (2015). A meta-analysis of soil erosion rates across the world. *Geomorphology, 239*, 160–173. https://doi.org/10.1016/j.geomorph.2015.03.008.

Gebrekirstos, A., van Noordwijk, M., Neufeldt, H., & Mitlöhner, R. (2011). Relationships of stable carbon isotopes, plant water potential and growth: An approach to asses water use efficiency and growth strategies of dry land agroforestry species. *Trees, 25*(1), 95–102. https://doi.org/10.1007/s00468-010-0467-0.

Gebrekirstos, A. (2025). The human factor. *Nature Plants, 11*, 1221. https://doi.org/10.1038/s41477-025-02075-w.

Gebremikael, A. T., Hassan, J. Y., Endris, G. S., & Aweke, C. S. (2024). Analysis of farmers' perceptions of bench terracing innovation in the eastern and southern Ethiopian highlands. *Sustainable Environment, 10*(1). https://doi.org/10.1080/27658511.2023.2293261.

Greene, L. (2022). *Their determination to remain: A Cherokee community's resistance to the trail of tears in North Carolina*. University of Alabama Press https://doi.org/10.2307/jj.30347513.

Habito, C. D., Sontillanosa, R. D., & Faurillo, D. L. C. (2023). *Analyzing the spatio-temporal changes of the biocultural landscape of Banaue, Philippines using GIS plant diversity in biocultural landscapes* (pp. 495–513). Springer Nature. https://doi.org/10.1007/978-981-19-8649-9_21.

Haq, S. M., Yaqoob, U., Calixto, E. S., Kumar, M., Rahman, I. U., Hashem, A., Abd-Allah, E. F., Alakeel, M. A., Alqarawi, A. A., Abdalla, M., Lone, F. A., Khan, M. A., Khan, U., & Ijaz, F. (2021). Long-term impact of transhumance pastoralism and associated disturbances in high-altitude forests of indian western himalaya. *Sustainability (Switzerland), 13*(22). https://doi.org/10.3390/su132212497.

Hock, R., Rasul, G., Adler, C., Cáceres, B., Gruber, S., Hirabayashi, Y., Jackson, M., Kääb, A., Kang, S., Kutuzov, S., & Milner, A. (2019). High mountain areas. In H.-O. Pörtner, D. C. Roberts, V. Masson-Delmotte, P. Zhai, M. Tignor, E. Poloczanska, K. Mintenbeck, A. Alegría, M. Nicolai, A. Okem, J. Petzold, B. Rama, & N. M. Weyer (Eds.), *IPCC special report on the ocean and cryosphere in a changing climate* (pp. 131–202). Cambridge University Press.

Imai, N., Otokawa, H., Okamoto, A., Yamazaki, K., Tamura, T., Sakagami, T., Ishizaka, S., & Shimojima, H. (2023). Abandonment of cropland and seminatural grassland in a mountainous traditional agricultural landscape in Japan. *Sustainability, 15*(10), 7742. https://doi.org/10.3390/su15107742.

Kadigi, W. R., Ngaga, Y. M., & Kadigi, R. M. (2021). Perceptions of smallholder farmers on nature-based income generating activities as potential livelihood and biodiversity conservation strategies in Uluguru Mountains, Tanzania. In *Journal of Development and Agricultural Economics: 13* (pp. 174–191).

Kimaro, A. A., Sererya, O. G., Matata, P., Uckert, G. ö, Hafner, J., Graef, F., Sieber, S., & Rosenstock, T. S. (2018). *Understanding the multidimensionality of climate-smartness: Examples from agroforestry in Tanzania* (pp. 153–162). Springer Science and Business Media LLC. https://doi.org/10.1007/978-3-319-92798-5_13.

Nepal, T. K. (2022). An overview of biodiversity in Bhutan. *Asian Journal of Research in Agriculture and Forestry*, 7–19. https://doi.org/10.9734/ajraf/2022/v8i130144.

Lavorel, S., Grigulis, K., Richards, D. R., Etherington, T. R., Law, R. M., & Herzig, A. (2022). Templates for multifunctional landscape design. *Landscape Ecology, 37*(3), 913–934. https://doi.org/10.1007/s10980-021-01377-6.

Lavorel, S., McIntyre, S., & Grigulis, K. (1999). Plant response to disturbance in a Mediterranean grassland: How many functional groups? *Journal of Vegetation Science, 10*(5), 661–672. https://doi.org/10.2307/3237081.

Law, J. (2022). The knotty politics of ginseng conservation and management in Appalachia. *Journal of Political Ecology, 29*(1), 36–50. https://doi.org/10.2458/JPE.2286.

Leceta, F., Binder, C., Mader, C., Mächtle, B., Marsh, E., Dietrich, L., Reindel, M., Eitel, B., & Meister, J. (2024). The impact of agriculture on tropical mountain soils in the western Peruvian Andes: A pedo-geoarchaeological study of terrace agricultural systems in the Laramate region (14.5° S). *Soil, 10*(2), 727–761. https://doi.org/10.5194/soil-10-727-2024.

Lentz, D. L., Slotten, V., Dunning, N. P., Jones, J. G., Scarborough, V. L., McCool, J. P., Owen, L. A., Fladd, S. G., Tankersley, K. B., Perfetta, C. J., Carr, C., Crowley, B., & Plog, S. (2021). Ecosystem impacts by the Ancestral Puebloans of Chaco Canyon, New Mexico, USA. *PLoS ONE, 16*(10). https://doi.org/10.1371/journal.pone.0258369.

Li, F., Gao, J., Xu, Y., Nie, Z., Fang, J., Zhou, Q., Xu, G., Shao, N., Xu, D., Xu, P., & Wang, M. (2021). Biodiversity and sustainability of the integrated rice-fish system in Hani terraces, Yunnan province, China. *Aquaculture Reports, 20*, 100763. https://doi.org/10.1016/j.aqrep.2021.100763.

Lüning, S., Gałka, M., Bamonte, F. P., Rodríguez, F. G., & Vahrenholt, F. (2019). The medieval climate anomaly in South America. *Quaternary International, 508*, 70–87. https://doi.org/10.1016/j.quaint.2018.10.041.

MacDonald, D., Crabtree, J. R., Wiesinger, G., Dax, T., Stamou, N., Fleury, P., Gutierrez Lazpita, J., & Gibon, A. (2000). Agricultural abandonment in mountain areas of Europe: Environmental consequences and policy response. *Journal of Environmental Management, 59*(1), 47–69. https://doi.org/10.1006/jema.1999.0335.

Mader, C., Godde, P., Hägele, E., Lyons, M., Weber, A. K., Odenthal, R., Stryjski, P., Binder, C., Leceta, F., Isla, J., Reindel, M., & Meister, J. (2025). Mapping and geospatial analysis of ancient terrace agricultural systems in Lucanas Province, Peruvian Andes, based on satellite imagery, high-resolution DSMs, and field surveys. *Geoarchaeology, 40*(2). https://doi.org/10.1002/gea.70002.

Manzano, P., Burgas, D., Cadahía, L., Eronen, J. T., Fernández-Llamazares, Á., Bencherif, S., Holand, Ø., Seitsonen, O., Byambaa, B., Fortelius, M., Fernández-Giménez, M. E., Galvin, K. A., Cabeza, M., & Stenseth, N. C. (2021). Toward a holistic understanding of pastoralism. *One Earth, 4*(5), 651–665. https://doi.org/10.1016/j.oneear.2021.04.012.

Martinoli, A., Regato, P., Samad, F. A., Kanso, L., Eddine, N. S., Panichi, M., Gagliardi, A., Sarkis, L., & Hani, N. (2022). Stone-walled terraces restoration: Conserving biodiversity and promoting economic functions of farmlands in Lebanon. *Journal of Agriculture and Environment for International Development (JAEID), 116*(1), 77–114. https://doi.org/10.36253/jaeid-13012.

Meddens, F. M., Aramburu Venegas, D. M., Branch, N. P., Fehren-Schmitz, L., Meddens, B. J., & Pomacanchari, C. V. (2025). The cloud people of Naupallacta: Middle horizon camelid management in the hinterland of the Chicha Soras Valley. *Nawpa Pacha, 45*(1), 21–54. https://doi.org/10.1080/00776297.2024.2399428.

Namgyel, U., Dorji, S., Lee, W.-K., & Wang, S. W. (2023). Farmers' perceptions of climate change and its socio-ecological consequences in Bhutan's biological corridor network. *Sustainability, 15*(19), 14517. https://doi.org/10.3390/su151914517.

Niguse, G., Iticha, B., Kebede, G., & Chimdi, A. (2022). Contribution of coffee plants to carbon sequestration in agroforestry systems of Southwestern Ethiopia. *The Journal of Agricultural Science, 160*(6), 440–447. https://doi.org/10.1017/s0021859622000624.

Pancorbo-Olivera, M., Parra-Rondinel, F., Torres-Guevara, J., Cruz-Soriano, A., & Casas, A. (2024). Gathering, agriculture, and exchange: An ethnoecological approach to the study of food patterns and feedstuff sources in communities of the Central Andes, Peru. *Journal of Ethnobiology and Ethnomedicine, 20*(1). https://doi.org/10.1186/s13002-024-00705-9.

Peters, M. K., Hemp, A., Appelhans, T., Becker, J. N., Behler, C., Classen, A., Detsch, F., Ensslin, A., Ferger, S. W., Frederiksen, S. B., Gebert, F., Gerschlauer, F., Gütlein, A., Helbig-Bonitz, M., Hemp, C., Kindeketa, W. J., Kühnel, A., Mayr, A. V., Mwangomo, E., & …, Steffan-Dewenter, I. (2019). Climate–land-use interactions shape tropical mountain biodiversity and ecosystem functions. *Nature, 568*(7750), 88–92. https://doi.org/10.1038/s41586-019-1048-z.

Platts, P. J., Schaafsma, M., Turner, R. K., Burgess, N. D., Fisher, B., Mbilinyi, B. P., Munishi, P. K. T., Ricketts, T. H., Swetnam, R. D., Ahrends, A., Ashagre, B. B., Bayliss, J., Gereau, R. E., Green, J. M. H., Green, R. E., Jeha, L., Lewis, S. L., Marchant, R., Marshall, A. R., & …, Balmford, A. (2023). Inequitable gains and losses from conservation in a global biodiversity hotspot. *Environmental and Resource Economics, 86*(3), 381–405. https://doi.org/10.1007/s10640-023-00798-y.

Rakocevic, M., Batista, E. R., Pazianotto, R. A. A., Scholz, M. B. S., Souza, G. A. R., Campostrini, E., & Ramalho, J. C. (2021). Leaf gas exchange and bean quality fluctuations over the whole canopy vertical profile of Arabic coffee cultivated under elevated CO_2. *Functional Plant Biology, 48*(5), 469–482. https://doi.org/10.1071/FP20298.

Röhrig, N., Hassler, M., & Roesler, T. (2021). Silvopastoral production as part of alternative food networks: Agroforestry systems in Umbria and Lazio, Italy. *Agroecology and Sustainable Food Systems, 45*(5), 654–672. https://doi.org/10.1080/21683565.2020.1835783.

Roy, M. K., Fort, M. P., Kanter, R., & Montagnini, F. (2025). Agroforestry: A key land use system for sustainable food production and public health. *Trees, Forests and People, 20*. https://doi.org/10.1016/j.tfp.2025.100848.

Rozaki, Z., Rahmawati, N., Wijaya, O., Fajriyah Mubarok, A., Senge, M., & Md. Kamarudin, F. (2021). A case study of agroforestry practices and challenges in Mt. Merapi risk and hazard prone area of Indonesia. *Biodiversitas Journal of Biological Diversity, 22*(6). https://doi.org/10.13057/biodiv/d220661.

Sa Rego, J., Cabo, P., & Castro, M. (2022). Pastoralism, multifunctionality, and environmental agency: Insights from mountain sheep pastoralists in Northern Portugal. *Journal of Agrarian Change, 22*(4), 766–786. https://doi.org/10.1111/joac.12480.

Santoro, A., Venturi, M., Bertani, R., & Agnoletti, M. (2020). A Review of the role of forests and agroforestry systems in the FAO globally important agricultural heritage systems (GIAHS) programme. *Forests, 11*(8), 860. https://doi.org/10.3390/f11080860.

Schlauderaff, S., Press, J., Huston, H., Su, C. H. J., & Tsai, C. H. K. (2022). Are we putting our favorite destinations in Peril? A no longer lost city - Machu Picchu case study. *Journal of Hospitality and Tourism Cases, 10*(1), 34–39. https://doi.org/10.1177/21649987220100010\4.

Scoones, I. (2021). Pastoralists and peasants: Perspectives on agrarian change. *The Journal of Peasant Studies, 48*(1), 1–47. https://doi.org/10.1080/03066150.2020.1802249.

Seid, G., & Kebebew, Z. (2022). Homegarden and coffee agroforestry systems plant species diversity and composition in Yayu Biosphere Reserve, southwest Ethiopia. *Heliyon, 8*(4). https://doi.org/10.1016/j.heliyon.2022.e09281.

Shaoliang, Y., Ismail, M., & Zhaoli, Y. (2012). Pastoral communities' perspectives on climate change and their adaptation strategies in the Hindukush-Karakoram-Himalaya. In *Pastoral practices in high Asia: Agency of'development'effected by modernisation, resettlement and transformation* (pp. 307–322). Springer. https://doi.org/10.1007/978-94-007-3846-1_17.

Showers, K. B. (1989). Soil erosion in the Kingdom of Lesotho: Origins and colonial response, 1830s-1950s. *Journal of Southern African Studies, 15*(2), 263–286. https://doi.org/10.1080/03057078908708200.

Sobala, M. (2023). Assessment of the traditional landscapes' state in mountain areas as the basis for their restoration (the Western Beskids, Poland). *Applied Geography, 161*, 103123. https://doi.org/10.1016/j.apgeog.2023.103123.

Sollen-Norrlin, M., Ghaley, B. B., & Rintoul, N. L. J. (2020). Agroforestry benefits and challenges for adoption in Europe and beyond. *Sustainability, 12*(17), 7001. https://doi.org/10.3390/su12177001.

Tappan, G., & McGahuey, M. (2007). Tracking environmental dynamics and agricultural intensification in southern Mali. *Agricultural Systems, 94*(1), 38–51. https://doi.org/10.1016/j.agsy.2005.07.011.

Tasser, E., Leitinger, G., & Tappeiner, U. (2017). Climate change versus land-use change—What affects the mountain landscapes more? *Land Use Policy, 60*, 60–72. https://doi.org/10.1016/j.landusepol.2016.10.019.

Thornton, J. M., Snethlage, M. A., Sayre, R., Urbach, D. R., Viviroli, D., Ehrlich, D., Muccione, V., Wester, P., Insarov, G., Adler, C., & Freeman, J. (2022). Human populations in the world's mountains: Spatio-temporal patterns and potential controls. *PLOS ONE, 17*(7), e0271466. https://doi.org/10.1371/journal.pone.0271466.

Tripathi, H. G., Kunin, W. E., Smith, H. E., Sallu, S. M., Maurice, S., Machera, S. D., Davies, R., Florence, M., Eze, S., Yamdeu, J. H. G., & Sait, S. M. (2022). Climate-smart agriculture and trade-offs with biodiversity and crop yield. *Frontiers in Sustainable Food Systems, 6*. https://doi.org/10.3389/fsufs.2022.868870.

Tumahai, L. (2024). Climate policy in Jacinda Ardern's New Zealand: A Ngāi Tahu perspective. In *New Zealand's foreign policy under the Jacinda Ardern government: Facing the challenge of a disrupted world* (pp. 139–158). New Zealand World Scientific Publishing Co.. https://doi.org/10.1142/9789811285165_0007.

Vanacker, V., Molina, A., Rosas, M. A., Bonnesoeur, V., Román-Dañobeytia, F., Ochoa-Tocachi, B. F., & Buytaert, W. (2022). The effect of natural infrastructure on water erosion mitigation in the Andes. *SOIL, 8*(1), 133–147. https://doi.org/10.5194/soil-8-133-2022.

Wang, N., Fang, M., Beauchamp, M., Jia, Z., & Zhou, Z. (2021). An indigenous knowledge-based sustainable landscape for mountain villages: The Jiabang rice terraces of Guizhou, China. *Habitat International, 111*, 102360. https://doi.org/10.1016/j.habitatint.2021.102360.

Wilson, M., & Lovell, S. (2016). Agroforestry—The next step in sustainable and resilient agriculture. *Sustainability, 8*(6), 574. https://doi.org/10.3390/su8060574.

Yin, C., Zhang, J., & Yu, X. (2023). Mountain valleys, alluvial fans and oases: Geomorphologic perspectives of the mixed agropastoral economy in Xinjiang (3000–200 BC). *Frontiers in Earth Science, 11*. https://doi.org/10.3389/feart.2023.1109905.

Zhang, Y., Zhang, A., & Ma, Y. (2024). An integrated mechanism and challenges of mountainous sustainable development: A review of Hani Terraces, China. *Sustainable Development, 32*(1), 101–118. https://doi.org/10.1002/sd.2651.

Zhu, G., Li, X., & Zhang, Y. (2021). Multi-stakeholder involvement mechanism in tourism management for maintaining terraced landscape in important agricultural heritage systems (IAHS) Sites: A case study of Dazhai Village in Longji Terraces, China. *Land, 10*(11), 1146. https://doi.org/10.3390/land10111146.

CHAPTER 6

Challenges faced by the world's mountain areas

6.1 Introduction

As detailed in previous chapters, mountains are amongst the world's most dramatic and most dynamic landscapes and are vital to the lives of people who depend, either directly or indirectly, on the many benefits that they provide. Mountain environments are particularly attractive due to their high diversity of natural resources, high agricultural productivity, reliable water supply, cooler climate (Ashagre et al., 2018), supply of fuel, and exceptional biodiversity (Capitani et al., 2019). Mountains determine many nations' sustainable development potential by underpinning food production, energy security, biocultural diversity, tourism income, and supplying timber and non-timber forest products (Capitani et al., 2019; Cuni-Sanchez et al., 2021). Human population pressure, and how this interacts with rapidly changing and more variable climates, represent immediate risks for ecosystem functioning, potential species extinctions, and loss of genetic diversity, with extreme consequences that extend far beyond the boundary of mountains. Additionally, mountain areas are the focus of restoration and afforestation initiatives aiming to both sequester carbon and conserve biodiversity to address the global climate change and biodiversity crisis, respectively. Partly due to their importance for underpinning the global development agenda, enhancing environmental resilience, and adaptability at scale, mountains and mountain people face numerous challenges. Mountains and mountain people are experiencing both slow-onset changes (e.g., in precipitation and temperature, habitat fragmentation, overgrazing, and glacier retreat) and sudden-onset events (e.g., hydrological risk, flash floods, landslides, and volcanic activity). This chapter will explore the geological, biophysical, ecological, and social challenges and hazards, and examine how these combine and cascade to develop potential impacts (Fig. 6.1). This chapter of 'mountain doom' is a precursor for a chapter of 'mountain hope', which explores how these challenges can be navigated through and mitigated against.

Mountain Diversity: Dynamics of High-Altitude Environments.
DOI: https://doi.org/10.1016/B978-0-443-22251-1.00006-X

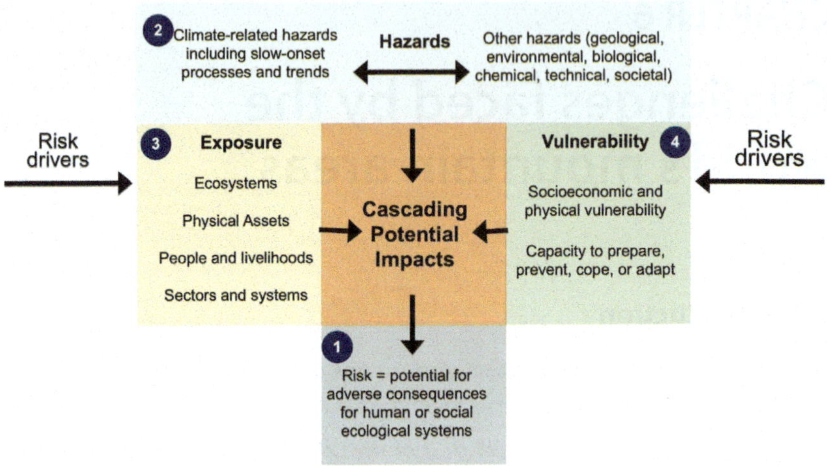

Figure 6.1 Risk is a factor complex arising from hazard, exposure, and vulnerability that combine and cascade to develop potential impacts.

6.2 Geological mountain hazards

Due to their steep terrain, dynamic geological activity, and extreme weather conditions, mountains are particularly prone to numerous geological hazards, including earthquakes, volcanic eruptions, landslides, and avalanches (Fig. 6.2). Geological hazards are one of the most common natural disasters in mountains, causing human casualties and economic losses each year. Their orogenesis and impact are closely related to the regional topography, landforms, lithology, vegetation, geological conditions, and meteorological conditions (Lin et al., 2022). Volcanic eruptions are one of the most devastating natural phenomena, posing multiple serious threats to human health, the ecosystem, and infrastructure. Volcanic hazards, such as pyroclastic lava flows and ashfall, can have widespread impacts on mountain communities and landscapes. Lahars pose a significant risk to communities, particularly those living near snow-capped volcanoes. These rapid flows of mud and debris, typically, but not necessarily, triggered by volcanic activity, can pose risks to human health, settlements, infrastructure, and ecosystems. For example, the eruption and ensuing Lahar flow at Nevado Del Ruiz, Colombia, in 1985 led to the loss of over 23,000 lives and destroyed an entire town (Vinnell et al., 2021).

Exposure to volcanic hazards can lead to a variety of health problems, with respiratory disorders being amongst the most common. Fine volcanic

Figure 6.2 Mountains are particularly prone to several types of geological hazards, including earthquakes, volcanic eruptions, landslides, and avalanches.

ash containing silica and sulphur dioxide can enter deep into the lungs and be highly toxic (Fernandez et al., 2025). The physical health effects are compounded by indirect effects from impact on critical services such as contaminated water and decreased food security, which, in turn, can cause secondary health problems (Fernandez et al., 2025). Monitoring and understanding people's readiness and intentions for adaptation, and the geological processes that trigger these hazards, is essential for hazard assessment, mitigation, and disaster preparedness in affected areas (Vinnell et al., 2021) (Fig. 6.3). For example, one area of significant progress has been the creation of a searchable Volcanic Hazard Maps Database, which has collected information from 1780 volcanic hazard maps to provide a warning system for volcanic hazards (Ogburn et al., 2023). In addition to collating information from volcanic observatories, this resource includes participatory community-based hazard maps designed by communities surrounding volcanoes. These direct understandings of risk and risk mitigation are created during collaborative workshops between NGOs, community members, and scientists. The resources focus on social aspects of risk mitigation displayed

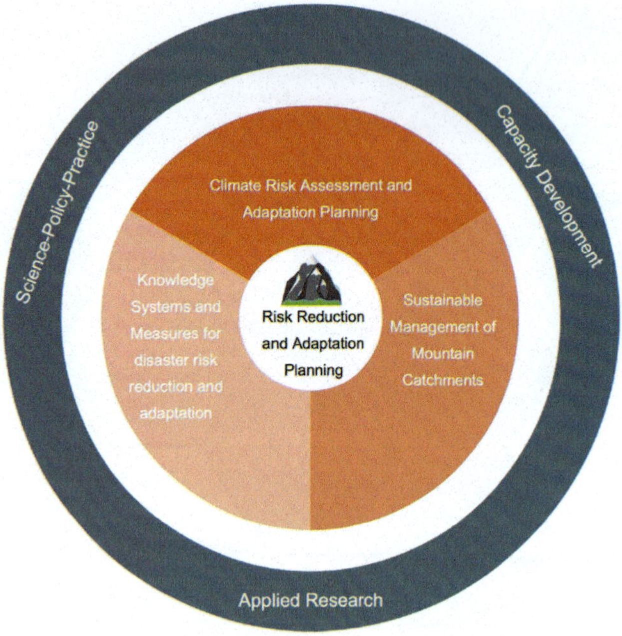

Figure 6.3 Monitoring and understanding people's readiness, and the geological processes that trigger hazards, is essential for hazard assessment and disaster mitigation.

on a locally useful map, such as evacuation routes, shelters, hospitals, and muster points (Ogburn et al., 2023).

Seismic hazards are often associated with tectonic activity along fault lines, and can result in earthquakes that cause surface rupture and secondary effects like landslides and tsunamis. Earthquakes can trigger landslides, avalanches, and even volcanic eruptions in mountainous regions, causing widespread destruction. Landslides in deglaciated and deglaciating mountains represent a major hazard. As mountain glaciers retreat and permafrost thaws, valley slopes can decrease in stability and fail (Pánek et al., 2022). This can lead to catastrophic landslides and is one of the most common natural hazards in rural areas, which can have a devastating impact. For example, on 28 May 2025, a disastrous landslide destroyed 90% of the village of Blatten in Switzerland (Esfahani et al., 2025) (Plate 6.1). The physical impact of paraglacial landslides can lead to a cascade of secondary hazards, such as glacier lake outburst floods and tsunamis, or a combination of these. In this way, landslides in deglaciating and deglaciated landscapes may pose a threat tens of kilometres downstream from the source area. Larger combined

Plate 6.1 Landslide covering the village of Blatten in Switzerland.

landslides can be triggered by factors such as heavy rainfall, earthquakes, volcanic activity, and human activities. Landslides can destroy infrastructure, bury communities, block rivers (leading to flooding), and cause loss of life.

Avalanches are rapid flows of snow, ice, and debris down a mountainside, often triggered by new snowfall, temperature changes, or human activity. Where these coincide with settlement, they can cause significant loss of life, especially in ski resorts and mountainous communities, and destroy structures and vegetation in their path (McClung & Schaerer, 2006). Avalanches can cause a considerable amount of damage to property and infrastructure, severely affecting socio-economic activities and livelihoods. In the past decades, Alpine countries such as Switzerland have experienced multiple catastrophic avalanches in the rapidly evolving mountain cryosphere (Ortner et al., 2023). Due to a lack of long-term observations, inaccessibility, severe weather conditions, and financial and logistical constraints, the complex relationship between climate and potential avalanche triggers is poorly understood. There is growing awareness of the avalanche dynamics as activity patterns are gradually shifting from low to high elevations under continued warming and increasing high altitude snow on steeper slopes (Eckert et al., 2024). These changes affect avalanche risk; however, risk is also influenced by factors such as land use and the growth or decline of human settlements. Indeed, the impact of these hazards is dependent on the presence of human activity – an avalanche in a valley where no one lives is much less hazardous than one in a heavily populated valley! Efforts by national agencies tasked with hazard monitoring should further invest

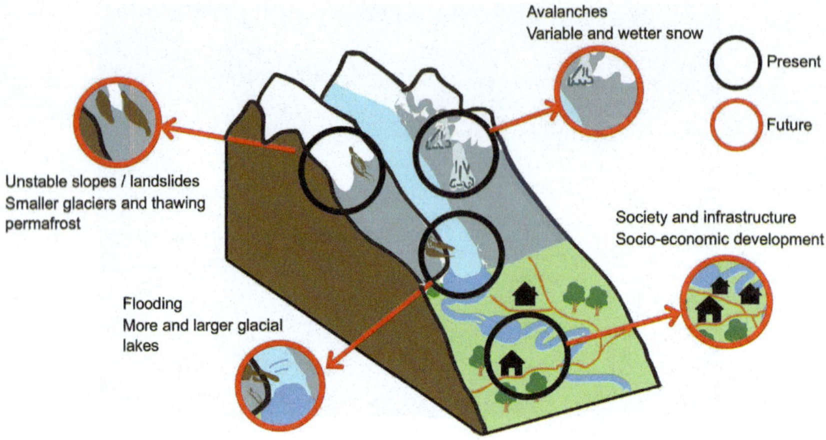

Figure 6.4 The future is likely to be characterised by a series of increasing challenges that combine increasing avalanches with more densely settled mountain environments.

in pairing scientific investigations of avalanche hazards with developing adaptation and mitigation solutions, an approach that has seen success in other mountain regions (Acharya et al., 2023). Avalanche risk maps (https://www.avalanchemap.com/) applied globally over large areas allow for the identification of the potential avalanche hazard and quantification of the associated risks at a local, regional, or even national level (Ortner et al., 2023). This combination of increasing avalanches impacting more densely settled mountain environments is likely to become more of a challenge in the future (Fig. 6.4).

Predicting geological hazards, formulating policies for disaster prevention and mitigation, and effectively reducing disaster losses, the risk assessment of geological hazards has become an intensive topic and an important part of international disaster prevention and reduction strategies (Lin et al., 2022; Pavlova et al., 2017). This can be challenging as mountain systems often undergo multiple, layered risks. For example, the Fujian Province in China, an area undergoing rapid economic development, is characterised by multiple geological hazards such as landslides, debris flows, and collapse, all limiting socio-economic development (Lin et al., 2022). Understanding the geological hazards in each mountainous area can be used to delineate key prevention and control areas and formulate targeted regional development strategies that can avoid the development of geological hazards caused by excessive land development (Lin et al., 2022). Tapping into valuable resources that document, map, and mitigate hazards in a joined-up manner opens the possibility for new kinds of understanding that can underpin planning in mountain regions (Ogburn et al., 2023).

6.3 Climate change impacts on mountains

Climate change is one of the main and rapidly growing challenges across the world's mountains. Observed changes in climate include increased temperatures, reduced fog, changes in rainfall amount and distribution, an increase in extreme droughts, fewer hailstorms, and increased wind strength during an increasingly unpredictable environment. Climate change is also aggravating social inequality, particularly for more vulnerable people, such as women and the landless (Cuni-Sanchez et al., 2025). It has long been known that the world is experiencing climate warming, with an increase in the global mean surface temperature over the last century by 0.6°C ± 0.2°C (Houghton et al., 2001). Subsequent IPCC reports (we are now approaching the seventh assessment report) continue to document increasing warming and a trend of increased climate variability and extremes (IPCC, 2022). Mountain areas are at the forefront of the global climate change challenge, partly as mountains are some of the most climate-sensitive systems in the world (Pepin et al., 2022; Seddon et al., 2016). It is well known that the rate of temperature change with increased levels of greenhouse gases in the atmosphere is amplified at high latitudes (Previdi et al., 2021), but there is also growing evidence that the rate of warming is amplified with elevation in mountain environments (Inouye, 2020; Mora et al., 2023). In recent decades, observed changes, including increasing temperatures, changing seasonal weather patterns, and reductions in snow cover extent and duration, are building a strong evidence base (Monteiro & Morin, 2023). However, this average increase is neither temporally nor spatially uniform. The increase in average temperature of 0.5°C/decade since 1971 in the European alpine area amounts to threefold that of the global trend (Pepin et al., 2022). Climate change impacts are increasingly frequent in Pakistan's northern mountain ranges, including the Pamir-Karakoram, Himalayas, and Hindu Kush, which have experienced rapidly increasing temperatures and decreasing monsoon precipitation (Norris et al., 2020). These increasing climate extremes have triggered extreme events, including floods, fires, glacial lake outbursts, glacial retreats, crop failures, landslides, heat waves, and cold waves. Clearly, climate-induced impacts are diverse and significant across various sectors, with a great impact on the economy, health, society, and biodiversity.

While climate change is undeniably happening across the world, it is the effects of enhanced climate variability and unpredictability (Cuni-Sanchez et al., 2025) that are much more impactful than a rise in global temperature or a reduction in precipitation. The disruption of historically predictable seasons, including droughts and floods, has major implications for crop yields and livestock, and hence for the livelihoods of pastoralists, farmers, and other small-scale producers (Kariuki et al., 2018). The recent Intergovernmental Panel on Climate Change (IPCC) Sixth Assessment report (AR6) illustrated how East African snow and glaciers have significantly decreased in recent decades, with high confidence that this trend will continue over the 21st century, where consequent changes in runoff will have downstream impacts, for example, impacting people across the Pangani River Basin (Adler et al., 2022).

Climate change impacts are still not well studied across tropical mountains, with many uncertainties remaining around the basic climatology and the nature of the changes in climate. The climatology and weather systems on mountains are highly complex due to the superimposition of major global convergence zones, with more localised regional feedbacks, which are in turn moderated by a diverse land surface characterised by very different albedos and surface properties (Chapter 2). Although climate change is a phenomenon felt across the world's mountains, the adverse impacts are experienced unevenly, based on the adaptive capacity of individual countries. Evidence shows that countries across the Global South are predominantly vulnerable to climate-induced impacts owing to characteristics like land-based economies, limited resources, low financial and technological capacity, high vulnerability, and low ability to lessen the undesirable impacts of climate change (Ngcamu, 2023). It is also about the nature of the impacts, for example, in Aspen or St Moritz, someone might not be able to ski, on Mt Elgon or in Nepal, a farmer may lose their livelihood!

Although there is significant global discourse around global warming and rising temperatures, it is the changing precipitation regimes and seasonal distribution of rainfall that are significant for many mountain regions. With a warmer atmosphere, annual precipitation is predicted to increase through the 21st century, although crucially, significant increases in precipitation extremes (around 75% for total precipitation on very wet days) occur together with an intensification of meteorological droughts caused by increased evapotranspiration, with relative changes being the greatest from June to August (Potter et al., 2023). For many mountain systems, there is a direct climatological link to ocean circulation systems. For example, future precipitation changes in the Andes will depend largely on changes in El Niño patterns (Potter et al., 2023). Across the Himalaya, projections of annual precipitation indicate increases on the order of 5%–20% over the century, particularly the frequency and intensity of extreme rainfall events during the summer monsoon periods, leading towards more episodic and intense monsoonal precipitation, especially in the easternmost part of the Himalayan chain (Dimri et al., 2021). This situation of increased rainfall and enhanced variability is not equally distributed across the world's mountains. For example, the Carpathian Mountains in Central and Eastern Europe, which span seven countries and are home to 18 million people, are characterised by less precipitation, leading to runoff reduction, threatening electricity generation capacities, agriculture production, increased concentration of pollutants, and eutrophication with irreversible damage to riparian ecosystems (Kato et al., 2021).

Quantifying climate change in mountain areas is challenging, partly because mountains are not defined by climate (Pepin et al., 2022). Although Global Climate Model (GCM) projections are currently the only available tool for understanding future temperature and precipitation change, and their use to drive hydrological models and project water resources can be insightful, GCMs fail to represent crucial processes that determine temperature and precipitation trends in mountains. There is great uncertainty in GCMs as they do not work well in mountain areas (Platts et al., 2013), particularly in terms of projecting precipitation. There are two main challenges required to build realistic climate models over mountain regions. Firstly, climate models need to be adapted to high-elevation areas that necessitate high-resolution ($<1\ km^2$) modelling grids that are appropriate for the topographical diversity; this is a far cry from the normal 300 km^2 or the current 25 km^2 regional climate models. To underline the climate modelling scale challenge, even at a high resolution of 1 km^2 in an area of steep terrain, there may be 3°C–4°C variation within a single climate model grid! The second major challenge is that climate models also need to be global, as the climate systems of mountains are part of a connected global climate system. Mountain climates are driven by climate teleconnections, particularly to coupled ocean systems (Lalande et al., 2021) such as the Himalaya to the north Atlantic (Hunt & Zaz, 2023), the Andes to the Pacific Ocean (Newell et al., 2022), and the East African Mountains to the Indian Ocean (Marchant et al., 2007). As computational capabilities improve, it will become possible to use high-resolution meteorological projections over complex mountain terrain that take into account the orographic flow patterns that respond uniquely to

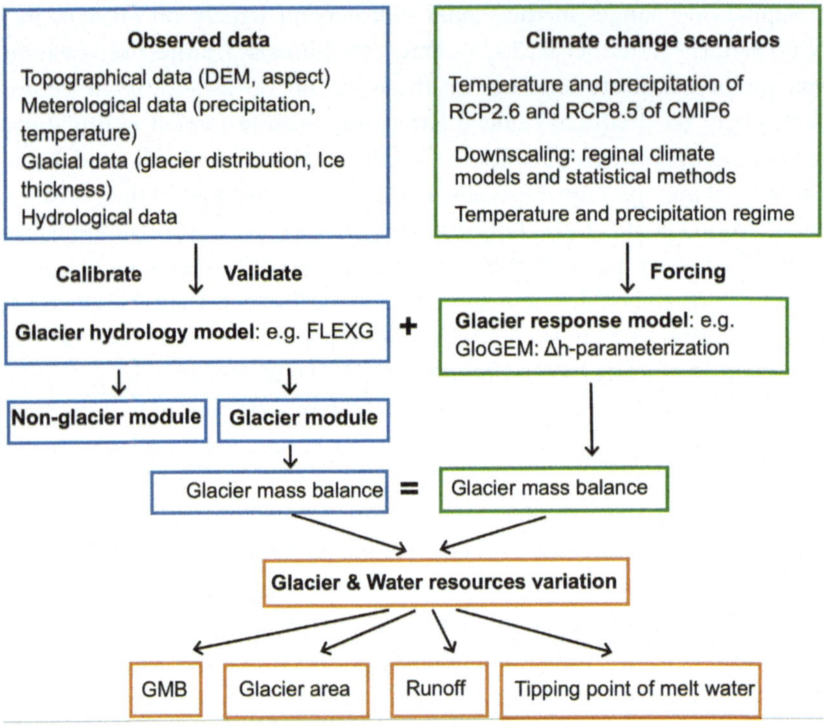

Figure 6.5 By combining climate, hydrology, and glacial response models with observation data, it is possible to make predictions on the future changes to glacial mass balance (GMB), runoff, and associated hazards. However, combining models with high amounts of uncertainty has to be used with caution and caveat.

regional climate change (Norris et al., 2020), but we remain a long way from climate models accurately reflecting mountain climates, and certainly from being able to use modelling outputs to feed into a planning and preparedness process (Fig. 6.5).

6.3.1 Climate change impacts on the distribution of animals and plants

As climatic conditions shape the ranges and distribution of species, and thus the composition of ecosystems, it is not surprising that climate change poses a significant threat to mountain communities; rising temperatures and changing precipitation patterns affecting the distribution of animals and plants are well documented (Parmesan & Yohe, 2003). How animal and plant habitats will change in response to climate change is a key question, and it is often assumed that habitats will shift in tandem with climate. However, habitats may change for other reasons, particularly due to human interventions, with such changes interacting with and being amplified by climate change (Mantyka-Pringle et al., 2012). Furthermore, differential shifts in climate and habitats may lead to a decoupling of suitable climatic and suitable habitat conditions (Platts et al., 2013). An understanding of how habitat and climate interact is therefore needed in order to forecast likely future impacts of changing climates on biodiversity (Platts et al., 2013). Unfortunately, although mountain ecosystems are predicted to be severely affected by environmental change, observations in these areas are typically sparse; we are not monitoring regions of the globe that are being impacted the most! Indeed, the information is incomplete, and relatively few mountains, and almost no tropical mountains, have been evaluated due to the lack of historical data on species distributions in these regions (Zu et al., 2022). Given the tremendous importance of plant life to Earth's ecosystems and humanity, a better understanding of these effects is needed to facilitate adaptation and mitigation planning (Rodman et al., 2020).

Typically, as a result of increased temperature, poleward and upward range shifts have been reported for a large number of organisms, including plants, birds, and butterflies (Settele et al., 2008). Recent environmental changes have pushed many ecosystems to the margins of their historic operating space (Gillson et al., 2025). A key question of current ecological research is thus to elucidate how ecosystems respond to increasing levels of warming, and quantify the relevant driver-state relationships (Albrich et al., 2020). However, this is not just a simple case of species moving uphill, as there may not be suitable habitat to move into. Furthermore, what happens as species move up? Based on a two-century dataset, Telwala et al. (2013) documented the impact of recent warmer winters in the Eastern Himalayas and reported increased species richness in the upper alpine zone since the 19th century, with a mean upward displacement rate of 28 ± 22 m/decade. Even if the current political climate targets are met by keeping global climate warming within a 1.5°C increase, fundamental changes in the characteristics of important forest ecosystems of the Alps are likely (Albrich et al., 2020). As this climate warming target seems to be woefully overshot, changes in mountains are locked in, and indeed are only going to accelerate – challenging times ahead for mountain biota.

Another important biogeographic aspect for montane flora is elevational range shifts, in which the majority of species are driven to higher elevations as the climate warms. Continued and unchecked global warming will present an imminent extinction risk as the scope for upward movement will be finite, especially since most of the available space for range expansion may not be accessible. For example, in the Himalayas, much of this 'adaptation space' falls behind the mountain range's rain shadow (Wambulwa et al., 2021). Future persistence of some species will be dependent upon their respective abilities to colonise new habitats, mostly via natural dispersal corridors and/or along ecological gradients. The effects of climate change will be even more pronounced if colonisation of higher unvegetated areas by grasslands is not possible due to constraints on soil formation (e.g., Freppaz et al., 2010). This future migration and enhancement of potential threats could be further exacerbated due to increased fragmentation of habitats. These natural dispersal routes may be interrupted, impeding movement and increasing the risk of extinction for many species in fragile ecosystems, such as the Himalayas (Wambulwa et al., 2021).

However, it is not a simple issue of species moving upslope! An understanding of the interplay of multiple drivers will be necessary to mitigate the far-reaching consequences of global change for tropical mountains and the human populations that rely on their ecosystem services. This change has multiple directions and is not a bidirectional situation of biodiversity loss, range contraction, and species extinction. For example, the warming-induced shifts to higher altitudes have increased the competition of cold-adapted alpine plants, which may then cause the decline of some high-altitude species (Zu et al., 2022). Elevational changes of mountain species in response to climate change differ across different regions; plant species in California have shifted their elevations downhill (Crimmins et al., 2011). Climate change may not only cause elevational shifts of mountain species but also contraction of their elevational ranges, which could increase the risk of local extinctions to endemic alpine species. The contraction of elevational ranges may be due to the upslope movement of lower limits, the downhill movement of upper limits, or both. There is a plethora of studies on changes of species elevational ranges, and it has been found that elevational ranges of mountaintop species from several plant and animal groups are significantly contracting. There are some 'inconsistencies' in the drivers of species redistribution in response to climate change due to the fact that different species have different climate adaptations, which may lead to different responses of elevational ranges to climate change. As mentioned previously, species are individuals with individual requirements for growth, and hence an ecosystem will respond in a myriad of ways.

Plate 6.2 Snow melt and overwintering plants. *(Photograph Rob Marchant).*

Snow cover is a major component in the alpine environment; it is a good insulator and can help protect the soil and its biota, for example, by overwintering plants (Plate 6.2), pupae, or larvae from extremely cold winter temperatures (Inouye, 2020). Consequences of earlier and more rapid snowmelt include advanced and perhaps longer growing seasons, more frequent frost damage, and increased summer drought (Inouye, 2020). One alarming study estimated a significant impact on European mountain plants with some 36%–55% of alpine species, 31%–51% of subalpine species, and 19%–46% of montane species losing more than 80% of their suitable habitat by 2070–2100 (Engler et al., 2011).

In mountain areas, a species' thermal niche range is an important trait determining response to climate change, and species with wider thermal tolerances are more likely to persist, moving both up and down (Engler et al., 2011; Mora et al., 2023; Rödder et al., 2021). Conversely, species with restricted ranges are likely to be less responsive. Species living at high elevations are particularly vulnerable to climate change impacts due to their fine-tuned adaptations to harsh environmental conditions and short periods

Plate 6.3 Shifts of endemic species in higher altitudes (e.g., Impatiens kiliminjari and Helichrysum spp.) may affect their ability to survive. *(Photographs Rob Marchant).*

of reproduction (Huber et al., 2013). Range-restricted and high-altitude species, such as some of the iconic *Scenecio* (Plate 6.3), are going to find the years ahead a 'fight for survival'. Mapping and monitoring the significant and constant shifts of distributions is crucial work. For example, butterfly populations in the eastern Alps have had an average altitudinal distributional shift of more than 300 m uphill from 1980 to 2020 (Rödder et al., 2021). Under predicted climate change, the Himalayan ibex will lose most of its current suitable habitat in the Himalayas and Hindu Kush (Zu et al., 2022), which will be outside of the confines of the current protected area network. Indeed, there is an urgent need to revisit protected areas management strategies in many mountain areas to enhance their effectiveness for the conservation of mountain species. There is clearly a mismatch between a static protected area network with well-defined borders and a dynamic and mobile population that is trying to keep pace with climate change. The situation is made more complicated as it is not just the impacts of climate change on the range of one species with multiple moving parts and interactions. These range modifications of mountain species result in a whole new factor-complex of abiotic and biotic interactions, having particular importance as many butterfly species depend on specific habitat types and larval food plants. For example, the fritillary *Euphydryas intermedia* is restricted to open mountain forest habitats supporting the only larval food plant, *Lonicera coerulea*, and the Small Apollo *Parnassius phoebus* is mostly found in habitats accompanying fast-flowing mountain streams (Rödder et al., 2021). These shifts in distribution will have cascading impacts across the system with intimate links to multiple interactions with other species.

High-mountain environments harbour ecosystems that are very sensitive and vulnerable to global changes (Gutiérrez-Girón et al., 2015). Threats to mountains are numerous, and the repercussions to humanity demand conservation and restoration of mountain ecosystems as they are an essential and highly sensitive part of the global life-support system (Schmeller et al., 2022). We are only beginning to understand the functional ecology of mountain ecosystems, but research already suggests that changing species communities will be detrimental to the environment and biodiversity, and therefore, a critical part of Earth's life-support system will be changed. Given these challenges, predicting the impacts of climate change on biodiversity has become a key research topic (Schmeller et al., 2022). Over recent decades, there have been steady declines in global biodiversity, both in terms of populations of wild species and the extent and condition of habitats. Although there are many drivers behind this, it has long been known that habitat destruction and degradation caused by human activities, and the subsequent fragmentation and loss of connectivity of ecosystems, are key factors in driving biodiversity loss (Platts et al., 2013).

6.3.2 Traits and phenology

Since species' responses to environmental change are mediated by their traits, vegetation studies frequently take advantage by classifying species distribution into life forms (Di Biase et al., 2021). Life forms are widely used in plant community ecology as they make it possible to compare plant assemblages with different taxonomic compositions. Functional classifications, based on morphological, phenological, and physiological features, have long been used to group plant species into classes based on similarity of function and structure (Di Biase et al., 2021). For example, Raunkiaer's classification system (Raunkiaer, 1905) grouped terrestrial plants into major groups according to the protection strategies during the unfavourable season and the height of the reproductive buds in relation to the soil surface.

Figure 6.6 While most species are migrating upslope to cooler climes, the average migration distances vary for different lifeforms.

Climate warming, shifting rainfall, and snow patterns are driving changes in vegetation phenology such as budburst, flowering times, or seed production (Piao et al., 2015), and increases in tree mortality rates (Van Mantgem et al., 2009). Redistribution of species across space in response to climate change may significantly influence the structures and functions of ecosystems, including the mismatch between pollinators and host plants (Zu et al., 2022). Species' functional traits also influence the redistribution of species in response to climate change (Fig. 6.6). However, how differences in climate adaptations and functional traits of species may influence species redistribution under climate change remains largely unknown (Zu et al., 2022). With range changes, contractions, and fragmentation, there is an increased likelihood of phenological mismatches that further complicate trying to get ahead of the climate change challenge.

6.4 Agricultural impacts

Agriculture is widely regarded as the most vulnerable sector to climate change (Ali et al., 2017), with the IPCC describing how climate shifts have severely impacted agricultural mountain systems (Adler et al., 2022; Pörtner et al., 2019). Climate change results in adverse impacts such as water scarcity, soil degradation, increased pest infestations, disease outbreaks, and declining crop yields (Cuni-Sanchez et al., 2025). Crop production, like any plant, is highly sensitive to changes in rainfall and temperature regimes, annual climate variations, shocks during different stages of growth, and extreme weather events (Srivastav et al., 2021). This high susceptibility challenges food security and rural livelihoods for billions of people worldwide, particularly as climate change challenges will be more impactful in the future (Jendritzki et al., 2023; Teklu et al., 2023). Snow-dominated and glacier-fed river basins are already experiencing changes in the amount and seasonality of water runoff, such as spring peaks set to occur earlier in the year; impacts that are projected to increase across the agriculture sector (Kato et al., 2021). These runoff changes are affecting water management, related hazards, and crop growth by reducing the productivity of irrigated agriculture far downstream. Consequently, these climate-induced impacts severely threaten rural livelihoods and global food security, leading to adaptation mechanisms at a range of different scales from the household to the nation, from the mountain to the wider landscape scale (Cuni-Sanchez et al., 2025).

Climate change gravely threatens economically disadvantaged nations' growth, more specifically, poor agrarian communities living in Global South countries are exceptionally susceptible to the adverse consequences of a changing climate, although they make a small contribution to global emissions of CO_2 and other greenhouse gases (Ali et al., 2017). There is a need to assess the on-the-ground realities of climate change impacts as perceived by farmers at the household level, particularly concerning their land-based livelihoods. For example, climate-induced events like flash floods and landslides have significantly affected agricultural practices in the high mountain Gilgit-Baltistan, decreasing food security (Ali et al., 2017; Geertsema & Alcántara-Ayala, 2023). Due to the diversity of mountain social ecological systems, the impact of climate change is having a differential impact across mountains, from coffee production to pastoralism, from staple crops to crop pests (Cuni-Sanchez et al., 2025). Women are often particularly vulnerable to the impacts of climate change due to multiple factors. First and foremost, women are often the key agricultural workforce in many mountain communities (Plate 6.4) (Kato et al., 2021). Reduced crop yields and impacts on food security are only predicted to increase (Mwalusepo et al., 2015).

Smallholder farmers' capacities need to be strengthened to withstand the climate change-related stresses, shocks, and impacts. These include the responsive capacity to already known threats and should also consider innovation, learning, and anticipation for the projected impacts of a changing climate on the agricultural system (Teklu et al., 2023). A collaborative and

Plate 6.4 Women are often particularly vulnerable as they are a key workforce in many mountain systems. *(Photographs Rob Marchant).*

localised climate adaptation framework that builds resilience in these agro-ecological systems can help to navigate climate change impacts. Climate-resilient agriculture can safeguard food security by enhancing smallholder farmers' productivity and transforming the current system to withstand present and future climate change effects on smallholder farmers' livelihoods (Teklu et al., 2023). Climate change-resistant crop development, expanding more traditional resilient crops, irrigation and water management, adapting climate-smart agriculture approaches, and promoting indigenous knowledge can ensure food security via sustaining agricultural yield (Srivastav et al., 2021). A comprehensive, multi-faceted approach is required to safeguard land-based livelihoods in the face of climate change and support climate change resilience and diversification. Governments, development NGOs, and the private sector are increasingly promoting nature-based solutions for mountain communities (Chapter 7), while at the same time making transitional infrastructure more climate resilient and using technology to help navigate the climate change challenge. For example, mountain communities are progressively using information and communication technologies to strengthen climate resilience, making use of mobile-based climate and market services (Kato et al., 2021). Policies that strengthen good governance, social cohesion, disaster communication and early warning systems, input supply of drought-resistant varieties, climate-smart extension services, and climate-resilient infrastructure can inform and develop climate-resilient agricultural strategies, particularly for remote and vulnerable mountain communities.

6.5 Challenges for mountain soils

Mountain soils are a vital component of mountain systems, supporting biodiversity, regulating water resources, and sustaining human livelihoods. Their unique characteristics, shaped by elevation, climate, and topography, make them distinct from lowland soils. However, mountain soils face significant challenges, including erosion, climate change, and human activities. Conservation efforts are essential to protect these soils and ensure their continued contribution to environmental and societal well-being. Mountain soils face several unique challenges: for example, warmer temperatures could lead to colder soils if snow cover, and its associated insulating properties, is reduced. Such an effect would prevent the development of vegetation, and in particular sward-forming grasses, at high altitudes (Cannone et al., 2022), yet at the same time, this habitat is likely to be lost at lower altitudes due to advancing treelines. Erosion is a significant threat to mountain soils, exacerbated by deforestation, overgrazing, and unsustainable agricultural practices. Soil loss can lead to reduced fertility, desertification, and the degradation of ecosystems. For example, the Loess Plateau in China has undergone extensive reforestation efforts to combat severe soil erosion and restore the landscape. Climate change poses a threat to mountain soils by altering precipitation patterns, thus increasing the frequency of extreme weather events that can destabilise soils and lead to increased erosion and loss of soil fertility. Global climate change is having a significant impact on the organic matter content and rates of decomposition, effects that are expected to be greater in soils at the highest altitudes (Cannone et al., 2022). Changes in organic matter decomposition rates could result in changes in biochemical cycles that would affect the structure and functioning of soil ecosystems. Indeed, changes in the structure and function of the ecosystems would alter carbon dioxide (CO_2) flux from the soil, provoking positive feedback that enhances the greenhouse effect (Gutiérrez-Girón et al., 2015).

6.6 Mountain land use and land cover change

One key challenge on mountains is the exploitation of natural resources that has led to marked changes in mountain landscapes. Habitats that have been disturbed or transformed by human land use have increasingly replaced natural vegetation along elevational gradients. Exploitation of natural resources has transformed mountain ecosystems across the world (Peters et al., 2019). The processes of land use and land cover change (LULCC) are determined by the interactions, in both space and time, between the biophysical factors (e.g., soil, climate, and topography) and the human factors (e.g., population, technology, and economic conditions) (Espinoza-Guzmán et al., 2023). Mountains have a long history of human-induced LULCC (Bucała-Hrabia, 2024). Nearly 60% of all mountainous land has been transformed by human activities to a state where intensive human pressure has reduced habitat connectivity (Elsen et al., 2020). Human pressure on montane landscapes is predominantly caused by resource extraction, infrastructure development, and habitat conversion for agriculture, pasture, and cropland (Körner, 2004). Conversion tends to be concentrated at lower elevations as higher elevation areas often offer fewer opportunities for human activities, such as agriculture or road building, due to being generally more topographically complex (Elsen et al., 2020). Because human pressure is typically biased towards low elevations and protection towards high elevations (Nogués-Bravo et al., 2008), higher elevations could provide refuge from human activities (Guo et al., 2021), particularly as it is these areas that are receiving the majority of the climate change migrants. The impact of LULCC on habitat quality in mountainous areas is of great significance for balancing regional ecological protection and the sustainable use of land resources (Yang, 2021). Spatial and temporal patterns of LULCC change can differ markedly between mountain regions due to geographical location, history, and political and socio-economic systems (Bucała-Hrabia, 2024).

Deforestation is one of the most obvious changes to land cover and can exacerbate the risk of flooding by destabilising slopes, reducing vegetation cover, and altering natural water flow patterns. For example, the rapid urbanisation and deforestation in the Himalayan foothills have been linked to increased flood risk, as seen in the 2024 Kathmandu floods. Flooding in mountainous regions has wide-ranging impacts, often exacerbated by the challenging terrain and isolation of these areas. Environmental degradation, driven by deforestation, resulting overgrazing, and unsustainable land use practices, poses a significant threat to the sustainability of mountain

Plate 6.5 Deforestation for agriculture and logging can leave the soils exposed and unprotected, leading to deep gully formation. *(Photographs Rob Marchant).*

environments. The loss of forests and the degradation of soil can lead to increased erosion, loss of biodiversity, and reduced agricultural productivity. For example, in the Eastern Arc Mountains of Tanzania, deforestation for agriculture and logging has led to significant soil erosion and loss of wildlife habitat (Plate 6.5). The degradation of these ecosystems ultimately threatens the livelihoods of local communities, who depend on the forests for food, medicine, and other resources.

Mining activities are another key direct and pervasive transformative activity in many mountains. For example, in the Peruvian Andes, the expansion of mining activities has led to significant environmental damage, including soil and water contamination. This has not only affected the health of local communities but also threatened their traditional livelihoods, particularly agriculture and livestock farming. Mountaintop removal mining, a practice common in the Appalachian Mountains, has caused significant environmental damage, including the destruction of entire ecosystems, contamination of water sources, and displacement of local communities.

The long-term impacts of this practice have led to a growing movement to restore and protect these vulnerable landscapes.

It must be emphasised that LULCC is not unidirectional. Indeed, LULCC is multidirectional, with people moving in and out of mountains with areas being deforested and afforested through natural or plantation forestry. While there had been a general expansion of LULCC, the progressive abandonment of agricultural land, and a parallel increase in forest cover, in the mountains of Western Europe, had already begun by the first half of the nineteenth century as a consequence of the Industrial Revolution, which caused migration from rural areas to urban centres. Across Europe, the upper transition zone between wooded and open landscapes, the so-called treeline ecotone, has been largely shaped by anthropogenic activities; initially characterised by grassland expansion and then followed by forest expansion (Barras et al., 2021; Stevens et al., 2022). Between the beginning of the 20th century and today, the share of the population employed in agriculture has decreased in most Alpine areas from 70% to below 5% as agriculturally used areas decreased by between −4% and −64% (Tasser et al., 2017).

LULCCs are considered to be the major driving forces for changes in biodiversity, ecosystem function, and dynamics, and ultimately for ecosystem services provision (Tasser et al., 2017). Land use change is increasingly impacting mountain social ecological systems and is considered an equally important, if not greater, threat than climate change in European cultural landscapes (Chamberlain et al., 2013). Given the increasing concern about the impacts of climate change, land-use changes are often discussed as secondary triggers of landscape transformation, acting indirectly through their effects on climate change (Tasser et al., 2017). Habitat conversion through historical LULCC, combined with the impacts of climate change on species distribution, is the main driver of the global biodiversity crisis (Rödder et al., 2021). Particularly in recent decades, where LULCC has been accelerating due to a combination of population growth and migration in and out of mountains, human activities have led to a widespread loss of species, habitat degradation and fragmentation, and the introduction of exotic species, which have negatively affected the well-being of local communities (Yang, 2021). However, LULCC has been so deep-rooted and so pervasive across the world's mountains that it is exacerbating the impacts of climate change and the ability of plants and animals to move across landscapes. With this increasing isolation, mountain species will not be able to track climate changes – the ability to migrate, which has worked so well in allowing species to adjust ranges through the glacial-interglacial cycles, is, and will continue to become, massively hindered!

Indeed, a limitation in many approaches to forecasting changes in available space for species to persist under climate change involves accounting for how LULCC limits species distributions and the movement required to shift their range to adapt (Elsen et al., 2020). Ultimately, accurate assessments of changes in habitable area for species undergoing upslope range shifts rely on integrating fine-scale topography with current land use (Elsen et al., 2020). Topography and current patterns of human pressure across the world's mountain ranges influence the extent of intact land area available to species undergoing elevational range shifts. While human pressure has undoubtedly reduced the amount of habitable area available, with potentially severe consequences for biodiversity, human pressure in mountains has functionally changed mountains compared to intact landscapes (Elsen et al., 2020).

Over the last two decades, an increasing number of studies have assessed LULCC across the world using various remote sensing products and land use modelling techniques (Ben-Said et al., 2025). Integrating LULCC into future scenarios is crucial to properly assess species vulnerability in the face of global change (Capitani et al., 2016) and thus make meaningful conservation and management recommendations (Kariuki et al., 2025), especially in ecosystems that are heavily modified by humans. Future studies should extend beyond a simple description of transformation towards a deeper understanding of LULCC impacts on biodiversity, ecosystem services, and population well-being under global changes, considering socio-economic, ecological, biophysical, and cultural characteristics in interpreting LULCC (Ben-Said et al., 2025). It is necessary to review the actions to avoid the transformation of mountain ecosystems to design new strategies to mitigate the effects of LULCC and the ensuing loss of biodiversity and ecosystem services. Information generated on LULCC can help design management strategies for critical protected areas and support the conservation of mountain ecosystems (Espinoza-Guzmán et al., 2023). Future land-use scenarios hold a high degree of uncertainty and depend on a host of factors, such as rates of agricultural intensification versus expansion, and fire management challenges (Capitani et al., 2016). Land use changes brought on by changing livelihoods can have cascading impacts that are driven by interactions with social, economic, and climate changes. It is these interactions that allow us to identify spaces where solutions can intervene to navigate through the complexities and support sustainable mountain livelihoods (Chapter 7).

6.7 Mountain glaciers

Glacier shrinkage and the development of post-glacial ecosystems related to anthropogenic climate change are some of the fastest and most dramatic changes in mountain regions across the world (Dangles et al., 2025). It is not just the loss of glaciers that is alarming; there are marked ecological, economic, and societal consequences that impact across the mountains and into the lowlands (Bosson et al., 2023). The progressive extinction of ice masses is having local to global cascading consequences. Ice melt affects hydrological and biogeochemical cycles, sea level, and the functioning of the ocean and climate systems, and thereby the distribution of life conditions and human activities. By 2100, the decline of all glaciers outside the Antarctic and Greenland ice sheets may produce new terrestrial, marine, and freshwater ecosystems over a wide-ranging area (Bosson et al., 2023; Dangles et al., 2025). The largest deglaciated areas will probably appear in Alaska, the Greenland periphery, and High Mountain Asia, where among the c. 210,000 glaciers, 148,100 ± 28,500 (71% ± 14%) to 33,900 ± 24,000 (16% ± 11%) are expected to barely subsist by 2100 (Bosson et al., 2023). With less than half of glaciated regions being located in protected areas, there is an urgent need for action to preserve the world's mountain glaciers, as echoed in the recent United Nations resolution declaring 2025 as the International Year of Glaciers' Preservation. The United Nations emphasises the need to urgently and simultaneously enhance climate-change mitigation and the in-situ protection of glaciers to secure their existence, functioning, and values (Bosson et al., 2023).

With decreasing surface albedo and the potential release of stored carbon as associated permafrost areas melt, deglaciation will feed back to further climate change impacts. Post-glacial ecosystems might, in favourable areas, capture and store carbon through biogeochemical processes (for example, pedogenesis, underwater weathering, sediment accumulation) and biomass (for example, plants, plankton, seabed communities), and contribute to mitigating climate change (Bosson et al., 2023). While glacial shrinkage and evapotranspiration intensify, emerging post-glacial ecosystems will become increasingly important components of the freshwater cycle. Not only will new lakes, wetlands, and river ecosystems be formed, but also terrestrial biomass, soils, and unconsolidated sediments that locally store, purify, regulate, and provide blue and green water will be created. However, because glaciers host various specialised species and influence downstream hydrosystems and, more broadly, climate and water cycles, their decline will affect geodiversity, biodiversity, and many ecosystems globally as they are pushed beyond their resilience capacities (Bosson et al., 2023).

Accelerated ice loss over the last four decades, particularly in the drier western Himalayas (Lutz et al., 2014), has significantly disrupted ecosystem services provision and delivery. The High Mountain Asia water tower is projected to have an initial increase in runoff due to glacier wastage, and continuous glacier retreat and subsequent disappearance will lead to a significant decrease in runoff in most regions (Huss & Hock, 2018). However, projections of future glacier area and volume, and their runoff yield, come with high uncertainties, as numerous gaps remain in our knowledge of regional glacier behaviour and the interaction of glaciers with a changing climate (Bhattacharya et al., 2021; Pandit et al., 2014). The spatial and temporal heterogeneity of mass loss from surge-type and lake-terminating glaciers in different parts of High Mountain Asia emphasises the need to look beyond climatic variables to understand variability in glacier behaviour into the future. A more detailed understanding of the current prevalence of surge-type glaciers and the likely future distribution of lake-terminating glaciers will help our understanding of glacier behaviour. Such understanding of the timing and magnitude of future glacier recession can only be achieved with observational data and improved spatial and temporal resolution (Bhattacharya et al., 2021).

6.7.1 Glacial lake outbursts and floods

Stores of mountain snow and ice are declining faster than previously assumed due to the combination of enhanced elevation-dependent warming and reduced elevation dependency of rainfall/snowfall (Pepin et al., 2022). Increasingly, glaciers in mountainous regions can pose hazards such as glacial lake outburst floods (GLOFs); when a glacier-fed lake breaches its ice or moraine dam, resulting in sudden, devastating floods that damage downstream communities and infrastructure (Richardson & Reynolds, 2000). For example, in 1985, the Dig Tsho glacial lake in Nepal burst, causing a flood that destroyed a hydroelectric project and killed several people (Khatri, 2025). Due to the unique topography and climatic conditions of the mountains, flooding events can be sudden, severe, and highly destructive. Increasingly intense rainfall, often associated with monsoons or extreme weather events, can lead to flash floods in mountainous areas. For example, in 2013, floods in North India, caused by exceptionally heavy rainfall, resulted in landslides and widespread devastation, including the loss of thousands of lives and severe damage to infrastructure (Geertsema & Alcántara-Ayala, 2023). Snowmelt in spring and early summer can significantly increase the volume of water in mountain streams and rivers. If the meltwater is rapid or combined with rain, it can lead to flooding. For example, the 2010 Indus River floods in Pakistan were exacerbated by snowmelt from the Himalayas, which, combined with heavy monsoon rains, led to one of the worst floods in the country's history (Atif et al., 2021). The rapid retreat of Himalayan glaciers also threatens the water supply for millions of people who depend on glacial melt for drinking water, irrigation, and hydropower.

It is not just the high magnitude, low frequency events of glacial melt that are impacting mountain communities; the continual decline from some mountain systems is having a pervasive and lasting impact. The retreat of the Quelccaya Ice Cap in Peru, the largest tropical ice cap in the world, has raised concerns about the long-term water supply for communities in the region (Rodda et al., 2023). The shrinking of glaciers and reduced snowpack in the Sierra Nevada due to climate change have raised concerns about water shortages for both local communities and downstream areas (Liu et al., 2021). Additionally, the increased frequency of wildfires in the region has devastated landscapes and displaced residents, highlighting the vulnerability of mountain communities to climate-related disasters. Hydrological flows from mountain water towers across East Africa have notably reduced as glaciers have retreated, particularly on Mount Rwenzori (Marchant & Taylor, 1998), Mount Kenya, and Mount Kilimanjaro (Dangles et al., 2025). This has had a negative impact on surface flows and underground aquifers, consequently affecting downstream hydropower generation and agro-production (Hock et al. 2019).

6.8 Economic marginalisation and declining indigenous knowledge

Many mountain communities are economically marginalised due to their remote locations, limited access to markets, and the challenges of modernising traditional livelihoods. This marginalisation often leads to poverty traps, outmigration, and a decline in traditional practices. For example, the inhabitants of the Rif Mountains in Morocco face economic challenges due to the region's isolation and limited agricultural productivity. Among the Berbers of the Atlas Mountains in Morocco, the younger generations are increasingly adopting urban lifestyles and modern values, leading to a decline in the use of the Berber language and traditional practices (De Haas, 2006).

Similarly, the Nuba people of the Nuba Mountains in Sudan have long faced economic marginalisation and conflict, which has exacerbated poverty and hindered development in the region (Barbour, 2023). This is made worse by limited access to education and healthcare, along with political instability, which has made it difficult for the Nuba to improve their living conditions and preserve their cultural heritage. Among the Naxi people of the Hengduan Mountains in China, the younger generations are increasingly adopting urban lifestyles and modern values, leading to a decline in the use of the Naxi language and traditional practices such as plant collecting and Dongba shamanism (Geng et al., 2017). In the nearby Gilgit-Baltistan area of the Himalayas, an ecologically sensitive protected area, only the older people in the valley know the local names, uses, grazing preferences, and effects of the plants. The reason for the decline of traditional knowledge of plant usage is the communication gap between old and young people. Only some herbalists, hakims, and elders know the uses and values of these plants; as these older people die, traditional knowledge is un-transferred and lost (Shedayi et al., 2016).

These shifts away from traditional farming practices in many mountain systems around the world have resulted in depopulation and the decline of small rural communities, raising concerns about the future of these mountain regions. Globalisation and the influence of external cultures can lead to the erosion of traditional customs, languages, and social structures in mountain communities. Younger generations may abandon traditional practices in favour of more modern lifestyles, leading to a loss of cultural heritage. Many young people migrate from mountains to urban areas in search of better opportunities, leading to the erosion of traditional knowledge, labour, innovation, and cultural practices. For example, many young Aymara from the Bolivian Altiplano have migrated to urban areas in search of better opportunities, leading to a decline in farming practices and a weakening of community ties (Flores-Palacios et al., 2023). The Carpathian Mountains, spanning several Central and Eastern European countries, have experienced significant outmigration as younger generations leave in search of better economic opportunities. This depopulation has led to the abandonment of farming and a decline in the local economy, contributing to the erosion of cultural traditions (Levytska et al., 2022). In the Alps, traditional alpine farming practices are in decline as younger generations move to urban areas for education and employment. This shift threatens the continuity of cultural traditions, such as the seasonal transhumance of livestock, which has shaped the landscape for centuries. However, this is not a single flow of traffic

Plate 6.6 Ski tourism has been a massive growth area in the past 50 years, but more marginal resorts at lower altitudes are having to invest in snow-making equipment, reservoirs, and energy infrastructure to buffer from increasingly unpredictable snow. *(Photographs Rob Marchant).*

and has led to many new arrivals into mountain areas who are seeking the high quality of environment and life and the ability to work remotely (Grüner, 2023).

6.9 Winter tourism and ski infrastructure

In the Alps, the retreat of glaciers and reduced snowfall due to climate change have raised concerns about the future of winter tourism, which is a major economic driver for the region. In New Zealand, the Franz Josef and Fox Glaciers in the Southern Alps have been retreating rapidly due to climate change. This not only affects the local environment but also threatens the tourism industry, which is a major source of income for the region. The skiing industry is of major economic importance to a number of alpine regions, and the industry experienced a period of great expansion from the 1960s (Abegg et al., 1997), particularly at increasingly higher altitudes as the reliability of the snow conditions at lower altitudes lessened. For high altitude alpine ecosystems, there has been significant investment in infrastructure such as artificial snow making and reservoirs (Plate 6.6). Meanwhile, those resorts at lower altitudes, where the energy and water costs of snow making are prohibitively expensive, are diversifying to different types of tourism such as mountain biking (Plate 6.7).

Plate 6.7 Ski tourism has been a massive growth area in the past 50 years, but increasingly, the more marginal resorts at lower altitudes are finding it challenging and having to diversify into mountain bike tourism. *(Photograph Rob Marchant)*.

Ski infrastructure also causes habitat alteration following the construction of ski-pistes and lifts, which significantly impact natural ecosystems by degrading soils, destroying natural habitats for high altitude grassland and forest animals (Caprio et al., 2011), fragmenting natural habitats by creating barriers to dispersal, and reducing the quality of adjacent natural habitats due to edge effects (Rolando et al., 2013). The removal of much of the vegetation and rocks, and disturbance to the soil, creates a habitat that is virtually unsuitable for nesting and is likely to provide very few food resources for high alpine species. Although snowfinches may derive some benefit from ski-pistes due to the creation of more open habitat, the abundances were still higher in natural prairies (Rolando et al., 2007). Ski-pistes can have a detrimental effect on alpine grassland bird communities at a landscape scale (Caprio et al., 2011). To reconcile the economic benefits of the skiing industry to human populations with its current clear negative impacts on biodiversity, a number of initiatives have been suggested for more sympathetic management of grassland on the piste (Braunisch et al., 2011; Caprio et al., 2011). A landscape-scale approach needs to be adopted in order to integrate biodiversity issues into land-use planning (Butchart et al., 2010), particularly for animals that typically range over wider areas and are more likely to be impacted.

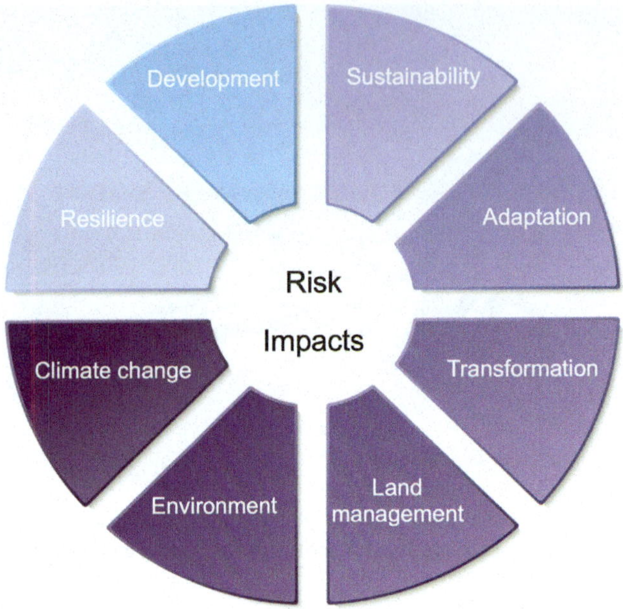

Figure 6.7 Risk impacts are a combination of multiple factors that combine in complex and sometimes unintended ways.

6.10 A combination of challenges

Mountains are the most climate-sensitive ecosystems worldwide (Seddon et al., 2016), and many mountain regions face unique problems that require bespoke solutions (Chapter 7). Many of the issues detailed in this chapter combine to form a complex layer of interacting challenges that can impact in an additive, multipurpose, or subtractive manner for mountain areas and their communities. Understanding the potential additive and interactive effects of climate and land use and land cover change on biodiversity and ecosystem functions is key in our attempts to predict and mitigate the consequences of global change on the world's mountains, along with other climatic gradients. The high population growth rate, growing poverty gap index, increasing pressure on water supplies (Ashagre et al., 2018), ecosystems (Marchant, 2021), rapidly changing climates, transforming models of socio-economic development, intensification of competing land uses, and ineffective national and international policies that sometimes seem to work at cross-cutting purposes, all threaten the future sustainability and resilience of mountain social-ecological systems (Archer et al., 2025;

Thorn et al., 2020). Risk impacts are a combination of multiple factors that interact in a series of complicated, often additive ways; such as climate change impacts being made worse by inappropriate land management and policy (Fig. 6.7)

Globally, nearly 60% of the mountainous areas are under intense human pressure with impacts that are felt far downstream, such as eutrophication and water supply. Mountains supply ecosystem services that much of the world relies on, not least since all major rivers have headwaters in highlands. Glaciers in many parts of the tropical Andes and Africa may disappear over the next few decades, with those in the Himalayas towards the later part of this century. Higher sensitivity and weaker resilience create an increasingly vulnerable system. Integrating all three aspects of vulnerability (exposure, sensitivity, and resilience) may offer more comprehensive and spatially explicit adaptation strategies to reduce the impacts of climate change on mountain systems and those downstream areas that rely on mountains as the engine that supplies water, food, and fuel.

References

Abegg, B., Koenig, U., Buerki, R., & Elsasser, H. (1997). Climate impact assessment in Tourismus. *Erde, 128*(2), 105–116.

Acharya, A., Steiner, J. F., Walizada, K. M., Zakir, Z. H., Ali, S., Caiserman, A., & Watanabe, T. (2023). Snow and ice avalanches in high mountain Asia–scientific, local and indigenous knowledge. *Natural Hazards and Earth System Sciences Discussions, 2023*, 1–35.

Adler, C., Wester, P., Bhatt, I., Huggel, C., Insarov, G., Morecroft, M., Muccione, V., Prakash, A., Alcántara-Ayala, I., Allen, S. K., & Bader, M. (2022). Cross-chapter paper 5: Mountains. In *Climate change 2022: Impacts, adaptation, and vulnerability* (pp. 2273–2318). Cambridge University Press.

Albrich, K., Rammer, W., & Seidl, R. (2020). Climate change causes critical transitions and irreversible alterations of mountain forests. *Global Change Biology, 26*(7), 4013–4027. https://doi.org/10.1111/gcb.15118.

Ali, S., Liu, Y., Ishaq, M., Shah, T., Abdullah, A. I., & Din, I. U. (2017). Climate change and its impact on the yield of major food crops: Evidence from Pakistan. *Foods, 6*(6), 1–19. https://doi.org/10.3390/foods6060039.

Archer, E., Arneth, A., Leadley, P., Mori, A., Obura, D., & Smith, P. (2025). Achieving the global biodiversity framework under climate change. *Global Change Biology, 31*(5). https://doi.org/10.1111/gcb.70249.

Ashagre, B. B., Platts, P. J., Njana, M., Burgess, N. D., Balmford, A., Turner, R. K., & Schaafsma, M. (2018). Integrated modelling for economic valuation of the role of forests and woodlands in drinking water provision to two African cities. *Ecosystem Services, 32*, 50–61. https://doi.org/10.1016/j.ecoser.2018.05.004.

Atif, S., Umar, M., & Ullah, F. (2021). Investigating the flood damages in Lower Indus Basin since 2000: Spatiotemporal analyses of the major flood events. *Natural Hazards, 108*(2), 2357–2383. https://doi.org/10.1007/s11069-021-04783-w.

Barbour, K. M. (2023). *Population, land and water in Central Sudan: Essays on African Population* (pp. 137–156). Taylor and Francis. https://doi.org/10.4324/9781003427643-10.

Barras, A. G., Braunisch, V., & Arlettaz, R. (2021). Predictive models of distribution and abundance of a threatened mountain species show that impacts of climate change overrule those of land use change. *Diversity and Distributions, 27*(6), 989–1004. https://doi.org/10.1111/ddi.13247.

Ben-Said, M., Chemchaoui, A., Etebaai, I., & Taher, M. (2025). Land use and land cover changes in Morocco: Trends, research gaps, and perspectives. *GeoJournal, 90*(1). https://doi.org/10.1007/s10708-024-11214-3.

Bhattacharya, A., Bolch, T., Mukherjee, K., King, O., Menounos, B., Kapitsa, V., Neckel, N., Yang, W., & Yao, T. (2021). High mountain Asian glacier response to climate revealed by multi-temporal satellite observations since the 1960s. *Nature Communications, 12*(1). https://doi.org/10.1038/s41467-021-24180-y.

Bosson, J. B., Huss, M., Cauvy-Fraunié, S., Clément, J. C., Costes, G., Fischer, M., Poulenard, J., & Arthaud, F. (2023). Future emergence of new ecosystems caused by glacial retreat. *Nature, 620*(7974), 562–569. https://doi.org/10.1038/s41586-023-06302-2.

Braunisch, V., Patthey, P., & Arlettaz, R. (2011). Spatially explicit modeling of conflict zones between wildlife and snow sports: Prioritizing areas for winter refuges. *Ecological Applications, 21*(3), 955–967. https://doi.org/10.1890/09-2167.1.

Bucała-Hrabia, A. (2024). Reflections on land use and land cover change under different socio-economic regimes in the Polish Western Carpathians. *Regional Environmental Change, 24*(1). https://doi.org/10.1007/s10113-024-02187-7.

Butchart, S. H. M., Walpole, M., Collen, B., Van Strien, A., Scharlemann, J. P. W., Almond, R. E. A., Baillie, J. E. M., Bomhard, B., Brown, C., Bruno, J., Carpenter, K. E., Carr, G. M., Chanson, J., Chenery, A. M., Csirke, J., Davidson, N. C., Dentener, F., Foster, M., Galli, A., & …, Watson, R. (2010). Global biodiversity: Indicators of recent declines. *Science, 328*(5982), 1164–1168. https://doi.org/10.1126/science.1187512.

Cannone, N., Guglielmin, M., Casiraghi, C., & Malfasi, F. (2022). Salix shrub encroachment along a 1000 m elevation gradient triggers a major ecosystem change in the European Alps. *Ecography, 2022*(2). https://doi.org/10.1111/ecog.06007.

Capitani, C., Garedew, W., Mitiku, A., Berecha, G., Hailu, B. T., Heiskanen, J., Hurskainen, P., Platts, P. J., Siljander, M., Pinard, F., Johansson, T., & Marchant, R. (2019). Views from two mountains: Exploring climate change impacts on traditional farming communities of Eastern Africa highlands through participatory scenarios. *Sustainability Science, 14*(1), 191–203. https://doi.org/10.1007/s11625-018-0622-x.

Capitani, C., Mukama, K., Mbilinyi, B., Malugu, I. O., Munishi, P. K. T., Burgess, N. D., Platts, P. J., Sallu, S. M., & Marchant, R. (2016). From local scenarios to national maps: A participatory framework for envisioning the future of Tanzania. *Ecology and Society, 21*(3). https://doi.org/10.5751/ES-08565-210304.

Caprio, E., Chamberlain, D. E., Isaia, M., & Rolando, A. (2011). Landscape changes caused by high altitude ski-pistes affect bird species richness and distribution in the Alps. *Biological Conservation, 144*(12), 2958–2967. https://doi.org/10.1016/j.biocon.2011.08.021.

Chamberlain, D. E., Negro, M., Caprio, E., & Rolando, A. (2013). Assessing the sensitivity of alpine birds to potential future changes in habitat and climate to inform management strategies. *Biological Conservation, 167*, 127–135. https://doi.org/10.1016/j.biocon.2013.07.036.

Crimmins, S. M., Dobrowski, S. Z., Greenberg, J. A., Abatzoglou, J. T., & Mynsberge, A. R. (2011). Changes in climatic water balance drive downhill shifts in plant species' optimum elevations. *Science, 331*(6015), 324–327. https://doi.org/10.1126/science.1199040.

Cuni-Sanchez, A., Aneseyee, A. B., Baderha, G. K. R., Batumike, R., Bitariho, R., Imani, G., Jha, N., Kaganzi, K. R., Kaplin, B. A., Klein, J. A., Leite, A., Marchant, R. A., Martin, E. H., Mcharazo, F., Mwangi, B., Ngute, A. S. K., Nkengurutse, J., Nkurunziza, A., Olaka, L., & …, Zafra-Calvo, N. (2025). Perceived climate change impacts and adaptation responses in ten African mountain regions. *Nature Climate Change, 15*(2), 153–161. https://doi.org/10.1038/s41558-024-02221-w.

Cuni-Sanchez, A., Sullivan, M. J. P., Platts, P. J., Lewis, S. L., Marchant, R., Imani, G., Hubau, W., Abiem, I., Adhikari, H., Albrecht, T., Altman, J., Amani, C., Aneseyee, A. B., Avitabile, V., Banin, L., Batumike, R., Bauters, M., Beeckman, H., Begne, S. K., & …, Zibera, E. (2021). High aboveground carbon stock of African tropical montane forests. *Nature, 596*(7873), 536–542. https://doi.org/10.1038/s41586-021-03728-4.

Dangles, O., Lana, S., Palomo, I., & Rabatel, A. (2025). *The voices of glaciers: Stories of grief and hope amidst shrinking glaciers in the tropics.* UNESCO Publishing https://doi.org/10.54677/PZZT2044.

De Haas, H. (2006). *The social and cultural impacts of international migration on Moroccan sending communities: A review. Morocco and the Netherlands: Society, economy, culture.* VU Publishers.

Di Biase, L., Pace, L., Mantoni, C., & Fattorini, S. (2021). Variations in plant richness, biogeographical composition, and life forms along an elevational gradient in a Mediterranean mountain. *Plants, 10*(10), 2090. https://doi.org/10.3390/plants10102090.

Dimri, A. P., Allen, S., Huggel, C., Mal, S., Ballesteros-Cánovas, J. A., Rohrer, M., Shukla, A., Tiwari, P., Maharana, P., Bolch, T., Thayyen, R. J., Stoffel, M., & Pandey, A. (2021). Climate change, cryosphere and impacts in the Indian Himalayan Region. *Current Science, 120*(5), 774–790. https://doi.org/10.18520/cs/v120/i5/774-790.

Eckert, N., Corona, C., Giacona, F., Gaume, J., Mayer, S., van Herwijnen, A., Hagenmuller, P., & Stoffel, M. (2024). Climate change impacts on snow avalanche activity and related risks. *Nature Reviews Earth & Environment, 5*(5), 369–389. https://doi.org/10.1038/s43017-024-00540-2.

Elsen, P. R., Monahan, W. B., & Merenlender, A. M. (2020). Topography and human pressure in mountain ranges alter expected species responses to climate change. *Nature Communications, 11*(1). https://doi.org/10.1038/s41467-020-15881-x.

Engler, R., Randin, C. F., Thuiller, W., Dullinger, S., Zimmermann, N. E., Araújo, M. B., Pearman, P. B., Lay, G. L., Piedallu, C., Albert, C. H., Choler, P., Coldea, G., De Lamo, X., Dirnböck, T., Gégout, J. C., Gómez-García, D., Grytnes, J. A., Heegaard, E., Høistad, F., & …, Guisan, A. (2011). 21st century climate change threatens mountain flora unequally across Europe. *Global Change Biology, 17*(7), 2330–2341. https://doi.org/10.1111/j.1365-2486.2010.02393.x.

Esfahani, R., Campillo, M., Seydoux, L., Nishida, K., & Favre-Bulle, G. (2025). Seismic precursors to the Blatten, Switzerland landslide revealed by unsupervised machine learning. https://doi.org/10.31223/X56X6V.

Espinoza-Guzmán, M. A., Aragonés Borrego, D., & Sahagún-Sánchez, F. J. (2023). Evaluation of recent land-use and land-cover change in a mountain region. *Trees, Forests and People, 11.* https://doi.org/10.1016/j.tfp.2023.100370.

Fernandez, M. A., Gaston, C. J., Cargo, L. C. M., & Estorico, G. C. (2025). A systematic review: Assessing organ system-specific health risks of volcanic eruptions across Asia with their corresponding volcanic hazard. *International Journal of Innovative Science and Research Technology*, 2284–2292. https://doi.org/10.38124/ijisrt/25mar1768.

Flores-Palacios, X., Ahmed, B., & Barbera, C. (2023). Micro-narratives on people's perception of climate change and its impact on their livelihood and migration: Voices from the indigenous aymara people in the Bolivian Andes. In *Rebuilding communities after displacement: Sustainable and resilience approaches* (pp. 29–57). Springer International Publishing. https://doi.org/10.1007/978-3-031-21414-1_2.

Freppaz, M., Godone, D., Filippa, G., Maggioni, M., Lunardi, S., Williams, M. W., & Zanini, E. (2010). Soil erosion caused by snow avalanches: A case study in the Aosta Valley (NW Italy). *Arctic, Antarctic, and Alpine Research, 42*(4), 412–421. https://doi.org/10.1657/1938-4246-42.4.412.

Geertsema, M., & Alcántara-Ayala, I. (2023). Mountain landslides: An overview of common types and future impacts. In *Montology palimpsest: A primer of mountain geographies* (pp. 187–209). Springer. https://doi.org/10.1007/978-3-031-13298-8_11.

Geng, Y., Hu, G., Ranjitkar, S., Shi, Y., Zhang, Y., & Wang, Y. (2017). The implications of ritual practices and ritual plant uses on nature conservation: A case study among the Naxi in Yunnan Province, Southwest China. *Journal of Ethnobiology and Ethnomedicine, 13*, 1–11. https://doi.org/10.1186/s13002-017-0186-3.

Gillson, L., Seddon, A., Mottl, O., Zhang, K., Kirsten, K., Gell, P., Marchant, R. A., Schwörer, C., Razanatsoa, E., Lane, P. J., Courtney-Mustaphi, C. J., & Dearing, J. (2025). Exploring the interface between planetary boundaries and palaeoecology. *Global Change Biology, 31*(1). https://doi.org/10.1111/gcb.70017.

Grüner, B. (2023). Two close-to-nature lifestyles, one benefit for the cultural landscape: Comparing Lifestyle movers and lifestyle farmers in the remote European Eastern Alps. *Mountain Research and Development, 43*(2), R1. https://doi.org/10.1659/mrd.2022.00033.

Guo, L., Zhao, H., Zhai, X., Wang, K., Liu, L., Wang, K., & Huang, D. (2021). Study on life history traits of Stellera chamaejasme provide insights into its control on degraded typical steppe. *Journal of Environmental Management, 291*, 112716.

Gutiérrez-Girón, A., Díaz-Pinés, E., Rubio, A., & Gavilán, R. G. (2015). Both altitude and vegetation affect temperature sensitivity of soil organic matter decomposition in Mediterranean high mountain soils. *Geoderma, 237–238*, 1–8. https://doi.org/10.1016/j.geoderma.2014.08.005.

Hock, R., Rasul, G., Adler, C., Cáceres, B., Gruber, S., Hirabayashi, Y., Jackson, M., Kääb, A., Kang, S., Kutuzov, S., Milner, A., Molau, U., Morin, S., Orlove, B., & Steltzer, H. (2019). High mountain areas. In H.-O. Pörtner, D. C. Roberts, V. Masson-Delmotte, P. Zhai, M. Tignor, E. Poloczanska, K. Mintenbeck, A. Alegriá, M. Nicolai, A. Okem, J. Petzold, B. Rama, & N. M. Weyer (Eds.), *IPCC special report on the ocean and cryosphere in a changing climate* (pp. 131–202). https://www.ipcc.ch/srocc/.

Houghton, J. T., Ding, Y. D. J. G., Griggs, D. J., Noguer, M., vanderLinden, P. J., Dai, X., Maskell, K., & Johnson, C. A. (Eds.). (2001). *Climate change 2001: The scientific basis: Contribution of working group I to the third assessment report of the Intergovernmental Panel on Climate Change*. Cambridge University Press.

Huber, R., Rigling, A., Bebi, P., Brand, F. S., Briner, S., Buttler, A., Elkin, C., Gillet, F., Grêt-Regamey, A., Hirschi, C., Lischke, H., Scholz, R. W., Seidl, R., Spiegelberger, T., Walz, A., Zimmermann, W., & Bugmann, H. (2013). Sustainable land use in mountain regions under global change: Synthesis across scales and disciplines. *Ecology and Society, 18*(3). https://doi.org/10.5751/ES-05499-180336Switzerland.

Hunt, K. M. R., & Zaz, S. N. (2023). Linking the North Atlantic Oscillation to winter precipitation over the western himalaya through disturbances of the subtropical jet. *Climate Dynamics, 60*(7-8), 2389–2403. https://doi.org/10.1007/s00382-022-06450-7.

Huss, M., & Hock, R. (2018). Global-scale hydrological response to future glacier mass loss. *Nature Climate Change, 8*(2), 135–140. https://doi.org/10.1038/s41558-017-0049-x.

Inouye, D. W. (2020). Effects of climate change on alpine plants and their pollinators. *Annals of the New York Academy of Sciences, 1469*(1), 26–37. https://doi.org/10.1111/nyas.14104.

IPCC. (2022). *Climate change 2022: Impacts, adaptation, and vulnerability. Contribution of working group II to the Sixth assessment report of the Intergovernmental Panel on Climate Change*. Cambridge University Press.

Jendritzki, I., Tonnang, H. E. Z., Calatayud, P. A., Borgemeister, C., Johansson, T., & Biber-Freudenberger, L. (2023). Uncertainties in the effectiveness of biological control of stem borers under different climate change scenarios in Eastern Africa. *Climatic Change, 176*(5). https://doi.org/10.1007/s10584-023-03514-3.

Kariuki, R. W., Thorn, J. P. R., Mfune, J. K. E., Marchant, R., & Capitani, C. (2025). Near and far future conservation, land use, and land cover interactions around the wider Etosha landscape, north-central Namibia. *Regional Environmental Change, 25*(2). https://doi.org/10.1007/s10113-025-02388-8.

Kariuki, R., Willcock, S., & Marchant, R. (2018). Rangeland livelihood strategies under varying climate regimes: Model insights from Southern Kenya. *Land,* 7(2), 47. https://doi.org/10.3390/land7020047.

Kato, T., Rambali, M., & Blanco-Gonzalez, V. (2021). *Strengthening climate resilience in mountainous areas.* OECD Development Co-operation Working Papers.

Khatri, B. (2025). Impacts of glacial lake outburst floods on Nepal's national security. *Unity Journal, 6*(1), 63–78. https://doi.org/10.3126/unityj.v6i1.75554.

Körner, C. (2004). Mountain biodiversity, its causes and function. *Ambio, 66,* 11–134.

Lalande, M., Ménégoz, M., Krinner, G., Naegeli, K., & Wunderle, S. (2021). Climate change in the High Mountain Asia in CMIP6. *Earth System Dynamics, 12*(4), 1061–1098. https://doi.org/10.5194/esd-12-1061-2021.

Levytska, O., Mulska, O., Shushkova, Y., Ivaniuk, U., Vasyltsiv, T., Krupin, V., & Ivanova, L. (2022). Modelling the impact of emigration upon social and economic development of the Carpathian region of Ukraine. *Bulletin of Geography. Socio-economic Series, 2022*(57), 141–154. https://doi.org/10.12775/bgss-2022-0028.

Lin, Q., Steger, S., Pittore, M., Zhang, J., Wang, L., Jiang, T., & Wang, Y. (2022). Evaluation of potential changes in landslide susceptibility and landslide occurrence frequency in China under climate change. *Science of The Total Environment, 850,* 158049. https://doi.org/10.1016/j.scitotenv.2022.158049.

Liu, Z., Herman, J. D., Huang, G., Kadir, T., & Dahlke, H. E. (2021). Identifying climate change impacts on surface water supply in the southern Central Valley, California. *Science of the Total Environment, 759.* https://doi.org/10.1016/j.scitotenv.2020.143429.

Lutz, A. F., Immerzeel, W. W., Shrestha, A. B., & Bierkens, M. F. P. (2014). Consistent increase in High Asia's runoff due to increasing glacier melt and precipitation. *Nature Climate Change, 4*(7), 587–592. https://doi.org/10.1038/nclimate2237.

Mantyka-pringle, C. S., Martin, T. G., & Rhodes, J. R. (2012). Interactions between climate and habitat loss effects on biodiversity: A systematic review and meta-analysis. *Global Change Biology, 18*(4), 1239–1252. https://doi.org/10.1111/j.1365-2486.2011.02593.x.

Marchant, R. (2021). *East Africa's human environment interactions.* Springer International Publishing.

Marchant, R., Mumbi, C., Behera, S., & Yamagata, T. (2007). The Indian Ocean dipole - The unsung driver of climatic variability in East Africa. *African Journal of Ecology, 45*(1), 4–16. https://doi.org/10.1111/j.1365-2028.2006.00707.x.

Marchant, R., & Taylor, D. (1998). Dynamics of montane forest in central Africa during the late Holocene: A pollen-based record from western Uganda. *The Holocene, 8*(4), 375–381. https://doi.org/10.1191/095968398672993971.

McClung, D., & Schaerer, P. A. (2006). *The avalanche handbook* (p. 342). The Mountaineers Books.

Monteiro, D., & Morin, S. (2023). Multi-decadal analysis of past winter temperature, precipitation and snow cover data in the European Alps from reanalyses, climate models and observational datasets. *The Cryosphere, 17*(8), 3617–3660. https://doi.org/10.5194/tc-17-3617-2023.

Mora, A., Wilby, A., & Menéndez, R. (2023). South European mountain butterflies at a high risk from land abandonment and amplified effects of climate change. *Insect Conservation and Diversity, 16*(6), 838–852. https://doi.org/10.1111/icad.12676.

Mwalusepo, S., Massawe, E. S., Affognon, H., Okuku, G. O., Kingori, S., Mburu, P. D., Ong'amo, G. O., Muchugu, E., Calatayud, P. A., Landmann, T., & Muli, E. (2015). Smallholder farmers' perspectives on climatic variability and adaptation strategies in East Africa: The case of mount Kilimanjaro in Tanzania, Taita and Machakos Hills in Kenya. *Earth Science & Climatic Change, 06,* 01–09.

Newell, F. L., Ausprey, I. J., & Robinson, S. K. (2022). Spatiotemporal climate variability in the Andes of northern Peru: Evaluation of gridded datasets to describe cloud forest microclimate and local rainfall. *International Journal of Climatology, 42*(11), 5892–5915. https://doi.org/10.1002/joc.7567.

Ngcamu, B. S. (2023). Climate change effects on vulnerable populations in the Global South: A systematic review. *Natural Hazards, 118*(2), 977–991. https://doi.org/10.1007/s11069-023-06070-2.

Nogués-Bravo, D., Araújo, M. B., Romdal, T., & Rahbek, C. (2008). Scale effects and human impact on the elevational species richness gradients. *Nature, 453*(7192), 216–219. https://doi.org/10.1038/nature06812.

Norris, J., Carvalho, L. M. V., Jones, C., & Cannon, F. (2020). Warming and drying over the central Himalaya caused by an amplification of local mountain circulation. *npj Climate and Atmospheric Science, 3*(1). https://doi.org/10.1038/s41612-019-0105-5.

Ogburn, S. E., Charlton, D., Norgaard, D., Wright, H. M., Calder, E. S., Lindsay, J., Ewert, J., Takarada, S., & Tajima, Y. (2023). The Volcanic Hazard Maps Database: An initiative of the IAVCEI Commission on Volcanic Hazards and Risk. *Journal of Applied Volcanology, 12*(1). https://doi.org/10.1186/s13617-022-00128-9.

Ortner, G., Bründl, M., Kropf, C. M., Röösli, T., Bühler, Y., & Bresch, D. N. (2023). Large-scale risk assessment on snow avalanche hazard in alpine regions. *Natural Hazards and Earth System Sciences, 23*(6), 2089–2110. https://doi.org/10.5194/nhess-23-2089-2023.

Pandit, M. K., Manish, K., & Koh, L. P. (2014). Dancing on the roof of the world: Ecological transformation of the himalayan landscape. *BioScience, 64*(11), 980–992. https://doi.org/10.1093/biosci/biu152.

Pánek, T., Břežný, M., Harrison, S., Schönfeldt, E., & Winocur, D. (2022). Large landslides cluster at the margin of a deglaciated mountain belt. *Scientific Reports, 12*(1). https://doi.org/10.1038/s41598-022-09357-9.

Parmesan, C., & Yohe, G. (2003). A globally coherent fingerprint of climate change impacts across natural systems. *Nature, 421*(6918), 37–42. https://doi.org/10.1038/nature01286.

Pavlova, I., Makarigakis, A., Depret, T., & Jomelli, V. (2017). Global overview of the geological hazard exposure and disaster risk awareness at world heritage sites. *Journal of Cultural Heritage, 28*, 151–157. https://doi.org/10.1016/j.culher.2015.11.001.

Pepin, N. C., Arnone, E., Gobiet, A., Haslinger, K., Kotlarski, S., Notarnicola, C., Palazzi, E., Seibert, P., Serafin, S., Schöner, W., Terzago, S., Thornton, J. M., Vuille, M., & Adler, C. (2022). Climate changes and their elevational patterns in the mountains of the world. *Reviews of Geophysics, 60*(1). https://doi.org/10.1029/2020rg000730.

Peters, M. K., Hemp, A., Appelhans, T., Becker, J. N., Behler, C., Classen, A., & ..., Steffan-Dewenter, I. (2019). Climate–land-use interactions shape tropical mountain biodiversity and ecosystem functions. *Nature, 568*(7750), 88–92. https://doi.org/10.1038/s41586-019-1048-z.

Piao, S., Yin, G., Tan, J., Cheng, L., Huang, M., Li, Y., Liu, R., Mao, J., Myneni, R. B., Peng, S., Poulter, B., Shi, X., Xiao, Z., Zeng, N., Zeng, Z. Z., & Wang, Y. (2015). Detection and attribution of vegetation greening trend in China over the last 30 years. *Global Change Biology, 21*(4), 1601–1609. https://doi.org/10.1111/gcb.12795.

Platts, P. J., Gereau, R. E., Burgess, N. D., & Marchant, R. (2013). Spatial heterogeneity of climate change in an Afromontane centre of endemism. *Ecography, 36*(4), 518–530. https://doi.org/10.1111/j.1600-0587.2012.07805.x.

(2019). IPCC special report on the ocean and cryosphere in a changing climate. In H.-O. Pörtner, D. C. Roberts, V. Masson-Delmotte, P. Zhai, M. Tignor, E. Poloczanska, K. Mintenbeck, A. Alegría, M. Nicolai, A. Okem, J. Petzold, B. Rama, & N. M. Weyer (Eds.). In *IPCC Intergovernmental panel on climate change: 1* (p. 255). Cambridge University Press. https://doi.org/10.1017/9781009157964.

Potter, E. R., Fyffe, C. L., Orr, A., Quincey, D. J., Ross, A. N., Rangecroft, S., Medina, K., Burns, H., Llacza, A., Jacome, G., Hellström, R., Castro, J., Cochachin, A., Montoya, N., Loarte, E., & Pellicciotti, F. (2023). A future of extreme precipitation and droughts in the Peruvian Andes. *npj Climate and Atmospheric Science, 6*(1). https://doi.org/10.1038/s41612-023-00409-z.

Previdi, M., Smith, K. L., & Polvani, L. M. (2021). Arctic amplification of climate change: A review of underlying mechanisms. *Environmental Research Letters, 16*(9), 093003. https://doi.org/10.1088/1748-9326/ac1c29.

Raunkiaer, C. (1905). Types biologiques pour la géographie botanique. *Oversigt over det Kgl. Danske Videnskabernes Selskabs Forhandlinger.* 5. Tryt hos Kgl. Hofbogtrykker Bianco Luno.

Richardson, S. D., & Reynolds, J. M. (2000). An overview of glacial hazards in the Himalayas. *Quaternary International, 65–66*, 31–47. https://doi.org/10.1016/S1040-6182(99)00035-X.

Rodda, H., Walsh, D., Ferreira, F., Meddens, F., Branch, N., Handley, J., Wade, A., Black, S., Herrera, A., Simmonds, M., & Lane, K. J. (2023). In *Adaptive capacity of farming communities to climate change in the Peruvian Andes: Past, present and future (preliminary findings of the ACCESS project): 8* (pp. 51–67). Revista de Glaciares y Ecosistemas de Montaña.

Rödder, D., Schmitt, T., Gros, P., Ulrich, W., & Habel, J. C. (2021). Climate change drives mountain butterflies towards the summits. *Scientific Reports, 11*(1). https://doi.org/10.1038/s41598-021-93826-0.

Rodman, K. C., Veblen, T. T., Battaglia, M. A., Chambers, M. E., Fornwalt, P. J., Holden, Z. A., Kolb, T. E., Ouzts, J. R., & Rother, M. T. (2020). A changing climate is snuffing out post-fire recovery in montane forests. *Global Ecology and Biogeography, 29*(11), 2039–2051. https://doi.org/10.1111/geb.13174.

Rolando, A., Caprio, E., & Negro, M. (2013). The effect of ski-pistes on birds and mammals. In *The impacts of skiing and related winter recreational activities on mountain environments* (pp. 101–122). Bentham Science Publishers. https://doi.org/10.2174/9781608054886113010009.

Rolando, A., Caprio, E., Rinaldi, E., & Ellena, I. (2007). The impact of high-altitude ski-runs on alpine grassland bird communities. *Journal of Applied Ecology, 44*(1), 210–219. https://doi.org/10.1111/j.1365-2664.2006.01253.x.

Schmeller, D. S., Urbach, D., Bates, K., Catalan, J., Cogălniceanu, D., Fisher, M. C., Friesen, J., Füreder, L., Gaube, V., Haver, M., Jacobsen, D., Le Roux, G., Lin, Y. P., Loyau, A., Machate, O., Mayer, A., Palomo, I., Plutzar, C., Sentenac, H., & …, Ripple, W. J. (2022). Scientists' warning of threats to mountains. *Science of the Total Environment, 853*. https://doi.org/10.1016/j.scitotenv.2022.158611.

Seddon, A. W. R., Macias-Fauria, M., Long, P. R., Benz, D., & Willis, K. J. (2016). Sensitivity of global terrestrial ecosystems to climate variability. *Nature, 531*(7593), 229–232. https://doi.org/10.1038/nature16986.

Settele, J., Kudrna, O., Harpke, A., Kühn, I., van Swaay, C., Verovnik, R., Warren, M., Wiemers, M., Hanspach, J., Hickler, T., Kühn, E., van Halder, I., Veling, K., Vliegenthart, A., Wynhoff, I., & Schweiger, O. (2008). Climatic risk atlas of european butterflies. *BioRisk, 1*, 1–712. https://doi.org/10.3897/biorisk.1.

Shedayi, A. A., Xu, M., Hussain, F., Sadia, S., Naseer, I., & Bano, S. (2016). Threatened plant resources: Distribution and ecosystem services in the world's high elevation park of the Karakoram ranges. *Pakistan Journal of Botany, 48*(3), 999–1012.

Srivastav, A. L., Dhyani, R., Ranjan, M., Madhav, S., & Sillanpää, M. (2021). Climate-resilient strategies for sustainable management of water resources and agriculture. *Environmental Science and Pollution Research, 28*(31), 41576–41595. https://doi.org/10.1007/s11356-021-14332-4.

Stevens, N., Bond, W., Feurdean, A., & Lehmann, C. E. R. (2022). Grassy ecosystems in the anthropocene. *Annual Review of Environment and Resources, 47*, 261–289. https://doi.org/10.1146/annurev-environ-112420-015211.

Tasser, E., Leitinger, G., & Tappeiner, U. (2017). Climate change versus land-use change—What affects the mountain landscapes more? *Land Use Policy, 60*, 60–72. https://doi.org/10.1016/j.landusepol.2016.10.019.

Teklu, A., Simane, B., & Bezabih, M. (2023). Effect of climate smart agriculture innovations on climate resilience among smallholder farmers: Empirical evidence from the Choke Mountain watershed of the Blue Nile highlands of Ethiopia. *Sustainability, 15*(5), 4331. https://doi.org/10.3390/su15054331.

Telwala, Y., Brook, B. W., Manish, K., & Pandit, M. K. (2013). Climate-induced elevational range shifts and increase in plant species richness in a himalayan biodiversity epicentre. *PLoS ONE, 8*(2). https://doi.org/10.1371/journal.pone.0057103.

Thorn, J. P. R., Klein, J. A., Steger, C., Hopping, K. A., Capitani, C., Tucker, C. M., Nolin, A. W., Reid, R. S., Seidl, R., Chitale, V. S., & Marchant, R. (2020). A systematic review of participatory scenario planning to envision mountain social-ecological systems futures. *Resilience Ecology and Society, 25*(3), 1–55. https://doi.org/10.5751/ES-11608-250306.

Van Mantgem, P. J., Stephenson, N. L., Byrne, J. C., Daniels, L. D., Franklin, J. F., Fulé, P. Z., Harmon, M. E., Larson, A. J., Smith, J. M., Taylor, A. H., & Veblen, T. T. (2009). Widespread increase of tree mortality rates in the Western United States. *Science, 323*(5913), 521–524. https://doi.org/10.1126/science.1165000.

Vinnell, L. J., Hudson-Doyle, E. E., Johnston, D. M., Becker, J. S., Kaiser, L., Lindell, M. K., Bostrom, A., Gregg, C., Dixon, M., & Terbush, B. (2021). Community preparedness for volcanic hazards at Mount Rainier, USA. *Journal of Applied Volcanology, 10*(1). https://doi.org/10.1186/s13617-021-00110-x.

Wambulwa, M. C., Milne, R., Wu, Z. Y., Spicer, R. A., Provan, J., Luo, Y. H., Zhu, G. F., Wang, W. T., Wang, H., Gao, L. M., Li, D. Z., & Liu, J. (2021). Spatiotemporal maintenance of flora in the Himalaya biodiversity hotspot: Current knowledge and future perspectives. *Ecology and Evolution, 11*(16), 10794–10812. https://doi.org/10.1002/ece3.7906.

Yang, Y. (2021). Evolution of habitat quality and association with land-use changes in mountainous areas: A case study of the Taihang Mountains in Hebei Province, China. *Ecological Indicators, 129*, 107967. https://doi.org/10.1016/j.ecolind.2021.107967.

Zu, K., Wang, Z., Lenoir, J., Shen, Z., Chen, F., & Shrestha, N. (2022). Different range shifts and determinations of elevational redistributions of native and non-native plant species in Jinfo Mountain of subtropical China. *Ecological Indicators, 145*, 109678. https://doi.org/10.1016/j.ecolind.2022.109678.

CHAPTER 7

Opportunities for the world's mountain areas

7.1 Introduction

In the same way that mountains and those communities that live in mountains face many challenges, there are also many ways to navigate these. The biggest opportunity comes from within mountain communities themselves: they are used to living in a dynamic and unpredictable environment with multiple adaptations having been developed, nurtured, and honed over millennia. Mountain communities are shaped by their unique geographical, ecological, environmental, and historical contexts. The challenging environment of mountain regions often fosters strong community ties and cooperative social structures. Mountain communities frequently rely on collective efforts for tasks such as farming, construction, and managing communal resources. For example, the Meru people of Mount Kenya maintain strong community bonds through their traditional land tenure system based on clan ownership. This system ensures that land is managed sustainably and that resources are distributed equitably among community members. The WaMeru have a rich oral tradition, with stories and songs passed down through generations that reinforce these social values and cultural identity. Tibetans have developed remarkable physiological adaptations to living at high-altitude and even have a unique genetic adaptation that allows them to thrive in lower levels of oxygen via a lower concentration of haemoglobin, reducing the risk of altitude sickness.

Many mountain communities rely on subsistence agriculture, animal husbandry, and the harvesting of forest products to sustain their livelihoods. These are characterised by resilient adaptation strategies and contributions to sustainable development way beyond the mountain border. Innovative technologies such as renewable energy systems, climate-resilient agricultural practices, and early warning systems can enhance the adaptive capacity of mountain communities. Mobile phone applications for weather forecasting, precision agricultural tools, access to new markets, and remote sensing technologies are being increasingly used by mountain farmers to improve

Mountain Diversity: Dynamics of High-Altitude Environments.
DOI: https://doi.org/10.1016/B978-0-443-22251-1.00007-1

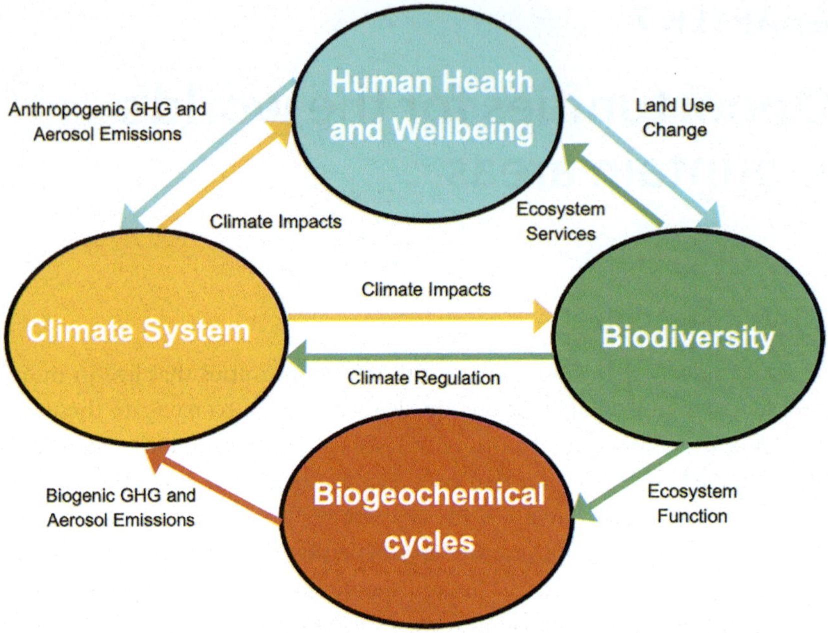

Figure 7.1 There are intricate and direct linkages between biodiversity, climate, biogeochemical, and human health and well-being.

decision-making and resource management. However, the adoption of these technologies must be context-specific and socially inclusive to ensure benefits that support society in a just and equitable way. By examining the challenges they face, the traditional knowledge they possess, and the innovative solutions they employ, this chapter will shed light on the importance of supporting resilient mountain communities in the face of global change so that these wider benefits of strong mountain social systems can be realised, not just for mountain regions but for the world that relies on them.

7.2 Climate change adaptation

Mountain communities are particularly vulnerable to the impacts of climate change, including glacier retreat, changes in precipitation patterns, and increased frequency of extreme weather events. These changes threaten water availability, agriculture, and biodiversity, exacerbating existing vulnerabilities with links between biodiversity, climate, biogeochemical, and human health and well-being (Fig. 7.1). Given these intricate connections, climate change

Plate 7.1 *Sorghum* (*top*) or Sunflower (*bottom*) can be more drought-resilient or bring higher value. *(Photographs Rob Marchant).*

adaptation strategies are crucial for building the resilience of mountain communities.

Mountain communities often rely on collective action and community solidarity to address shared challenges. This cooperative working includes initiatives such as community-managed irrigation systems, disaster preparedness committees, and indigenous land management practices. Social structures can support improving water management, promoting climate-resilient agriculture that focuses on growing more drought-resistant crops such as Sorghum, higher-value crops such as Sunflower (Plate 7.1), and implementing early warning systems for natural disasters.

Many systems for adapting to climate variability are deep-rooted and have a long history of use. For example, in the Andes, the use of ancient

Plate 7.2 Traditional water management features store water, such as the Ndivo. *(Photograph Daryl Stump).*

water management systems such as 'Amunas' (a traditional method of water harvesting), or the 'Ndiva' community-managed water reservoirs in the Pare Mountains of northern Tanzania (Plate 7.2), has been constructed to buffer water variability. While many traditional management systems have fallen into disrepair, particularly through colonial administration and the growing reliance on technologically engineered 'solutions', they are having a revival to improve water security in the face of unprecedented climate change and variability. These systems help capture and store water during the rainy season, ensuring a steady supply during dry periods. In Nepal, the revival of ancient water management systems such as 'Guthi' (traditional irrigation networks) has helped improve water security in the face of climate change. These traditional systems capture and store water during the monsoon season, ensuring a steady supply during the dry season and reducing the risk of drought. Adaptation responses within irrigation systems include the adoption of new irrigation technologies or upgrading existing technologies, adopting water conservation measures, water rationing, constructing water storage infrastructure, and changes in cropping patterns.

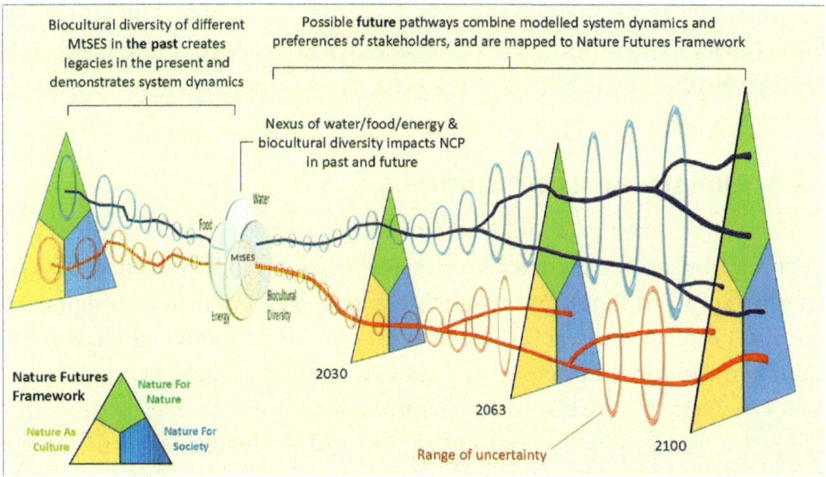

Figure 7.2 Timeline that will be used to explore how water, energy and food systems have evolved in mountain systems from the deep past to present to futures (2030, 2063, and 2100). Our visions of the past and the future are clearly more uncertain, as shown through the increasing rings of uncertainty.

Past signals of climate change impact the future in two fundamental ways: firstly, models that forecast future changes at decadal and longer time scales are validated by reference to data on past changes (e.g., Gillson et al., 2025); and secondly, because our perception of the rate and degree of past change affects our assessment of the plausibility of future changes. In other words, predictions of environmental or climate change can be rejected because they fall outside an individual's lived experience, or outside the wider memories of their community (e.g., Soga & Gaston, 2024). A more complete synthesis of data on past changes within socio–ecological systems can thus document how contemporary landscapes, ecosystems, and human land-use strategies are legacies of past changes, improving forecasts and models of future change. Such an approach can also enable a diverse range of stakeholders from community members to policy makers to envision complex system dynamics in landscapes they know well, thus increasing trust in future forecasts and allowing the co-production of plausible and mutually acceptable pathways to sustainable futures (Fig. 7.2). Given the value of these past perspectives for navigating future challenges, more cross-disciplinary work should be done on integrating deep time perspectives from archaeological and palaeoecological knowledge into predictive models of the future, one of the spear points of the emerging discipline of conservation paleobiology and validating outputs from species distributions models for past climate conditions with palaeoecological records (Hooghiemstra & Flantua, 2019).

7.2.1 Climate-smart agriculture

Climate change is challenging current food production systems with wide-reaching impacts for societies; challenges which are heightened when considering some of the more vulnerable and highly populated mountainous regions. Fostering a greater understanding of climate-induced changes will require meaningful cooperation between farmers, suppliers, markets, and policy-makers. Climate-smart agricultural systems continue to develop, as a concept and in practice, but enhancing and facilitating clear implementation pathways could be the difference between forest communities either thriving or declining as threats change (Bowditch et al., 2020). Sustainable agricultural development in mountain regions requires ecosystem-based approaches that promote biodiversity conservation, soil fertility, and watershed management, and enhance agricultural productivity and resilience, while some areas are seeing a resurgence of terracing. Governments and non-governmental organisations can play a crucial role in promoting and supporting terracing practices. This includes providing financial assistance, technical support, and training to farmers, as well as implementing policies that encourage sustainable land management practices. In some regions, terracing is being integrated into broader soil and water conservation programs, helping to address environmental challenges and improve agricultural productivity.

Mountains have supported agricultural livelihoods for centuries. Farmers are adopting climate-resilient farming techniques, such as conservation agriculture, crop diversification, and rainwater harvesting, to mitigate the impacts of climate change and enhance agricultural productivity. The climate-smart concept originated with agriculture (FAO, 2010) and recognises the need for 'an approach that helps to guide actions needed to transform and reorient agricultural systems to effectively support development and ensure food security in a changing climate'. In the Ethiopian Highlands, farmers have adapted to changing climatic conditions by diversifying their crops and adopting drought-resistant varieties or returning to traditional crops. For example, there has been a growth in reviving traditional varieties such as millet on many African mountains, versus the more water-thirsty maize that has come to dominate over the past 400 or so years, since its introduction from Latin America. Similarly, farmers in northwest India have increased production of lentils and vegetables, which provide important nutrients to the local diet, with support from government watershed improvement programs, which help address decreased availability of irrigation water (Hock et al., 2019). Farmers who rely on irrigation in the Naryn River basin in Kyrgyzstan have shifted from the water-intensive fruits and vegetables to fodder crops such as barley and alfalfa, which are more profitable (Hill et al., 2017).

Rural communities are dependent on adequate levels of soil moisture at planting time, derived in many cases from irrigation water, which includes glacier and snowmelt water. As a result, communities can be exposed to risk stemming from cryosphere changes (Hock et al., 2019). The challenging terrain combined with limited arable land often necessitates innovative agricultural practices, such as terracing and agroforestry (Chapter 4). Sustainable terraced agriculture requires careful consideration of ecological factors, such as biodiversity conservation, water quality protection, and ecosystem services enhancement. Agroecological principles, landscape planning tools, and adaptive management practices can help optimise the environmental performance of terraced landscapes. The Quechua people of the Peruvian Andes have adapted to high-altitude living by developing a diet rich in high-altitude crops like quinoa and potatoes, which are both nutritious and resilient to cold temperatures. Additionally, the practice of chewing coca leaves is widespread among Andean communities, as it helps to alleviate the effects of altitude sickness and provides energy for labour-intensive activities. In Rwanda, the government has implemented a nationwide program to promote terracing as part of its efforts to improve food security and reduce poverty. The program provides financial incentives and technical assistance to farmers, encouraging them to adopt terracing and other sustainable agricultural practices. As a result, terracing has become a key component of Rwanda's agricultural development strategy, contributing to increased crop yields and improved livelihoods. Ongoing research and innovation are essential for improving the efficiency and sustainability of terracing practices. This includes developing new techniques and technologies for terrace construction and maintenance, as well as exploring ways to integrate terracing with other sustainable agricultural practices. Research is also being conducted to assess the long-term impacts of terracing on soil health, water management, and biodiversity. In Nepal, the government, along with international organisations, has implemented programs to support terracing as part of its broader efforts to improve food security and promote sustainable agriculture. These programs across the Global South clearly show the importance of encouraging people to adopt and maintain terracing as a key component of their farming systems, which has benefits far outside the terrace.

At a local scale, farmers adapt by using mostly on-farm adaptation strategies, including changing planting dates, adopting new crop varieties, investing in water harvesting infrastructure, increasing soil conservation techniques, diversifying livelihoods, the use of inputs (fertiliser, pesticides), and the increased use of veterinary products. However, wealth is often an important driver of adaptation, with wealthier households using a greater number of adaptation strategies. Local socio-cultural and political contexts are also of key importance. Cuni-Sanchez et al. (2025) found fewer adaptation strategies implemented in conflict-affected studied sites of Cameroon and the Democratic Republic of Congo than in more stable African countries. Access to land and credit are important constraints to adaptation, as are improving market access and strengthening value chains for mountain agricultural products – these can provide economic opportunities for farmers, reduce post-harvest losses, and enhance food security in remote mountain areas. Cuni-Sanchez et al. (2025) identified three key priorities for moving forward with locally driven and culturally appropriate farmers' climate change adaptation in African mountain regions that apply to mountain systems across the world. (1) Improving access to credit and markets; (2) increasing knowledge exchange among farmers; and (3) carefully considering the legacy of conflicts and non-intended consequences of national policies.

Part of the solution for mountain food producers is to add value to their produce and focus on quality over quantity. The Slow Food movement in Italy's Piedmont region has helped to promote sustainable agriculture and preserve traditional food practices. By supporting local farmers and artisanal producers, the movement has revitalised rural economies and helped to preserve the cultural heritage of mountain communities. Access to markets is not just related to physical infrastructure (e.g., roads [Plate 7.3]); other actors, such as NGOs or extension services, can help increase awareness of market prices using mobile phone apps. Product certification (e.g., organic coffee certification) can increase economic benefits and compensate for lower yields, contributing to farmers' climate change adaptation. Diversifying livelihoods can also support agricultural resilience. For example, the Gamo populations of Ethiopia engage in crafts such as weaving and pottery (Plate 7.4), which are important for both cultural expression and economic sustenance with diversified income streams.

Plate 7.3 One of the continued challenges for mountain communities is the road infrastructure that can impact access to the market. *(Photographs Rob Marchant)*.

Plate 7.4 The Gamo Highlands in Ethiopia also engage in traditional crafts, such as weaving and pottery, which are important for both cultural expression and economic sustenance. *(Photograph Rob Marchant)*.

7.2.2 Climate-smart forestry

Forestry has been placed at the forefront of action to mitigate climate change through afforestation and carbon sequestration efforts. In recent years, climate-smart forestry has become increasingly common as organisations like the Intergovernmental Panel on Climate Change place great emphasis on forests as being crucial for mitigating the global impacts of climate change (IPCC, 2022). Indeed, in order to meet the Paris Agreement goals, considerable contributions will be required from forests, as nations are simultaneously aiming to reduce forest degradation and enhance carbon sinks (Rockström et al., 2017). However, many forests may be in vulnerable climatic zones with concentrated climate-induced changes impacting species mixes, habitat suitability, and ultimately shifting distribution ranges (Bowditch et al., 2020). Large-scale restoration efforts, such as China's Grain-for-Green Program, have resulted in significant additional forest cover in mountainous landscapes (Piao et al., 2015). Though the net benefits do not equal those provided by naturally managed forests, there are some benefits associated with more tree cover, with healthy and widely beneficial forests, the ultimate goal of sustainable mountain development. Appropriate management strategies have the potential to regulate emissions, as well as produce a wider range of ecosystem services that will support and enhance communities. To maximise the potential benefits, careful establishment and subsequent management are required to support the forests, communities, and economies under the banner of the 'right tree in the right place'. In conjunction with a series of well-defined indicators, afforestation can support climate-smart decision-making within the context of the complex socio–ecological systems to achieve sustainable development goals (Bowditch et al., 2020; Denton et al., 2015).

Developing climate-smart systems, where contemporary choices are embedded within the historical context of what is appropriate, drawing on experienced forest and land use managers, is central to shaping best practice. On the international level, forestry is increasingly viewed as one of the most effective ways to mitigate climate change (Chazdon & Brancalion, 2019); however, the discussion between forest accounting being based upon past management activities and prospective management is ongoing (Grassi et al., 2018). Each country and region will interpret definitions, principles, and indicators differently, as diverse forest types and tree species, ecosystems, socio–economic conditions, and people will vary. These will be especially prevalent in critical or more sensitive areas such as mountains (Lexer & Bugmann, 2017). As the need for action against climate change heightens, greater guidance will be needed at all scales, from international to national policy to individual forest managers. A key development to bridging the gap between policy and practice will be effective engagement with forest managers, forest communities, and different forest owner types to assess the accessibility of climate-smart forestry indicators. Implementing locally adapted climate-smart forestry, through interactive and participatory methods, can help communities develop alongside a productive forest resource (Capitani et al., 2016).

To maintain the potential for future cultivation and to protect water resources, it is important for any further development in mountain regions to involve dedicated efforts to avoid soil loss (Harden, 2001). The most effective way to control erosion due to rainfall and wind is to ensure that the land is densely vegetated, such as through appropriate agroforestry systems (Chapter 5). These changes, if effectively implemented, can favour and provide new production opportunities and important revenue sources for local economies, which support community development, especially for communities suffering from depopulation or a lack of services and infrastructure (Bowditch et al., 2020). These efforts need to not only help preserve the forested environment but also support the livelihoods of local communities by enhancing biodiversity and improving sustainability. Reforestation and agroforestry initiatives, if properly implemented, can improve soil fertility and reduce the risk of flooding, landslides, and soil erosion.

7.2.3 Climate-smart pastoralism

Upslope shifts in treelines are mainly ascribed to woody vegetation growth, following pastoral abandonment (Barras et al., 2021; Dirnböck et al., 2011). Declining transhumance and the use of high-altitude pasture can lead to shrub encroachment into mesic grasslands. In the Pyrenean mountains, this is resulting in an increase in soil organic carbon stocks offset by greater stocks (Gutiérrez-Girón et al., 2015). At first glance, this temporary habitat gain may provide a short-term window of opportunity for several treeline species. However, treeline upward shifts can also be considered as a major threat (Barras et al., 2021) since open Alpine grasslands, which harbour a very rich biodiversity, would decrease in area as a result (Dirnböck et al., 2011). Because of the pyramid shape of mountains, grasslands risk getting squeezed between a moving treeline and rocky substrates where soil development is especially slow (Hagedorn et al., 2019). An elegant solution to solve this apparent conservation dilemma of declining biodiversity is to maintain pasturing in mountain ecosystems, often via financial incentives to farmers. To many upland pastoralists, this can provide additional income streams and support biodiversity (Bani et al., 2019). Keeping both biodiverse-rich open Alpine grasslands and semi-open wooded pastures allows habitat heterogeneity and is key for birds such as the Ringed Ouzel and a number of associated treeline species of birds and mammals (Barras et al., 2021). This upland management practice can be implemented at scale and can be facilitated by the 'Alp

Plate 7.5 Switzerland's high-altitude pastures have exceptionally high plant diversity, partly due to a long history of land management. *(Photograph Rob Marchant).*

co-operatives'. For example, in Switzerland, groups of farmers share the use and management of high-altitude pastures during the summer months (Plate 7.5). By pooling resources and labour, these communities are able to maintain their traditional way of life while ensuring the sustainable use of their land.

In the arid lands of northern Kenya, pastoralists and agro-pastoralists use the mountain forests of Mount Marsabit as grazing areas during droughts. The leaves and branches from the trees found in these montane forests collect water droplets from low clouds (or fog), maintaining the forest's humidity and keeping the grass underneath green. However, with increasing global temperatures, fog is formed at higher elevations, and these montane forests collect less fog water, slowly drying up. Pastoralists cut tree branches to feed the hungry cows during major droughts, which translates into less branches available to collect fog water, further drying up the forests. Increased human and cattle populations and increasingly recurrent drought events mean widespread 'branch-cutting for cow feeding', further degrading these vital montane forests. The water supply to the summit of Mount Marsabit has diminished over the last few decades, with a 60% reduction in hours of fog per year since 1981, and rainfall decreased from 1961 to 2010 by about 10 mm year^{-1} (Los et al., 2019). To mitigate the challenges and impacts on the water tower, pastoral populations try to control access to the forest, while also diversifying livelihoods, with beekeeping and small-scale agriculture. Tree planting (including the stimulant khat, *Catha edulis*) and shifting to camel herding, from traditional cattle, sheep, and goats, are some of the adaptive strategies used by these pastoralists. However, the lack of water for everyone to farm at lower elevations (where it needs to be piped from the mountain), and the high cost of acquiring a camel, hamper the widespread adoption of such strategies (Cuni-Sanchez et al., 2019).

7.3 Species response to climate change

There are many knowledge gaps and areas of uncertainty on how species will respond to current and future climate change. In particular, we know very little about how biotic interactions (e.g., between plants and their competitors, herbivores, pollinators, pathogens, and soil microbes) will be affected by altered climates and how these interactions will influence individual species and communities (Tito et al., 2020). Experimental studies are needed to empirically examine the effects of climate change on terrestrial organisms and to serve as the basis for predictions and management practices. However, designing and implementing experiments that can fully simulate multifactorial changes in the natural environment remains a daunting problem (Tito et al., 2020). The use of elevational gradients can be an especially powerful and logistically feasible tool for studying how climate and other factors influence species' distributions and physiologies (Malhi et al., 2010). One of the most important advantages of field experiments is the possibility of detecting changes in biotic interactions and their effects (Alexander et al., 2018). Although the topography of mountains changes considerably from one region to another, there is a universal and steady decrease in mean air temperatures as elevation increases (Chapter 3) (Körner, 2007). To explore how the responses of plants to climate change depend on the migration of other species, Alexander et al. (2018) transplanted alpine plant communities along an elevational gradient in the Swiss Alps to better understand plant community responses to altered competitive interactions under climate change and address additional areas of uncertainty, such as priority effects. Indeed, for plant species with relatively poor dispersal abilities, artificial dispersal strategies and translocations may be needed (e.g., Gaitán-Espitia & Hobday, 2021).

Over the last decades, biodiversity losses have escalated as a result of increasing biotic homogenisation, overexploitation, pollution, habitat fragmentation, and climate and land use changes (Chauvier et al., 2022). These biodiversity losses, particularly acute in mountain regions, have altered ecosystem functioning, stability, and services provided upon which modern society depends. Conservation initiatives and adaptive planning are rising to counteract this decay of our natural heritage. Accurately monitoring rare plant populations is critical to identifying threats to occurrences and establishing long-term population trends. To fulfil the monitoring mandates, conservation biologists are applying new approaches to traditional plot-based long-term demographic studies. With advances in image recognition and interaction with species distribution models, it is possible to use machine learning to plan targeted drone flights for rare plant monitoring in harder-to-reach mountain areas. A recent study mapping *Geum radiatum*, a rare plant in the Blue Ridge Mountains ecoregion, used a combination of drone survey and machine learning to enhance current conservation planning and track the status of the monitored plants over time (Reckling et al., 2021).

Species distribution models (SDMs) can be used for improved species and biodiversity management, provided that good modelling practices are followed (Chauvier et al., 2022; Platts et al., 2013; Tang et al., 2018). Using 21 understory bamboo species occurring within the mountainous geographic range of the Giant Panda in southwest China, found that due to high variations in local and regional gradients of temperature and precipitation, neglecting uncertainty can lead to misinformed research agendas, policies, and decisions. Being sensitive to the choice of baseline climate information, especially in mountainous environments with complex climatic gradients, is vital if these potential powerful tools are to maximise their utility. In order to address such issues and therefore better understand the actual demographic processes underlying species' distributions, the adoption of a process-based dynamic range modelling framework with SDMs can be ideal. Continued monitoring and research should be prioritised for potentially threatened habitats, in particular at the ecological interfaces, for example, between the grassland and bare rock habitats, across the treelines, or at higher altitudes crucial for vegetation communities and consequently possible distributional shifts of alpine fauna (Chamberlain et al., 2013; Lu et al., 2021).

For the potential utility of SDMs to be realised, there are a series of research priorities. Critically, there is a need to fill observational data gaps by working towards a comprehensive list of species for mountain regions, increasing species occurrence data, installing more climate stations, particularly above the upper forest line, and sharing data widely. Secondly, advanced climate modelling that represents key features in climate models (e.g., sharp rainfall gradients) and generates high-resolution climate data that better represent the complex mountain topography. Thirdly, improved plant distribution models by including key ecological/biological processes, data on plant traits, and available palaeoecological data to understand the momentum of ecological change and knowledge, but also by accounting for

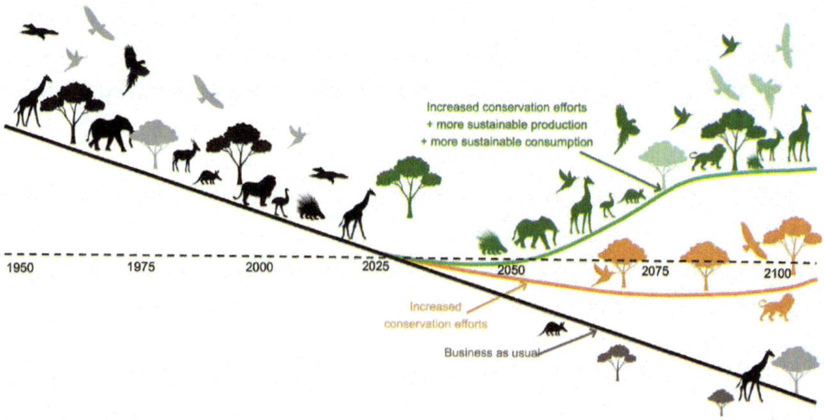

Figure 7.3 It is possible to stem the loss of biodiversity and bend the curve to a more positive future.

disturbance regimes and land-use changes. Lastly, an increased understanding of the locations and conditions which promote species diversification to support the integration of evolutionary refugia into conservation planning. Increasingly joined-up and concerted effort between botanists, ecologists, and climatologists working in mountain regions is helping to achieve better insight into how species are responding to changing mountain climates, and they are using this information to plan for the future – allowing us to 'get ahead of the challenge' and stem the loss of biodiversity and bend the curve to a more positive future (Fig. 7.3).

7.4 Biodiversity conservation, environmental restoration, and protected areas

In the Anthropocene, which is increasingly characterised by human-induced challenges, mountains may play a crucial role as biodiversity reservoirs. In many regions, there is a disproportionate amount of protection afforded to montane areas relative to lowlands (Rouget et al., 2003). This is largely due to the recognition of mountains as important centres of diversity (Chapter 3) (Perrigo et al., 2020). The biodiversity of mountain areas has also been disproportionately spared from human land use and land cover change, which has been so transformative across the more habitable and workable lowlands. There are some benefits from mountains being peripheral and relatively isolated (Sandel & Svenning, 2013).

Protected areas remain an essential tool to support conservation efforts globally (Naidoo et al., 2019). Protected areas and their expansion could play a significant role in the conservation of species under future climate scenarios. As mountains are predicted to be extremely sensitive to the impacts of climate change (Platts et al., 2013), protected areas can play a critical role in conserving biodiversity, maintaining viable populations of threatened species, and providing additional co-benefits. For example, the designation of protected areas can have differentiated management strategies that optimise resource allocation (for example, establishing more plants with less water consumption in areas rich in water resources) to coordinate the development of mountain resources and the environment (Yu et al., 2021). However, due to legal constraints, terrestrial protected areas are usually designed to be static and do not account for ecosystem changes or shifts in species' ranges, potentially limiting their effectiveness under climate or land-use change (Vincent et al., 2019). Implementing an optimum protected area network, although difficult to achieve, may be more effective and allow for biotic connections between the protected areas. Such establishment of ecological corridors can accelerate species' dispersal, migration, and exchange, contributing to ecosystem integrity (Du et al., 2022).

In the last decade, global conservation organisations have started to recognise that protected areas should address aspects of geodiversity and that geodiversity is part of natural diversity (Lausch et al., 2020). Indeed, global priorities for biodiversity conservation have heavily targeted mountain landscapes for additional protection and restoration. For example, a recent global synthesis identified that the highest global and regional priorities for protected area expansion are located primarily in mountainous regions, including in the Neotropics (Central America and along the Andes and the Brazilian coast), Africa (Madagascar, the Eastern Arc Mountains, and west African forests), and South and Southeast Asia (the Himalayan slopes, Indonesia, Papua New Guinea, and the Philippines) (Montesino Pouzols et al., 2014). Where additional protection of intact land cannot meet current conservation targets, restoration can meet shortfalls in several montane ecoregions, like the Sulaiman Range alpine meadows on the border of Afghanistan and Pakistan, the Madagascar subhumid forests along the Ankaratra Range, and the Huon Peninsula montane rain forests spanning the Finisterre and Saruwaged Ranges in Papua New Guinea (Mappin et al., 2019). Protecting and restoring mountain landscapes will also provide important benefits for lowland biodiversity in their future mountain habitats – rates of plant richness have already increased due to warming-driven upslope range shifts (Elsen et al., 2020). These mountain refuges will provide a lifeline for lowland species, enabling them to persist before their expected migration to lower latitudes under the next ice age. Geologically speaking, such a shift is imminent, providing a lifeline for lowland species, enabling them to persist before they are expected to migrate to lower latitudes under the next ice age. Geologically speaking, such a shift is imminent.

Ecological, demographic, and political processes that facilitate landscape recovery in mountainous ecosystems provide opportunities for species retreating upslope that conservation practitioners must integrate to assess species vulnerability to climate change (Elsen et al., 2020). Mountains provide a fantastic natural experimental set-up to answer some of these questions, because climatic conditions, soil properties, human settlement, plant species, and plant productivity change drastically within a short distance. Future studies can take advantage of elevational gradients as a 'natural experiment' to mechanistically understand diversity patterns and the associated ecosystem functioning in mountain regions from theoretical and evolutionary perspectives and further improve our predictions of the species, community, and ecosystem level consequences under global change (Platts et al., 2013). For example, the biota of the Qinghai-Tibet Plateau, like many mountain areas, is sensitive to ongoing global climate warming and human activities; these regional species-rich areas should be prioritised for conservation. A combined strategy that includes establishing more reserves, enlarging others, and increasing connectivity among them will better protect biodiversity, locally and regionally. Improving connectivity and migration corridors of key species, structure, and genetic variation will support resilience and adaptability of native forest diversity (Jump et al., 2014). A regional approach is essential for preventing a species' risk of extinction. Given this urgency, the broader conservation community is working together to ensure a more aligned regional approach, such as in Central Asia (Wilson et al., 2021) or the mountains of Equatorial Africa.

Environmental restoration and protection efforts are safeguarding and supporting ecosystems that mountain communities depend on. This includes reforestation, habitat restoration, and sustainable land management practices that mitigate the impacts of environmental degradation. However, setting appropriate restoration goals requires detailed information on former and current climates, forest extent, ecosystem composition and structure, and the heterogeneity of mountain socio–ecological landscapes. Without such information, we cannot assess the true impact of initiatives such as afforestation on biodiversity, or its effect on other ecosystem services. In the Pyrenees, reforestation projects have been implemented to restore degraded landscapes and protect watersheds (Nadal-Romero et al., 2021) to preserve the natural environment, but also support the livelihoods of local communities by enhancing biodiversity and improving the sustainability of agricultural practices. Effective conservation of mountain biological diversity will be achieved only by recognition of, and response to, the critical role of environmental drivers in shaping ecosystems. For example, managing woody biomass may reduce the risk of large catastrophic wildfires and alter the flow rates of water courses during extreme weather conditions (droughts, heat waves, rainstorms, and strong winds) to reduce flooding risk. Innovative science and new partnerships across a range of sectors are helping to navigate decisions about new and changing ecosystems – whether it be consideration of fire in the context of meeting global biodiversity targets, safeguarding regional ecosystem services, or protecting homes and habitats. Placing the increasingly important role of people, and their relationships with biodiversity, at the forefront of efforts to understand and adapt to changes, is central to these endeavours (Kelly et al., 2020).

7.5 Adaptive mountain social–ecological systems

There is a need to recognise mountains as entwined and deeply embedded social–ecological systems that have evolved over multiple spatial and temporal scales. Such a shift in perspective allows us to view mountain systems with a more nuanced and realistic lens. The multiple challenges mountains are facing can be made less challenging by recognising, documenting, and working with the valuable insights that communities living in mountain regions have accrued over aeons, as they adapt to the multiple environmental consequences of a dynamic climate. Mountain communities have to be at the heart of any problem-based solution and can provide the insights needed to design effective adaptation responses that are relevant to the, and work in the local context – one size does not fit all! Indeed, a 'one size fits all' approach to anything, be it reducing wildfire hazards in the Rocky Mountains (Schoennagel et al., 2004), or agricultural adaptation to climate change pressure (Cuni-Sanchez et al., 2025), is unlikely to be effective, and may even lead to maladaptation and exacerbate the challenge. For example, in a number of locations, community-based water management systems are replaced by an engineered solution, such as a pump by a well-meaning NGO; however, once the pump stops working, there is no one to pay for the fix, or the part is not available locally, and the traditional water management structures are abandoned.

The nature of mountain systems has often fostered strong community ties and cooperative social structures. Mountain communities frequently rely on collective efforts for tasks such as farming, construction, managing communal resources, and building communal infrastructure. Creating and maintaining social capital, such as knowledge exchange, upskilling, technological development, and use, as well as better monitoring and reporting to aid adaptive learning (Lawrence, 2017), should be key components to support mountain life (Bowditch et al., 2020). The rapidly changing nature of mountain social–ecological systems to a diversity of external effects, such as climate and land use change, can require shedding legacy social–ecological systems to be replaced by new social–ecological systems capable of responding to current and emerging challenges that threaten human well-being (Bowditch et al., 2020). The Ute's intimate knowledge of the landscape has been crucial for their survival in the harsh conditions of the Rockies (Ciocco et al., 2024), where they have developed extensive knowledge of the mountain environment, including seasonal migration patterns to follow game and the use of natural materials for shelter and clothing. Clearly, this knowledge has been created and implemented at a very different point in history and has to be reimagined and reimplemented in the context of contemporary challenges in North America. Similarly, the Māori people of New Zealand have a deep understanding of their mountainous regions. Their traditional knowledge includes an intimate understanding of the land, weather, and natural cycles, which has been passed down through generations, embodied in the concept of 'kaitiakitanga', which refers to the guardianship and sustainable management of natural resources, ensuring that the land is preserved for future generations (McAllister et al., 2023). The concept of 'ayni', a system of mutual aid and reciprocity, is deeply ingrained in Andean culture (Yana-Salluca et al., 2024). This system involves the exchange of labour and resources among community members, ensuring that everyone has access to what they need, and this can be adjusted to individuals' or communities' needs, which obviously change with the challenges in the Andes. In the Pamir Mountains of Tajikistan, the Aga Khan Development Network has worked to empower local communities through education, healthcare, and economic development programs. Pairing these deep-rooted and hefted concepts, that have worked through generations, with contemporary planning laws and ownership, remains a challenge, but if navigated correctly, can yield multiple benefits to societies and to mountains. These initiatives have strengthened the resilience of mountain people and improved their quality of life. Although the national political systems of protected area governance mean that communities are often excluded from managing land within the confines of a National Park, they can be instrumental in promoting sustainable land management practices. Indigenous communities should be front and centre in driving the agenda-setting around environmental management programmes so meaningful partnerships are created between communities and national agencies (Moyo, 2023). These efforts have helped to empower local people and ensure that their voices are heard in discussions about land use and conservation, and ultimately lead to stronger management of the protected area.

7.6 Mountain geodiversity

Nature conservation in many parts of the world is shifting from a focus on plants and animals towards a climate change adaptation approach, in which expected species range shifts are increasingly considered. Facilitating species' movements requires improving ecological connectivity across landscapes, and for these assessments, new and powerful approaches are emerging that can be crucial for supporting mountain systems into the future. One such approach is the use of 'geodiversity' as a proxy for biodiverse, resilient, and dispersal-enhancing areas (Määttänen et al., 2024) (Fig. 7.4). Geodiversity is defined as 'the natural range (diversity) of geological (rocks, minerals, fossils), geomorphological (landforms, topography, physical processes), soil, and hydrological features' (Gray, 2013). By taking a more holistic approach that goes beyond the intrinsic properties of individual species and ensures preservation of the complex web of biotic co-dependencies that are critical for healthy ecosystem function (Spicer et al., 2020), we can scale up efforts toward more effective ecosystem conservation and minimise the potential impacts of climate change. To better conserve geodiversity at a global scale, there is a need for a future assessment of geodiversity, not only at global but also local, national, and regional scales (Polman et al., 2024). Ultimately, a representative network of geoparks across the world where mountain systems are connected at scale will allow for enhanced resilience and the ability to buffer future shocks.

Figure 7.4 Geodiversity, climate diversity, and biodiversity underpin the resilient and effective function of mountain social–ecological systems.

Heterogeneous geophysical settings also support species adaptation and community changes, thereby offering resilience against ecological disturbances (Anderson et al., 2023). The highly heterogeneous geophysical nature of mountains is both advantageous and precarious, as it provides diverse opportunities yet also leaves mountains prone to rapid change. Despite some initial scepticism, geodiversity is becoming a more widely accepted concept with many research groups working on geodiversity topics around the world (Gray, 2024). Georestoration, where some of the engineering interventions that have taken place in the past century are removed, is starting to shape a more positive future. For example, river restoration, in which dams, weirs, straightened reaches, and concrete channels are removed and rivers allowed to operate naturally once again, seems to be less prone to flooding and less impactful on society (Gray, 2024). River restoration brings

Plate 7.6 'Hush', a temporary installation in one of the old lead mining valleys in Teesdale, North Pennines National Geopark in northern England. The valley was created during the mining process of using water to carve out the hillsides and access the mineral-bearing ore. *(Photograph Rob Marchant)*.

both geodiversity and biodiversity benefits as illustrated by Swindale Beck in Cumbria, England, where salmon and trout have returned to spawn in the redeposited gravels in the re-meandered reaches where river flows have been slowed. The concept of geodiversity has flourished and is leading to paradigm shifts in shaping ways of working with nature, using ecosystem engineers such as beavers, wolves, or lamas to help create resilient and connected mountain landscapes.

The underpinning geodiversity concept is being celebrated within the UNESCO Global Geoparks, which are identifiable geographical areas where sites and landscapes of international geological significance are managed with a holistic concept of protection, education, and sustainable development (Plate 7.6). The Geopark approach, which combines conservation with sustainable development while involving local communities, is becoming increasingly popular. At present, there are 229 UNESCO Global Geoparks in 50 countries, and a growing number of these make the string link from understanding the geodiversity and then using this understanding to inform the management. The importance of geodiversity has to be celebrated. There is a strong need to connect geodiversity to the community in diverse ways to emphasise the importance of mountains and their social–ecological history. For example, the North Pennines National Geopark in northern England is a region that has had multiple transformations from an agricultural landscape, to one driven by the lead mining industry, to one of upland farming and increasingly a region of tourism and recreation. These changing landscapes offer opportunities and can be the catalyst for new economies and connections that support the mountain social–ecological system, such as between communities and artists, or between hoteliers and farmers.

7.7 Ecosystem-based approaches

Mountains provide countless valuable ecosystem services that not only contribute to social capital, support rural economies, and generate significant income for communities that live in mountains, but also contribute to wider global society (Platts et al., 2023). Mountains supply freshwater for an estimated half of humanity (Ives, 1997) and are also vital centres for food production, clean energy, medicines, and protection against soil degradation. Traditional medicines, such as those derived from alpine plants, have been used for generations, and certain plants have economic value as sources of timber, food, and other products. Healthy, functioning, and conserved mountain ecosystems help mitigate the impacts of climate change, cool temperatures, maintain or even increase water retention, provide a vital carbon store, and reduce the risk of erosion and landslides. Mountain ecosystem services could improve the financial and spiritual well-being of people in mountainous areas and achieve 'win-win development' of ecosystem services and human welfare (Peng et al., 2017). Indeed, the 'social and ecological' are inextricably intertwined, shaped across space and time to influence current opinion and behaviour (Bowditch et al., 2020). Importantly, ecosystem services can form a mechanism that links stakeholders across the highlands to the lowlands with management practices that can lead to societal benefits. Payment for ecosystem services, conservation partnerships, and bridging organisations with local stakeholder expertise should play an increasingly important role in building the capacity of mountain systems (Platts et al., 2023).

Nature-based foci to restore ecosystems can help local communities build resilience and reduce dependencies on unsustainable practices. Nature-based solutions for restoration in mountains can diversify livelihoods and strengthen resilience, if conducted in line with guidelines from the UN Decade on Ecosystem Restoration 2021–2030. Strengthening regional and transboundary cooperation can ensure the long-term integrity of mountain ecosystems and their vital services throughout their catchments. Incorporating the existing wealth of knowledge – including traditional, indigenous, and experiential local knowledge – into the development of effective policies and activities can really underpin a sustainable future. Investing in restoration through long-term, nature-positive funding and community participation using incentive-based mechanisms will help to ensure the sustainability of mountain social–ecological systems.

Linking palaeoecological proxies to ecosystem services (the provisioning, supporting, and regulating benefits provided by nature) provides an interface for engagement between palaeoecologists and those working in the sustainable development arena. Such holistic perspectives will expand our comprehension of the range of national and regional interventions feasible for landscape management today, and lay foundations for informed strategies in the future, which include adaptation (Gillson et al., 2025). Moreover, a comprehensive understanding of the interlinked relationships between habitat structure, biodiversity, and key ecosystem services, although representing a great challenge because of the complexity of its analysis, may help to uncover important information to better guide our management of natural resources.

7.7.1 Sustainable carbon resource management in mountains

Although mountain forests provide various ecosystem services, such as wood production, protection against natural hazards, and carbon storage, it is important to understand such nuances and address the trade-offs among these diverse sets of ecosystem services. While carbon storage and sequestration are clearly one of the key services from intact forests (Cuni-Sanchez et al., 2021) and their underlying soils, an over-dependency on carbon sequestration and storage as a panacea to climate-induced changes can be a negative as much as a positive. Externally driven afforestation programmes fixated on maximum carbon storage with fast-growing tree species, such as seen with the expansion of *Eucalyptus* across a number of African countries, seriously overshadow adaptive capacity on regional and local scales, inhibiting small innovations and planning strategies. Protected areas, particularly mountain ecosystems, represent an important reservoir of terrestrial organic carbon, and reliable quantifications of their soil organic carbon stocks are essential to have a real estimate of their contribution to carbon storage and inform baseline soil carbon assessments. Similarly, soil organic carbon stocks of mountain soils are an amazing resource that can be maximised to provide a global benefit.

7.7.2 Sustainable water resource management in mountains

One of the main services produced and delivered from the mountains is their role in water provision. As the human population continues to increase worldwide, and thus demands on water resources increase, the protection of mountain catchments will be more important than ever (Perrigo et al., 2020). Hydropower comprisesabout 16% of electricity generation globally, but close to 100% in many mountainous countries. For example, Antioquia contributes 35% to Colombia's total electricity generation (López-Bermeo et al., 2022). The provision of water and power is changing due to climate change and global warming. In some glacier-fed rivers, summer and annual runoff have increased due to intensified glacier melt. However, other areas are decreasing where glacier meltwater has lessened as glacier area shrinks. Decreases are observed in regions dominated by small glaciers, such as the European Alps. Glacier retreat and snow cover changes have also contributed to localised declines in agricultural yields in some high mountain regions, including the Hindu Kush Himalayas and the tropical Andes (medium confidence). High Mountain Asia contains the largest number of glaciers (>10,000), glacier lakes, and dense annual snow cover outside the polar region, and functions as the 'Water Tower of Asia' (Banerjee et al., 2024; Immerzeel et al., 2020).

More than 70% of the population downstream of mountains directly depends on primary activities that rely on this source of water (Plate 7.7). Montane forests play a crucial role in water regulation, particularly in maintaining dry season flows and smoothing out floods and droughts. They capture moisture from the air (especially in cloud forests), slow down surface runoff, and release water gradually, ensuring a steady supply to lower elevations and nearby watersheds, thus playing a critical role in water regulation with downstream impacts (Cuni-Sanchez et al., 2025). They capture and store moisture from snowmelt, releasing it gradually into lower watersheds. This slow release helps maintain water flow in rivers and streams during the dry season. The forests act as a buffer, moderating temperatures and precipitation regimes by providing shade during the day and retaining heat at night. This helps to maintain a more stable climate, both within the forest and in surrounding areas.

Temperature changes dramatically alter mountain hydrological regimes by reducing water storage in snow and glaciers, increasing evapotranspiration

Plate 7.7 Water is the key commodity that makes mountains key for the world, and at the heart of future sustainability plans. *(Photographs Rob Marchant)*.

and permafrost degradation (Somers & McKenzie, 2020). Mountains are climate 'hotspots' where change can amplify what is occurring elsewhere; accelerating mountain climate change has extensive environmental impacts, including depletion of snow/ice reserves, critical for the world's water supply (Pepin et al., 2022). To cope with seasonal water scarcity at critical times for irrigation, villagers in the Himalayas have developed four types of artificial ice reservoirs: basins, cascades, diversions, and a form known locally as ice stupas. Ice reservoirs capture water in the autumn and winter, allowing it to freeze and hold it until spring, when it melts and flows down to fields, supplementing any decreased flow in the following spring (Clouse, 2016) (Plate 7.8). Frozen basins are formed from water which is conveyed across a slope through channels and check dams to shaded surface depressions near the villages. Cascades and diversions direct water to pass over stone walls, slowing its movement and allowing it to freeze. Ice stupas direct water through pipes into fountains, where it freezes into conical shapes. These locally crafted solutions to the challenges of water variability again exemplify the creative nature of mountain societies that have to be supported. Lower down in the Himalayas in Nepal, women's groups are helping to maintain springs, protect the surrounding catchment forest, ensure water use is restricted to drinking, and prohibit

Plate 7.8 Water trapped in ice and snow can mitigate seasonal water scarcity through relatively slow release into streams. *(Photographs Rob Marchant).*

construction close to springs. The sustainability of many centres of human populations, from Andean cities to isolated communities, depends on water. Quantifying and understanding precipitation processes is key to their sustainability.

7.7.3 Sustainable biodiversity management in mountains

Dealing with the biodiversity crisis is one of the main focuses of ecological restoration. Ecosystem functioning fundamentally depends on biodiversity, with soil microbial diversity supporting processes essential for both human health and ecosystem sustainability. Endemicity is indeed an essential component of biodiversity because endemic species are at a higher risk of extinction globally, given their small and or restricted ranges. One key element around biodiversity is that there are links between ecosystem diversity and society. Several international agreements and goals, including the UN Declaration on the Rights of Indigenous Peoples and the Sustainable Development Goals (SDGs), highlight the importance of Indigenous peoples' rights and emphasise that their participation is key to sustaining and improving biodiversity. Granting forest-dwelling Indigenous communities basic rights is the first step towards sustainable natural resource management and can strengthen the conservation regime of the forests. Forest Rights holders must underpin the sustainable use, conservation of biodiversity, and maintenance of ecological balance. For example, the Kurichya Indigenous population, along with the Kannavam Reserve Forest, Western Ghats, India, have been crucial in guiding sustainable forest management, conserving forest biodiversity, enhancing ecosystem resilience, and promoting sustainable forest management under a changing climate.

Active revegetation is the most used NbS in Europe and is increasingly applied to mountains in other parts of the world, sometimes without due care and attention. There is a need to focus on the effectiveness of biodiversity conservation and restoration and how the results deliver both societal and biodiversity benefits, thus leading to evidence-based conservation for mountains. For example, it often remains unclear whether NbS focuses on improving carbon capture or protection against hazards, and whether these approaches are still effective in biodiversity conservation. In areas where there is a clear mandate, there is often a need to help the restoration process, for example, with the implementation of tree guards to reduce the impact of grazers on restoring forest cover (Plate 7.9).

7.8 Community empowerment and participation in tourism

Given the often remote and rugged location of mountain communities, outdoor recreation and tourism can provide an opportunity to aid economic diversification, generating employment and income for residents while simultaneously supporting local cross-sector industries, which collectively help combat local population losses (Steiger et al., 2024). Mountain tourism has the advantage of attracting diverse year-round visitors through a range of activities, including mountaineering, hiking, and biking, and snow sports such as skiing, snowboarding, snowshoeing, and sledging. There has been a massive growth in mountain tourism in the past decades. However, Mountain tourism destinations are increasingly vulnerable to the impacts

Plate 7.9 Tree guards and infrastructure may be needed to reduce the impact of grazers on restoring forest cover, such as on Mount Teide (*left*) and on the Tramuntana mountains, Mallorca. *(All Photographs Rob Marchant).*

of climate change, including rising temperatures, reduced snowfall, more frequent extreme weather events, and environmental degradation. These changes threaten not only the natural landscapes that attract visitors but also the socio-economic sustainability of mountain communities that depend on tourism. Promoting sustainable mountain tourism often involves supporting traditional livelihoods, while also introducing new opportunities for income generation. This can include eco-friendly tourism, organic farming, homestays, and the production of high-value niche products such as artisanal crafts and food that can make tourism more diverse. This diversification also makes it more resilient to change, as the destination has a greater ability to absorb shocks without losing functionality, leveraging its adaptive capacity to reorganise, recover, and enhance long-term sustainability.

Building resilience in mountain tourism requires a multifaceted and integrated approach (Fig. 7.5), addressing both environmental and socio-economic dimensions, and involving a wide range of stakeholders and different approaches. For example, for the ski industry, this can include investment in snow cannons, snow-making systems (Plate 6.6), and hybrid or indoor ski facilities to extend the winter sports season. However, such a hard-engineered approach can be both expensive and energy-demanding, and a short-term solution to the challenge of declining snow levels. Reducing dependence on snow-based activities by expanding cultural wellness,

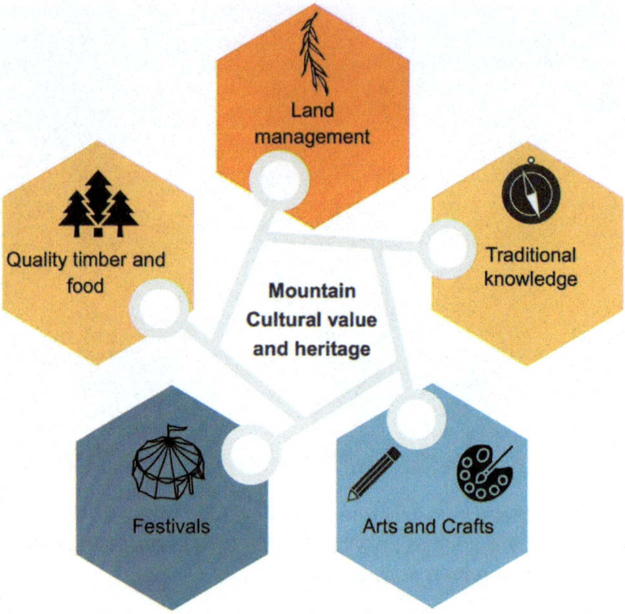

Figure 7.5 Building resilience in mountain communities and supporting heritage tourism requires a multifaceted and integrated approach that links environmental and socio-economic dimensions, and mountain natural resources with communities.

walking, cycling, agritourism, spa retreats, conventions, and meetings, and nature-based experiences can lead to a much more resilient tourist economy that benefits a wider segment of society. By aligning resilience strategies with climate adaptation, mountain destinations can safeguard their future whilst offering sustainable tourism experiences in a changing climate. To be most effective, this needs to combine top-down and bottom-up approaches that meet in the mountain social–ecological system. Bottom-up adaptiveness – community-based measures and knowledge-sharing initiatives empower residents and strengthen local resilience. Top-down adaptiveness – raising awareness among policymakers, investors, and senior stakeholders to climate change impacts on mountain tourism, while promoting the development of strategic plans and policies for effective adaptation – has to connect with bottom-up implementation and knowledge. Strengthening collaboration among policymakers, tourism stakeholders, and local communities ensures adaptive strategies are practical, widely supported, and bridge gaps between policy and local needs.

Plate 7.10 The Quiraing in the north of Skye has attracted so many visitors in the past decades that they are leading to footpath erosion, blocking of infrastructure, and an overload of visitor numbers to the iconic mountains, bringing unintended conflict. *(Photographs Rob Marchant).*

Mountain tourism facilities and activities rely on alpine climate, topography, scenery, and seasonal cycles, where climate change is having an impact. The growth of tourism in mountain regions has not been matched by the expansion of infrastructure such as roads, parking, and hotels with associated impacts on the quality of life for residents in tourism-dependent mountain communities (Steiger et al., 2024). For example, there has been a massive increase in the number of visitors to the mountains of Skye, with an associated impact on footpath erosion, blocking of infrastructure, and an overload of camper vans and visitor numbers to the iconic mountains such as the Quiraing (Plate 7.10).

These interrelated changes emphasise the linked climate and biodiversity crises that directly impact summer nature-based activities in mountain destinations. Climate change is affecting alpine species' ranges, habitats, and behaviour, for example, invasive species, resulting in increasingly unfamiliar ecological contexts within which tourism operates (Knowles & Scott, 2021). Considering these rare, raw, and remote environments are what draw tourists to the mountains to begin with, the tourism industry and all its stakeholders have a vested interest in being agents for positive change. As ecosystems naturally adapt to the climate, tourism systems will require new knowledge and strategies to adapt alongside nature, dealing with changing water

Plate 7.11 Cultural festivals are a vital part of mountain communities, such as the Rollate carnival in Sapadda, Italy. *(Photographs Rob Marchant).*

availability, seasonality of species, weather, and hazards. Again, considering the formidable stature of mountains is a key tourist attraction, a response that aligns with the natural setting and not only mitigates negative impact to natural resources, views and vistas, or nature-based tourism assets but uses these as nature-based climate adaptations, may see an advantage in drawing new tourists to destinations (Steiger et al., 2024). It seems clear that continued growth of tourism does not align with a decarbonised economy, considering all the travel and high energy demand inherent in tourism – from snow-making infrastructure to transport of food into mountain resorts. Therefore, in looking forward, we must look for ways that ensure mountain tourism gets better but not bigger, and continues to serve the needs of the local communities and ecosystems upon which it is built (Steiger et al., 2024). Sustainable tourism initiatives that engage local communities can promote cultural exchange, economic opportunities, and cultural preservation in mountain regions. Community homestays, cultural festivals (Plate 7.11), and ecotourism ventures empower communities to showcase their heritage and generate income. Mountain communities engage in diverse livelihood activities, ranging from subsistence agriculture and pastoralism to tourism and handicrafts. However, these livelihoods are often vulnerable to environmental degradation, climate change, and fluctuating market conditions. Adaptation to cryosphere change for ski tourism focuses on snowmaking and is expected to be moderately effective for many locations in the near term (Hock et al., 2019).

7.8.1 Cultural tourism

Tourism plays a significant role in many mountain economies, offering opportunities for income generation and employment. However, unchecked tourism can also pose threats to the environment and local cultures, necessitating sustainable tourism practices and community-based ecotourism initiatives. Investing in cultural revitalisation programs, language preservation efforts, and indigenous education initiatives can strengthen cultural identity and pride among mountain communities. Intergenerational knowledge exchange, storytelling, and traditional arts and crafts revival contribute to cultural resilience. To address the challenges faced by mountain people, various strategies are being implemented to preserve their cultural heritage, promote sustainable development, and empower these communities. Efforts to document and preserve the languages, customs, and traditional knowledge of mountain communities are essential for maintaining cultural diversity. This includes supporting indigenous education, promoting cultural tourism, and recognising the value of traditional practices. In Ladakh, India, community-based tourism initiatives have provided an alternative source of income for local people, helping to reduce poverty and incentivise the preservation of natural and cultural resources. These initiatives offer visitors the opportunity to experience the region's natural beauty and learn about the traditional practices of Ladakhi communities, creating a sustainable model for economic development. For example, the government of Bhutan has implemented policies to preserve its cultural heritage by promoting Dzongkha (the national language), traditional arts, and architecture. Bhutan's focus on Gross National Happiness also emphasises the importance of cultural preservation alongside economic development. In the Ethiopian Highlands, government and non-governmental organisations have worked to preserve traditional knowledge through the promotion of indigenous education and the documentation of oral histories. Cultural tourism initiatives, such as community-based eco-tourism, also provide an opportunity for visitors to learn about and support the preservation of traditional practices. In Peru, organisations such as the Quechua-Aymara Association for Sustainable Livelihoods (ANDES) work to preserve traditional knowledge and promote biocultural heritage. This includes the conservation of native crops, the promotion of traditional farming practices, and the documentation of oral histories. In the Rocky Mountains, community-based ecotourism initiatives have provided an alternative source of income for local people while promoting the conservation of natural and cultural resources. These initiatives offer visitors the opportunity to experience the region's natural beauty and learn about the traditional practices of indigenous communities, creating a sustainable model for economic development. The Ladin Cultural Institute in the Dolomites works to preserve and promote Ladin culture through language courses, cultural events, and the documentation of oral histories. This initiative is crucial for passing on Ladin cultural practices to future generations and ensuring the survival of their unique heritage.

7.9 Tools to 'get ahead' of the challenges

In recent decades, there has been an increasing reliance on Earth Observation data to supplement gauging and monitoring data. While the increased range of sensors and resolution of remotely sensed products is really quite exciting, there are a number of ongoing caveats that make the value of ground observation vital, and indeed more valuable. One of the key characteristics and challenges for mountain areas comes from precipitation, seasonal changes to this, and the extremely high local variability (Cuni-Sanchez et al., 2025). One useful remotely sensed data set is the Climate Hazards Group Infrared Precipitation with Stations (CHIRPS) dataset, a rainfall product that can be useful for analysing broad-scale seasonal and interannual variability and spatial precipitation patterns, and can provide good spatio-temporal coverage, which is particularly suitable for areas with few rainfall gauges. However, its accuracy is minimal for fine-scale daily analysis (López-Bermejo et al., 2022), or indeed the fine spatial scale that is ecologically important or relevant at the household level.

Similarly, there is great uncertainty on snow measurement; like many phenomena, it is incredibly variable in space and time. Satellites such as Sentinel have the potential to provide measurements, but deep snow, wet snow, or snow under forests curtail sensor ability. LiDAR can be used to map snow at high resolution, but again, deep snow, snow under forest and on steep slopes can be challenging to sense. There are advances in sensor and machine learning that can help with image processing and correcting, but these need good quality, repeatable data to calibrate and improve, particularly important due to changing signals. One of the keys is remotely sensed data sets, particularly high-quality digital elevation models. The two TanDEM-X satellites operated by the German Aerospace Centre have been acquiring X-band radar data for the generation of digital elevation models since late 2010. This wealth of data can be used to track annual elevation change rates, such as glacier thickness. In the first decade of the 21st century, Fedchenko Glacier in the Pamir Mountains experienced a typical elevation decrease in the lower parts and an elevation increase in the upper accumulation zone. In contrast, the glacier has thinned over its whole extent since the first years of TanDEM-X acquisitions. Understanding the interacting effects of climate drivers that constrain different species and life stages, and how those drivers will vary over both space and time, is critical to predicting species distributions and responses to climate change in climatically diverse landscapes.

Current challenges in mountain social–ecological systems continue to be dealt with in a reactionary way – be it food insecurity, water rationing during recurrent droughts, or fences and culling to mitigate human-wildlife conflict. The importance of mountains and mountain ecosystems for societies is often only fully realised when there are disasters, and we only react to change once change has happened and the intervention space has passed. This is not a new phenomenon – there is something in human psychology that people are very good at not dealing with a problem until they have to! For example, the landslide and destruction of the Swiss mountain village of Blatten in 2025 are now resulting in policies and planning for the future. Similarly, in Switzerland, the massive flood event in 1868 that impacted large areas went on to enact changes in Swiss forest law, leading to replanting large areas of forest and halting loss and conversion to agriculture and pasture. Similarly, in 2023, Cyclone Gabriel led to severe flooding that resulted in regulations on woody debris. Despite these multiple examples of action after the event, it is possible to get ahead of the challenges. Even with the often seemingly insurmountable challenges, given the right context and support, mountain social–ecological systems can identify pathways toward sustainable human development under multiple and interacting drivers of change arising from environmental (e.g., climate and vegetation change), demographic (e.g., population growth and migration), socio-economic (e.g. market integration and foreign investment flows), and cultural (e.g., loss of traditional ecological knowledge) aspects. For instance, the Middle Drâa valley in the Atlas Mountains of Morocco is one of many semi-arid to arid mountainous areas that have been struggling with increasing water scarcity for the last 50 years, threatening self-sufficient animal husbandry and

Plate 7.12 The area surrounding Kilimanjaro has seen a massive expansion of groundwater-fed intensive agriculture. *(Photograph Rob Marchant).*

consequent food security (Nsengiyumva, 2019). The area has a long history and an indigenous population which has been able to adapt and optimise scarce natural resources, but in the last decade, the Zagora region has been amongst the regions worst affected by desertification – with further impacts on palm groves or oases, salinity, road infrastructure, irrigation networks, and groundwater upstream and downstream. Similarly, the area surrounding Kilimanjaro has seen a massive expansion of more intensive agriculture that is groundwater-fed (Plate 7.12). The transformation of previously open grazing land into more intensive agricultural use, coupled with increased water abstraction and the application of fungicides and pesticides, is clearly unsustainable. As water resources become harder to access, short-term gains will give way to long-term consequences. An integrated approach to water resource management is essential to prevent sedimentation and mitigate environmental and economic risks—for example, through participatory irrigation management, crop diversification, improved coordination between farmers and planners, and the adoption of integrated water management practices.

To address these challenges and confront uncertainty in areas where the social-economic opportunities mountain ecosystems provide, we need to synergise new fertile niches between and across, different disciplines, between engineers, anthropologists, ecologists, biologists, zoologists, political scientists, social scientists, and others to reduce risk and support adaptation (Fig. 7.6). However, this is much more than another call for interdisciplinary science: this requires a synergy between those already working within interdisciplinary space to apply novel scientific approaches across time. For this integrated approach to happen, we must foster non-typical knowledge

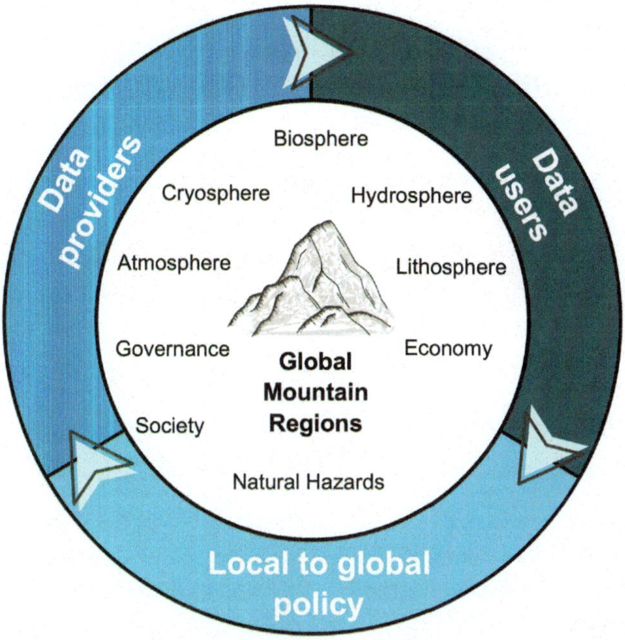

Figure 7.6 An integrated perspective that links data generators and providers to their user community and policy can allow us to navigate the mountain challenges.

exchanges between diverse fields and communities in a process of collaborative learning, facilitated by innovative exchanges of researchers between institutions – through knowledge exchanges, field visits, researchers working together on international collaborative research programmes, and building capacities in both the Global North and South in terms of transdisciplinary practice. This will add value and lead to breakthroughs from a social–ecological perspective that can be applied through time to explore alternative future scenarios. By combining landscape history and regionally adapted dynamic models that reflect ecological reality, with stakeholders' values that acknowledge socio-economic and cultural complexity, it is possible to produce useful scenarios of mountain land use futures. Such integrative approaches allow researchers, policy makers, and implementers to get ahead of the challenges rather than continually reacting from one crisis to the next.

An integrated perspective linking data generators and providers to their user community and policy that is in a continuing interactive loop, which allows them to strengthen and navigate through the challenges in global mountain regions (Fig. 7.7). A joined-up approach will allow us to deal with contemporary challenges such as climate and ecosystem change,

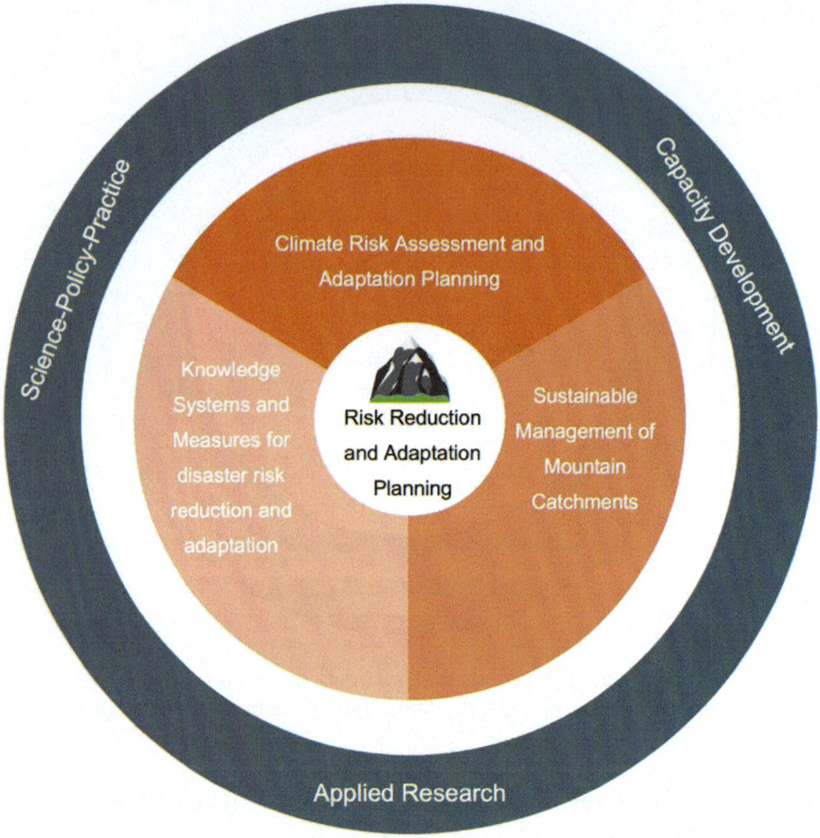

Figure 7.7 Combining research, linking to policy, and supporting capacity is possible and can be used to reduce risk and support adaptation.

conservation strategies, biocultural heritage protection, and the achievement of nature-based solutions for inclusive development in mountain social–ecological systems. An integrative approach is critical to modelling pathways for sustainability and will provide a better perspective for the implementation of large-scale targeted initiatives such as Africa100, the UN's decade of restoration, and the near-term 2030 Sustainable Development Goal agenda, and to match the far-term (2063) African Union Development Goals.

Appreciation of land use futures and getting ahead of projected climate changes, be they societal, environmental, or economic, can be achieved through the 'lens of land'. Land is where we have choice and agency and can make a difference. Participatory scenario planning has been a key tool in recent decades to understand the multifaceted nature of land use change and the drivers of this, be it societal, economic, or environmental (Capitani et al., 2019). Participatory scenario planning engages participants from diverse sectors and levels and involves them across a process to collaboratively co-create credible outcomes that support long-term policy and decision-making. Participatory scenario planning can promote an inclusive management of multifaceted challenges, foster a discourse for integrating science-based knowledge with indigenous wisdoms, link science to policy, balance top-down and bottom-up policies, and identify risks and opportunities (Butler et al., 2020; Capitani et al., 2016; Kariuki et al., 2018, 2025). While participatory scenario planning has received increasing attention across diverse disciplines, including; natural capital assessments and ecosystem service management (Saito et al., 2019), protected area management (Palomo, 2017; Palomo et al., 2017), climate-resilient agriculture and food security (Olabisi et al., 2020), land use and land cover changes (Capitani et al., 2016), risks and adaptation (Olabisi et al., 2020), its progress and effectiveness in fostering sustainable land management and climate resilience require further investigation due to some critical challenges. These challenges include: resource-intensive processes, power dynamics and stakeholder politics, meaningful participant engagement, integrating diverse knowledge and interests, transforming scenario outcomes into actionable policies and strategies, and systematic monitoring and evaluation (Oteros-Rozas et al., 2015). Participatory scenario planning is a key tool for supporting sustainable land management and climate resilience across diverse social–ecological systems worldwide. Participatory scenario planning fosters stakeholder engagement and acts as a social learning process, building trust and co-producing knowledge that navigates inclusive, equitable future scenarios and actionable pathways to tackle complex, emerging social–ecological challenges. Embedding Participatory scenario planning within national planning and policy frameworks, alongside targeted capacity-building, can enhance multi-scale coordination, translate co-developed pathways into tangible policies, and maximise practical impact, thereby enabling transformative, sustainable, and resilient futures.

7.10 Policy implications and recommendations

Mountain policies should aim to be comprehensive and support joined-up action (Fig. 7.8). Such policies should be flexible enough to include

Figure 7.8 Climate change adaptation has to be supported in multiple dimensions, from policies that support joined-up action through to sustainable forest and soil management.

broad actions and strategies with locally tailored solutions for regions and individual countries. Mismatches in communication and failed initiatives often originate from a narrow scope and a lack of vital information. Combining social sciences with more traditional areas of forest science could help avoid issues that perpetuate problems (Duckett et al., 2016). Recognising and integrating novel factors such as social territories and scales, based upon core social–ecological systems, could identify key intervention points that help enhance our ability to manage forests and avoid future spatial, temporal, and political mismatches. This underlines the need to develop resilience with local community sustainable development goals, as well as balance long-term policy goals with short-term needs of the locality (Bull et al., 2018). An approach that operates through co-design and production may have a greater chance of creating a climate-smart framework with bespoke set-ups for different regions, ownership, and forest types (Bowditch et al., 2020). Improved communication and use of expansive participatory methods to engage and interact with forest managers and wider stakeholders of forest communities will be central to developing, testing, and refining further steps that aim to produce a valuable resource for forest practitioners to enhance the management of their forests in response to future uncertainty and growing demands on these forests. Focusing on such concepts as bioeconomy is likely to be less effective as these are dominated by economic, resource use, and commodity concerns, which have the tendency to neglect social and ecological considerations (Bowditch et al., 2020).

Integrated Development Policies should prioritise and address the multidimensional challenges faced by mountain communities, including poverty alleviation, environmental conservation, and cultural preservation. Strengthening local governance structures and institutions is essential for empowering mountain communities to participate in decision-making processes and implement sustainable development initiatives. Facilitating knowledge exchange and capacity building initiatives between mountain communities, researchers, and policymakers can foster mutual learning and enhance the resilience of mountain ecosystems. Setting appropriate restoration goals requires detailed information on current and former ecosystem extent, and on the climate resilience, composition variability, and heterogeneity of mountain social–ecological systems; without such information, we cannot assess the true impact of afforestation on biodiversity or its effect on other ecosystem services. To provide this, we must understand the entangled environmental and human histories of mountains better, to define the varied social–ecological dynamics and legacies of this biocultural heritage, and to combine these data with historical and observational studies of the political ecology of environmental change. More longitudinal studies are needed, as well as research comparing different communities not only in the same mountain range and communities from different mountain areas worldwide, but also communities from highlands and lowlands (Schneiderbauer et al., 2021). In-depth participatory work with local stakeholders will both inform and be informed by these syntheses, to refine and develop solutions for challenging issues across highly dynamic, but inherently resilient, landscapes that have evolved, are evolving, and will continue to do so under increasing multiple pressures of a changing and variable climate, rapid development, and increasing demands of a growing population.

7.11 Mountain living labs—Joining people, science, and monitoring

Mountain areas are important geographical units in terms of their biological diversity, natural resources, and cultural values. Mountains can form unique open-air laboratories to study biogeography, evolution, and climate change impacts. Long-term ecological research programs provide unique opportunities to study, on a global scale, the community dynamics of biodiverse ecosystems, including montane ecosystems and how they are responding to global changes. The establishment of permanent plots dedicated to the provision, documentation, and monitoring of long-term data, with a time horizon of decades to centuries, is the most reliable way to track how global-change drivers are affecting natural ecosystems (Silveira et al., 2019). Long-term biomonitoring in mountains is producing relevant data for predicting the effects of global change on species range shifts, local extinctions, alterations in species richness and composition, changing connections, and the rewiring of species interactions. Raising environmental awareness among local communities and involving them in the process through nature education can underpin sustainable development processes. Taking an approach that encompasses experiential insights from long-term living labs can be powerful in supporting sustainable mountain social–ecological futures. Such activities can build long-term, in-depth, and nuanced collaboration to imagine new ideas for what mountains and mountain futures could look like. This knowledge generation on how the system functions must combine local and scientific knowledge and link with issues of sustainable land management, while being supported by, and working within, the sustainable governance structures (Fig. 7.9). Choices need to be made on what could be implemented to create the desired type of landscape, or at least to avoid the landscapes that are not. There is a need to harmonise different perspectives from hydrologists, energy companies, farmers, tour operators, and dwellers. The Southern Andes Natural Laboratory is an articulation node based in Chile and funded by the National Agency for Research and Development (ANID). It brings together researchers, decision-makers, and members of organised civil society to contribute to the development of public policies for the management of these territories, while also promoting scientific knowledge and the training of specialised human capital. They aim to investigate, produce, integrate, and share scientific and technical knowledge in an adaptive, coordinated, and transdisciplinary manner to enhance the present and future well-being of communities in the Southern Andes of Chile.

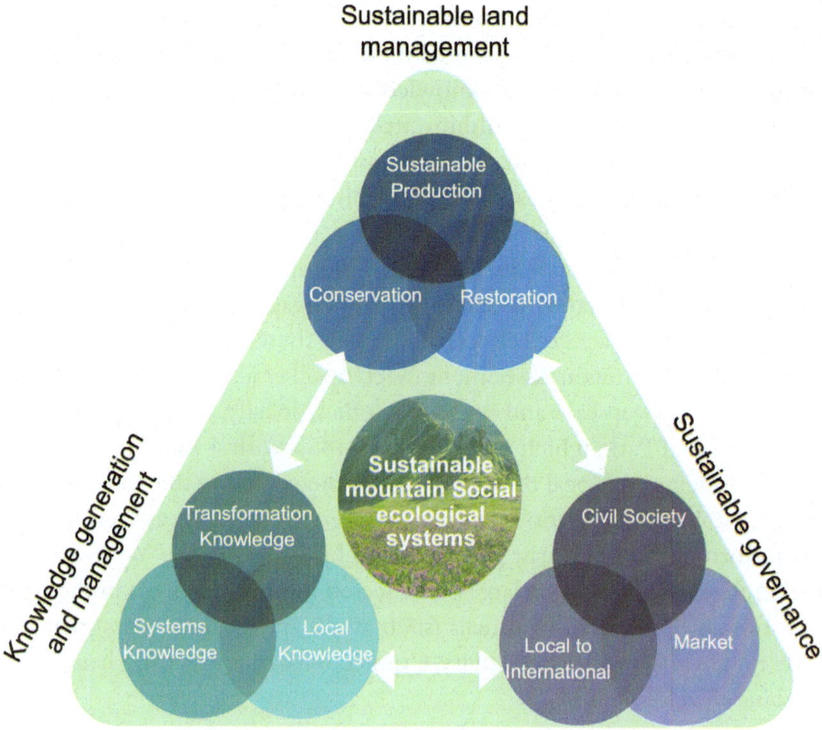

Figure 7.9 Understanding mountain social–ecological systems must be driven by generation on how the system functions and must combine local and scientific knowledge and link with issues of sustainable land management.

Transformations toward desirable futures in mountain regions remain challenged by the complexity of the social–ecological systems. Central to this complexity is recognising the diversity of actors' perspectives on what constitutes desirable futures and how to achieve them. For example, the Andes is home to 130 million people with a number of key cities such as Quito, Bogota/ ANDEX is aiming to improve hydroclimatic processes and the need to integrate knowledge. The local knowledge is often strong, but really needs to be connected globally and regionally in nested and impactful ways. Key themes are around water security and managing high-impact events – possible with multi-disciplinary teams. This is achievable with data sharing platforms such as Observatorio Andino on snow cover and the continued development of young scientists. These regional initiatives are linked to global programs such as WCRP and GEWEX (Global Energy and Water Exchanges Program).

Many people live in mountains – some 80% of the Canadian population lives within or close to (100 km) of mountains. Indigenous people possess distinct, valid, and insightful knowledge. One way of bringing people together is through a 'stewardship circle' that can link different kinds of knowledge. There are five main areas of issues around service to mountains and mountain communities: (1) reaping culturally protected knowledge, (2) creating space for multiple ways of knowing, (3) knowledge sharing within ways of knowing, (4) respectful braiding of multiple ways of knowing, and (5) lastly, but very importantly, is that long term engagement is crucial. A recent example of good practice that connects across these multiple spheres is from the Canadian Mountain Assessment (McDowell et al., 2023). This approach has been able to upscale and manage conflicts locally, such as those within Banff National Park, which was created in 1885 with 4 million visitors and 11,000 residents and local development conflicts, such as the increasing cost of living for local populations and the lack of opportunities, as their territory is transformed into a tourist playground. This is part of the Yellowstone to Yukon region, which is the largest connected mountain system in the world and still has many large mammals (such as elk and Grizzly bears) that have wide ranges and need connectivity. Connectivity that is getting harder with ongoing development.

7.12 Conclusions

The prominent role of human activity in shaping mountains, with impact at planetary scales, is the hallmark of the Anthropocene and sets the context for emerging strategies and actions to navigate the challenges faced by mountains and societies that live within and rely on them. The enormity of the challenges demands that scientists, stakeholders, and decision-makers confront the diverse and often synergistic changes to mountain systems that are occurring worldwide to create new, bolder conservation initiatives. Such an approach places an increasingly important role for people at the forefront of efforts to understand and adapt to ecosystem changes. By linking people and local land uses with ecosystems, there is a greater likelihood of finding effective, place-based solutions to suit species and ecosystems (Kelly et al., 2020).

Mountain people are stewards of biodiversity. Indigenous and local populations in mountains have unique and valuable local knowledge, traditions, and cultural practices that must inform effective land management strategies and sustainable food systems. Despite the resilience and adaptability of these communities, they face significant challenges from climate change, economic marginalisation, environmental degradation, and cultural erosion. Efforts to preserve their heritage, promote sustainable development, and empower these communities are essential for ensuring their survival and well-being. By recognising the value of mountain cultures and supporting their continued vitality, we can help safeguard the rich diversity of human life on our planet. Mountain communities, characterised by their geographic isolation, rugged terrain, and harsh climatic conditions, have long been at the forefront of environmental and socio-economic challenges. Despite these challenges, these communities have developed intricate systems of resilience and adaptation, rooted in centuries-old traditions and local knowledge. Understanding the dynamics of mountain communities is crucial for devising effective policies and interventions that promote their sustainable development while preserving their unique cultural heritage and ecological integrity. Mountain communities often comprise multiple ethnic groups, each with its own language, customs, and traditions. Mountain communities have developed rich repositories of traditional knowledge and practices, honed over generations to navigate their challenging environments, and they often exhibit vibrant cultural diversity, shaped by their geographical isolation and historical interactions. These cultures are closely intertwined with the natural environment, influencing traditional livelihoods, rituals, and belief systems, sustainable agricultural techniques, water management systems, and indigenous medicinal knowledge, which are going to be tested in a future that is unprecedented in terms of the pace of change and the ability of the ecosystem to adapt through migration. There is a positive future for mountains that must place communities at the heart of creating an enabling environment for a transformative, sustainable, and resilient future.

References

Alexander, J. M., Chalmandrier, L., Lenoir, J., Burgess, T. I., Essl, F., Haider, S., Kueffer, C., McDougall, K., Milbau, A., Nuñez, M. A., Pauchard, A., Rabitsch, W., Rew, L. J., Sanders, N. J., & Pellissier, L. (2018). Lags in the response of mountain plant communities to climate change. *Global Change Biology, 24*(2), 563–579. https://doi.org/10.1111/gcb.13976.

Anderson, M. G., Clark, M., Olivero, A. P., Barnett, A. R., Hall, K. R., Cornett, M. W., Ahlering, M., Schindel, M., Unnasch, B., Schloss, C., & Cameron, D. R. (2023). A resilient and connected network of sites to sustain biodiversity under a changing climate. *Proceedings of the National Academy of Sciences of the United States of America, 120*(7). https://doi.org/10.1073/pnas.2204434119.

Banerjee, A., Kang, S., Guo, W., Meadows, M. E., Wang, W., Sengupta, D., & Zhang, T. (2024). Glacier retreat and lake outburst floods in the central Himalayan region from 2000 to 2022. *Natural Hazards, 120*(6), 5485–5508. https://doi.org/10.1007/s11069-024-06415-5.

Bani, L., Luppi, M., Rocchia, E., Dondina, O., & Orioli, V. (2019). Winners and losers: How the elevational range of breeding birds on the Alps has varied over the past four decades due to climate and habitat changes. *Ecology and Evolution, 9*(3), 1289–1305. https://doi.org/10.1002/ece3.4838.

Barras, A. G., Braunisch, V., & Arlettaz, R. (2021). Predictive models of distribution and abundance of a threatened mountain species show that impacts of climate change overrule those of land use change. *Diversity and Distributions, 27*(6), 989–1004. https://doi.org/10.1111/ddi.13247.

Bowditch, E., Santopuoli, G., Binder, F., del Río, M., Porta, N. L., Kluvankova, T., Lesinski, J., Motta, R., Pach, M., Panzacchi, P., Pretzsch, H., Temperli, C., Tonon, G., Smith, M., Velikova, V., Weatherall, A., & Tognetti, R. (2020). What is climate-smart forestry? A definition from a multinational collaborative process focused on the mountain regions of Europe. *Ecosystem Services, 43*, 101113. https://doi.org/10.1016/j.ecoser.2020.101113.

Bull, G. Q., Boedhihartono, A. K., Bueno, G., Cashore, B., Elliott, C., Langston, J. D., Riggs, R. A., & Sayer, J. (2018). Global forest discourses must connect with local forest realities. *International Forestry Review, 20*(2), 160–166. https://doi.org/10.1505/146554818823767573.

Butler, J. R. A., Bergseng, A. M., Bohensky, E., Pedde, S., Aitkenhead, M., & Hamden, R. (2020). Adapting scenarios for climate adaptation: Practitioners' perspectives on a popular planning method. *Environmental Science and Policy, 104*, 13–19. https://doi.org/10.1016/j.envsci.2019.10.014.

Capitani, C., Garedew, W., Mitiku, A., Berecha, G., Hailu, B. T., Heiskanen, J., Hurskainen, P., Platts, P. J., Siljander, M., Pinard, F., Johansson, T., & Marchant, R. (2019). Views from two mountains: Exploring climate change impacts on traditional farming communities of Eastern Africa highlands through participatory scenarios. *Sustainability Science, 14*(1), 191–203. https://doi.org/10.1007/s11625-018-0622-x.

Capitani, C., Mukama, K., Mbilinyi, B., Malugu, I. O., Munishi, P. K. T., Burgess, N. D., Platts, P. J., Sallu, S. M., & Marchant, R. (2016). From local scenarios to national maps: A participatory framework for envisioning the future of Tanzania. *Ecology and Society, 21*(3). https://doi.org/10.5751/ES-08565-210304.

Chamberlain, D. E., Negro, M., Caprio, E., & Rolando, A. (2013). Assessing the sensitivity of alpine birds to potential future changes in habitat and climate to inform management strategies. *Biological Conservation, 167*, 127–135. https://doi.org/10.1016/j.biocon.2013.07.036.

Chauvier, Y., Descombes, P., Guéguen, M., Boulangeat, L., Thuiller, W., & Zimmermann, N. E. (2022). Resolution in species distribution models shapes spatial patterns of plant multifaceted diversity. *Ecography, 2022*(10). https://doi.org/10.1111/ecog.05973.

Chazdon, R., & Brancalion, P. (2019). Restoring forests as a means to many ends. *Science, 365*(6448), 24–25. https://doi.org/10.1126/science.aax9539.

Ciocco, T. W., Miller, B. W., Tangen, S., Crausbay, S. D., Oldfather, M. F., & Bamzai-Dodson, A. (2024). Indigenous knowledge in climate adaptation planning: Reflections from initial efforts. *Frontiers in Climate, 6*. https://doi.org/10.3389/fclim.2024.1393354.

Clouse, C. (2016). Frozen landscapes: Climate-adaptive design interventions in Ladakh and Zanskar. *Landscape Research, 41*(8), 821–837. https://doi.org/10.1080/01426397.2016.1172559.

Cuni-Sanchez, A., Aneseyee, A. B., Baderha, G. K. R., Batumike, R., Bitariho, R., Imani, G., Jha, N., Kaganzi, K. R., Kaplin, B. A., Klein, J. A., Leite, A., Marchant, R. A., Martin, E. H., Mcharazo, F., Mwangi, B., Ngute, A. S. K., Nkengurutse, J., Nkurunziza, A., Olaka, L., & …, Zafra-Calvo, N. (2025). Perceived climate change impacts and adaptation responses in ten African mountain regions. *Nature Climate Change, 15*(2), 153–161. https://doi.org/10.1038/s41558-024-02221-w.

Cuni-Sanchez, A., Omeny, P., Pfeifer, M., Olaka, L., Mamo, M. B., Marchant, R., & Burgess, N. D. (2019). Climate change and pastoralists: Perceptions and adaptation

in montane Kenya. *Climate and Development, 11*(6), 513–524. https://doi.org/10.1080/17565529.2018.1454880.
Cuni-Sanchez, A., Sullivan, M. J. P., Platts, P. J., Lewis, S. L., Marchant, R., Imani, G., Hubau, W., Abiem, I., Adhikari, H., Albrecht, T., Altman, J., Amani, C., Aneseyee, A. B., Avitabile, V., Banin, L., Batumike, R., Bauters, M., Beeckman, H., Begne, S. K., & …, Zibera, E. (2021). High aboveground carbon stock of African tropical montane forests. *Nature, 596*(7873), 536–542. https://doi.org/10.1038/s41586-021-03728-4.
Denton, F., Wilbanks, T. J., Abeysinghe, A. C., Burton, I., Gao, Q., Lemos, M. C., & …, Wright, S. B. (2015). Climate-resilient pathways: Adaptation, mitigation, and sustainable development. In *Climate change 2014 impacts, adaptation and vulnerability: Part A: Global and sectoral aspects* (pp. 1101–1131). Cambridge University Press. https://doi.org/10.1017/CBO9781107415379.025.
Dirnböck, T., Essl, F., & Rabitsch, W. (2011). Disproportional risk for habitat loss of high-altitude endemic species under climate change. *Global Change Biology, 17*(2), 990–996. https://doi.org/10.1111/j.1365-2486.2010.02266.x.
Du, W. B., Jia, P., & Du, G. Z. (2022). Current patterns of plant diversity and phylogenetic structure on the Kunlun Mountains. *Plant Diversity, 44*(1), 30–38. https://doi.org/10.1016/j.pld.2021.04.007.
Duckett, D., Feliciano, D., Martin-Ortega, J., & Munoz-Rojas, J. (2016). Tackling wicked environmental problems: The discourse and its influence on praxis in Scotland. *Landscape and Urban Planning, 154*, 44–56. https://doi.org/10.1016/j.landurbplan.2016.03.015.
Elsen, P. R., Monahan, W. B., & Merenlender, A. M. (2020). Topography and human pressure in mountain ranges alter expected species responses to climate change. *Nature Communications, 11*(1). https://doi.org/10.1038/s41467-020-15881-x.
Food and Agriculture Organisation of the United Nations (FAO). (2010). *Climate Smart Agriculture: Policies, Practices and Financing for Climate Smart Agriculture*: 43. Rome, Italy: FAO.
Gaitán-Espitia, J. D., & Hobday, A. J. (2021). Evolutionary principles and genetic considerations for guiding conservation interventions under climate change. *Global Change Biology, 27*(3), 475–488. https://doi.org/10.1111/gcb.15359.
Gillson, L., Seddon, A., Mottl, O., Zhang, K., Kirsten, K., Gell, P., Marchant, R. A., Schwörer, C., Razanatsoa, E., Lane, P. J., Courtney-Mustaphi, C. J., & Dearing, J. (2025). Exploring the interface between planetary boundaries and palaeoecology. *Global Change Biology, 31*(1). https://doi.org/10.1111/gcb.70017.
Grassi, G., Pilli, R., House, J., Federici, S., & Kurz, W. A. (2018). Science-based approach for credible accounting of mitigation in managed forests. *Carbon Balance and Management, 13*(1). https://doi.org/10.1186/s13021-018-0096-2.
Gray, M. (2013). *Geodiversity: Valuing and conserving abiotic nature*. Wiley-Blackwell.
Gray, M. (2024). Case studies associated with the 10 major geodiversity-related topics. *Philosophical Transactions of the Royal Society A: Mathematical, Physical and Engineering Sciences, 382*(2269). https://doi.org/10.1098/rsta.2023.0055.
Gutiérrez-Girón, A., Díaz-Pinés, E., Rubio, A., & Gavilán, R. G. (2015). Both altitude and vegetation affect temperature sensitivity of soil organic matter decomposition in Mediterranean high mountain soils. *Geoderma, 237-238*, 1–8. https://doi.org/10.1016/j.geoderma.2014.08.005.
Hagedorn, F., Gavazov, K., & Alexander, J. M. (2019). Above- And belowground linkages shape responses of mountain vegetation to climate change. *Science, 365*(6458), 1119–1123. https://doi.org/10.1126/science.aax4737.
Harden, C. P. (2001). Soil erosion and sustainable mountain development. *Mountain Research and Development, 21*(1), 77–83. https://doi.org/10.1659/0276-4741(2001)021[0077:seasmd]2.0.co;2.

Hill, A., Minbaeva, C., Wilson, A., & Satylkanov, R. (2017). Hydrologic controls and water vulnerabilities in the Naryn River Basin, Kyrgyzstan: A socio-hydro case study of water stressors in Central Asia. *Water, 9*(5), 325. https://doi.org/10.3390/w9050325.

Hock, R., Rasul, G., Adler, C., Cáceres, B., Gruber, S., Hirabayashi, Y., Jackson, M., Kääb, A., Kang, S., Kutuzov, S., & Milner, A. (2019). High mountain areas. In H.-O. Pörtner, D. C. Roberts, V. Masson-Delmotte, P. Zhai, M. Tignor, E. Poloczanska, K. Mintenbeck, A. Alegría, M. Nicolai, A. Okem, J. Petzold, B. Rama, & N. M. Weyer (Eds.), *IPCC special report on the ocean and cryosphere in a changing climate* (pp. 131–202). Cambridge University Press.

Hooghiemstra, H., & Flantua, S. G. A. (2019). Colombia in the Quaternary: An overview of environmental and climatic change. In J. Gomez-Tapias (Ed.), *The geology of Colombia Book: Vol. Volumen 4 Quaternary* (pp. 33–57). Servicio Geológico Colombiano.

Immerzeel, W. W., Lutz, A. F., Andrade, M., Bahl, A., Biemans, H., Bolch, T., Hyde, S., Brumby, S., Davies, B. J., Elmore, A. C., Emmer, A., Feng, M., Fernández, A., Haritashya, U., Kargel, J. S., Koppes, M., Kraaijenbrink, P. D. A., Kulkarni, A. V., Mayewski, P. A., & …, Baillie, J. E. M. (2020). Importance and vulnerability of the world's water towers. *Nature, 577*(7790), 364–369. https://doi.org/10.1038/s41586-019-1822-y.

IPCC. (2022). Climate change 2022: Impacts, adaptation, and vulnerability. *Contribution of working group II to the sixth assessment report of the Intergovernmental Panel on Climate Change.* Cambridge University Press https://doi.org/10.1017/9781009325844.

Ives, J. (1997). *Mountains of the world: A global priority, 17*(4), 375–376.

Jump, A. S., Carr, M., Ahrends, A., & Marchant, R. (2014). Genetic divergence during long-term isolation in highly diverse populations of tropical trees across the Eastern Arc mountains of Tanzania. *Biotropica, 46*(5), 565–574. https://doi.org/10.1111/btp.12139.

Kariuki, R., Willcock, S., & Marchant, R. (2018). Rangeland livelihood strategies under varying climate regimes: Model insights from Southern Kenya. *Land,* 7(2), 47. https://doi.org/10.3390/land7020047.

Kariuki, R. W., Thorn, J. P. R., Mfune, J. K. E., Marchant, R., & Capitani, C. (2025). Near and far future conservation, land use, and land cover interactions around the wider Etosha landscape, north-central Namibia. *Regional Environmental Change, 25*(2). https://doi.org/10.1007/s10113-025-02388-8.

Kelly, L. T., Giljohann, K. M., Duane, A., Aquilué, N., Archibald, S., Batllori, E., Bennett, A. F., Buckland, S. T., Canelles, Q., Clarke, M. F., Fortin, M. J., Hermoso, V., Herrando, S., Keane, R. E., Lake, F. K., McCarthy, M. A., Morán-Ordóñez, A., Parr, C. L., Pausas, J. G., & …, Brotons, L. (2020). Fire and biodiversity in the Anthropocene. *Science, 370*(6519). https://doi.org/10.1126/science.abb0355.

Knowles, N. L. B., & Scott, D. (2021). Media representations of climate change risk to ski tourism: A barrier to climate action? *Current Issues in Tourism, 24*(2), 149–156. https://doi.org/10.1080/13683500.2020.1722077.

Körner, C. (2007). The use of 'altitude' in ecological research. *Trends in Ecology and Evolution, 22*(11), 569–574. https://doi.org/10.1016/j.tree.2007.09.006.

Lausch, A., Schaepman, M. E., Skidmore, A. K., Truckenbrodt, S. C., Hacker, J. M., Baade, J., Bannehr, L., Borg, E., Bumberger, J., Dietrich, P., Gläßer, C., Haase, D., Heurich, M., Jagdhuber, T., Jany, S., Krönert, R., Möller, M., Mollenhauer, H., Montzka, C., & …, Thiel, C. (2020). Linking the remote sensing of geodiversity and traits relevant to biodiversity—Part II: Geomorphology, terrain and surfaces. *Remote Sensing, 12*(22), 1–61. https://doi.org/10.3390/rs12223690.

Lawrence, A. (2017). Adapting through practice: Silviculture, innovation and forest governance for the age of extreme uncertainty. *Forest Policy and Economics, 79,* 50–60. https://doi.org/10.1016/j.forpol.2016.07.011.

Lexer, M. J., & Bugmann, H. (2017). Mountain forest management in a changing world. *European Journal of Forest Research, 136*(5-6), 981–982. https://doi.org/10.1007/s10342-017-1082-z.

López-Bermeo, C., Montoya, R., Caro-Lopera, F. J., & Díaz-García, J. (2022). Validation of the accuracy of the CHIRPS precipitation dataset at representing climate variability in a tropical mountainous region of South America. *Physics and Chemistry of the Earth, Parts A/B/C, 127*, 103184. https://doi.org/10.1016/j.pce.2022.103184.

Los, S. O., Street-Perrott, F. A., Loader, N. J., Froyd, C. A., Cuní-Sanchez, A., & Marchant, R. A. (2019). Sensitivity of a tropical montane cloud forest to climate change, present, past and future: Mt. Marsabit, N. Kenya. *Quaternary Science Reviews, 218*, 34–48. https://doi.org/10.1016/j.quascirev.2019.06.016.

Lu, X., Liang, E., Wang, Y., Babst, F., & Camarero, J. J. (2021). Mountain treelines climb slowly despite rapid climate warming. *Global Ecology and Biogeography, 30*(1), 305–315.

Määttänen, A. M., Kemppinen, J. and Hällfors, M., 2024. Geodiversity is an inseparable but underutilized part of connectivity in nature conservation amid climate change. Authorea.https://doi.org/10.22541/au.172977173.33452815/v1

Malhi, Y., Silman, M., Salinas, N., Bush, M., Meir, P., & Saatchi, S. (2010). Introduction: Elevation gradients in the tropics: Laboratories for ecosystem ecology and global change research. *Global Change Biology, 16*(12), 3171–3175. https://doi.org/10.1111/j.1365-2486.2010.02323.x.

Mappin, B., Chauvenet, A. L. M., Adams, V. M., Marco, M. D., Beyer, H. L., Venter, O., Halpern, B. S., Possingham, H. P., & Watson, J. E. M. (2019). Restoration priorities to achieve the global protected area target. *Conservation Letters, 12*(4). https://doi.org/10.1111/conl.12646.

McAllister, T., Hikuroa, D., & Macinnis-Ng, C. (2023). Connecting science to indigenous knowledge. *New Zealand Journal of Ecology, 47*(1), 1–13.

McDowell, G., Stevens, M., Marshall, S., Higgs, E., Jacob, A., Johnson, G., Johnson, L., Dicker, M., Inkpen, D., Koppes, M., Lightning, K., Parlee, B., Pascal, W., Shea, J., Sims, D., Smiles, N. D., Andrew, L., Aubry-Wake, C., & ..., Vamosi, S. M. (2023). *Canadian mountain assessment: Walking together to enhance understanding of mountains in Canada*. University of Calgary Press https://doi.org/10.2307/jj.8816123.

Montesino Pouzols, F., Toivonen, T., Di Minin, E., Kukkala, A. S., Kullberg, P., Kuusterä, J., Lehtomäki, J., Tenkanen, H., Verburg, P. H., & Moilanen, A. (2014). Global protected area expansion is compromised by projected land-use and parochialism. *Nature, 516*(7531), 383–386. https://doi.org/10.1038/nature14032.

Moyo, I. (2023). Beyond a tokenistic inclusion of indigenous knowledge systems in protected area governance and management in Okhahlamba-Drakensberg. *African Geographical Review, 42*(2), 141–156. https://doi.org/10.1080/19376812.2021.1997613.

Nadal-Romero, E., Juez, C., Khorchani, M., Peña-Angulo, D., Lana-Renault, N., Regüés, D., Lasanta, T., & García-Ruiz, J. M. (2021). Impacts of land abandonment on flood mitigation in Mediterranean mountain areas. In *Nature-based solutions for flood mitigation: Environmental and socio-economic aspects* (pp. 189–214). Springer International Publishing. https://doi.org/10.1007/698_2021_772.

Naidoo, R., Gerkey, D., Hole, D., Pfaff, A., Ellis, A. M., Golden, C. D., Herrera, D., Johnson, K., Mulligan, M., Ricketts, T. H., & Fisher, B. (2019). Evaluating the impacts of protected areas on human well-being across the developing world. *Science Advances, 5*(4). https://doi.org/10.1126/sciadv.aav3006.

Nsengiyumva, P. (2019). African mountains in a changing climate: Trends, impacts, and adaptation solutions. *Mountain Research and Development, 39*(2), P1. https://doi.org/10.1659/MRD-JOURNAL-D-19-00062.1.

Olabisi, L. S., Onyeneke, R. U., Choko, O. P., Chiemela, S. N., Liverpool-Tasie, L. S. O., Achike, A. I., & Aiyeloja, A. A. (2020). Scenario planning for climate adaptation in agricultural systems. *Agriculture, 10*(7), 274. https://doi.org/10.3390/agriculture10070274.

Oteros-Rozas, E., Martín-López, B., Daw, T. M., Bohensky, E. L., Butler, J. R. A., Hill, R., Martin-Ortega, J., Quinlan, A., Ravera, F., Ruiz-Mallén, I., Thyresson, M., Mistry, J.,

Palomo, I., Peterson, G. D., Plieninger, T., Waylen, K. A., Beach, D. M., Bohnet, I. C., Hamann, M., & ..., Vilardy, S. P. (2015). Participatory scenario planning in place-based social-ecological research: Insights and experiences from 23 case studies. *Ecology and Society, 20*(4). https://doi.org/10.5751/ES-07985-200432.

Palomo, I. (2017). Climate change impacts on ecosystem services in high mountain areas: A Literature Review. *Mountain Research and Development, 37*(2), 179–187. https://doi.org/10.1659/mrd-journal-d-16-00110.1.

Palomo, I., Múgica, M., Piñeiro, C., Martín-López, B., & Montes, C. (2017). Envisioning protected areas through participatory scenario planning: Navigating coverage and effectiveness challenges ahead. *PARKS, 23*(1), 29–44. https://doi.org/10.2305/IUCN.CH.2017.PARKS-23-1IP.en.

Peng, J., Hu, X., Zhao, M., Liu, Y., & Tian, L. (2017). Research progress on ecosystem service trade-offs: From cognition to decision-making. *Dili Xuebao/Acta Geographica Sinica, 72*(6), 960–973. https://doi.org/10.11821/dlxb201706002.

Pepin, N. C., Arnone, E., Gobiet, A., Haslinger, K., Kotlarski, S., Notarnicola, C., Palazzi, E., Seibert, P., Serafin, S., Schöner, W., Terzago, S., Thornton, J. M., Vuille, M., & Adler, C. (2022). Climate changes and their elevational patterns in the mountains of the world. *Reviews of Geophysics, 60*(1). https://doi.org/10.1029/2020rg000730.

Perrigo, A., Hoorn, C., & Antonelli, A. (2020). Why mountains matter for biodiversity. *Journal of Biogeography, 47*(2), 315–325. https://doi.org/10.1111/jbi.13731.

Piao, S., Yin, G., Tan, J., Cheng, L., Huang, M., Li, Y., Liu, R., Mao, J., Myneni, R. B., Peng, S., Poulter, B., Shi, X., Xiao, Z., Zeng, N., Zeng, Z. Z., & Wang, Y. (2015). Detection and attribution of vegetation greening trend in China over the last 30 years. *Global Change Biology, 21*(4), 1601–1609. https://doi.org/10.1111/gcb.12795.

Platts, P. J., Gereau, R. E., Burgess, N. D., & Marchant, R. (2013). Spatial heterogeneity of climate change in an Afromontane centre of endemism. *Ecography, 36*(4), 518–530. https://doi.org/10.1111/j.1600-0587.2012.07805.x.

Platts, P. J., Schaafsma, M., Turner, R. K., Burgess, N. D., Fisher, B., Mbilinyi, B. P., Munishi, P. K. T., Ricketts, T. H., Swetnam, R. D., Ahrends, A., Ashagre, B. B., Bayliss, J., Gereau, R. E., Green, J. M. H., Green, R. E., Jeha, L., Lewis, S. L., Marchant, R., Marshall, A. R., & ..., Balmford, A. (2023). Inequitable gains and losses from conservation in a global biodiversity hotspot. *Environmental and Resource Economics, 86*(3), 381–405. https://doi.org/10.1007/s10640-023-00798-y.

Polman, E. M. N., Seijmonsbergen, A. C., Versteegh, H., & Kissling, W. D. (2024). Global geodiversity components are not equally represented in UNESCO Global Geoparks. *Philosophical Transactions of the Royal Society A: Mathematical, Physical and Engineering Sciences, 382*(2269). https://doi.org/10.1098/rsta.2023.0054.

Reckling, W., Mitasova, H., Wegmann, K., Kauffman, G., & Reid, R. (2021). Efficient drone-based rare plant monitoring using a species distribution model and AI-based object detection. *Drones, 5*(4), 110. https://doi.org/10.3390/drones5040110.

Rockström, J., Gaffney, O., Rogelj, J., Meinshausen, M., Nakicenovic, N., & Schellnhuber, H. J. (2017). A roadmap for rapid decarbonization. *Science, 355*(6331), 1269–1271. https://doi.org/10.1126/science.aah3443.

Rouget, M., Richardson, D. M., & Cowling, R. M. (2003). The current configuration of protected areas in the Cape Floristic Region, South Africa – Reservation bias and representation of biodiversity patterns and processes. *Biological Conservation, 112*(1-2), 129–145. https://doi.org/10.1016/S0006-3207(02)00396-8.

Saito, O., Kamiyama, C., Hashimoto, S., Matsui, T., Shoyama, K., Kabaya, K., Uetake, T., Taki, H., Ishikawa, Y., Matsushita, K., Yamane, F., Hori, J., Ariga, T., & Takeuchi, K. (2019). Co-design of national-scale future scenarios in Japan to predict and assess natural capital and ecosystem services. *Sustainability Science, 14*(1), 5–21. https://doi.org/10.1007/s11625-018-0587-9.

Sandel, B., & Svenning, J. C. (2013). Human impacts drive a global topographic signature in tree cover. *Nature Communications, 4*. https://doi.org/10.1038/ncomms3474.

Schneiderbauer, S., Fontanella Pisa, P., Delves, J. L., Pedoth, L., Rufat, S., Erschbamer, M., Thaler, T., Carnelli, F., & Granados-Chahin, S. (2021). Risk perception of climate change and natural hazards in global mountain regions: A critical review. *Science of the Total Environment, 784*. https://doi.org/10.1016/j.scitotenv.2021.146957.

Schoennagel, T., Veblen, T. T., & Romme, W. H. (2004). The interaction of fire, fuels, and climate across Rocky Mountain forests. *BioScience, 54*(7), 661–676. https://doi.org/10.1641/0006-3568(2004)054[0661:TIOFFA]2.0.CO;2.

Silveira, F. A. O., Barbosa, M., Beiroz, W., Callisto, M., Macedo, D. R., Morellato, L. P. C., Neves, F. S., Nunes, Y. R. F., Solar, R. R., & Fernandes, G. W. (2019). Tropical mountains as natural laboratories to study global changes: A long-term ecological research project in a megadiverse biodiversity hotspot. *Perspectives in Plant Ecology, Evolution and Systematics, 38*, 64–73. https://doi.org/10.1016/j.ppees.2019.04.001.

Soga, M., & Gaston, K. J. (2024). Do people who experience more nature act more to protect it? A meta-analysis. *Biological Conservation, 289*, 110417. https://doi.org/10.1016/j.biocon.2023.110417.

Somers, L. D., & McKenzie, J. M. (2020). A review of groundwater in high mountain environments. *Wiley Interdisciplinary Reviews: Water, 7*(6). https://doi.org/10.1002/wat2.1475.

Spicer, R. A., Farnsworth, A., & Su, T. (2020). Cenozoic topography, monsoons and biodiversity conservation within the Tibetan Region: An evolving story. *Plant Diversity, 42*(4), 229–254. https://doi.org/10.1016/j.pld.2020.06.011.

Steiger, R., Knowles, N., Pöll, K., & Rutty, M. (2024). Impacts of climate change on mountain tourism: A review. *Journal of Sustainable Tourism, 32*(9), 1984–2017. https://doi.org/10.1080/09669582.2022.2112204.

Tang, Y., Winkler, J. A., Viña, A., Liu, J., Zhang, Y., Zhang, X., Li, X., Wang, F., Zhang, J., Zhao, Z., & Bosso, L. (2018). Uncertainty of future projections of species distributions in mountainous regions. *PLOS ONE, 13*(1), e0189496. https://doi.org/10.1371/journal.pone.0189496.

Tito, R., Vasconcelos, H. L., & Feeley, K. J. (2020). Mountain ecosystems as natural laboratories for climate change experiments. *Frontiers in Forests and Global Change, 3*. https://doi.org/10.3389/ffgc.2020.00038.

Vincent, C., Fernandes, R. F., Cardoso, A. R., Broennimann, O., Di Cola, V., D'Amen, M., Ursenbacher, S., Schmidt, B. R., Pradervand, J. N., Pellissier, L., & Guisan, A. (2019). Climate and land-use changes reshuffle politically-weighted priority areas of mountain biodiversity. *Global Ecology and Conservation, 17*. https://doi.org/10.1016/j.gecco.2019.e00589.

Wilson, B., Dolotbakov, A., Burgess, B. J., Clubbe, C., Lazkov, G., Shalpykov, K., Ganybaeva, M., Sultangaziev, O., & Brockington, S. F. (2021). Central Asian wild tulip conservation requires a regional approach, especially in the face of climate change. *Biodiversity and Conservation, 30*(6), 1705–1730. https://doi.org/10.1007/s10531-021-02165-z.

Yana-Salluca, M., Arapa, J. J. P., Vilca-Apaza, H. M., Yana-Salluca, N., Adco-Valeriano, H., Argollo, K. P., & Turpo, W. S. (2024). Management of knowledge breeding in the Andean cosmovision in the ayllu of Amantani, Peru. *Journal of Social and Environmental Management, 18*(6). https://doi.org/10.24857/RGSA.V18N6-036.

Yu, Y., Li, J., Zhou, Z., Ma, X., & Zhang, X. (2021). Response of multiple mountain ecosystem services on environmental gradients: How to respond, and where should be priority conservation? *Journal of Cleaner Production, 278*, 123264. https://doi.org/10.1016/j.jclepro.2020.123264.

CHAPTER 8

Sustainable futures for the world's mountains

8.1 Introduction

As showcased throughout this book, mountains are special environments with unique biophysical and cultural character that connect the globe. Indeed, the Alpine zone is the only ecosystem that unifies the world and is found across all continents and across all latitudes. Despite covering a strikingly small area of the globe, mountains house most of the terrestrial biodiversity (Kreft & Jetz, 2007); their island-like nature and heterogeneous landscape support diversity at multiple levels from the gene to geodiversity (Cortés et al., 2018). Underpinning mountains is the diversity of mountain environments, communities, geologies, climate systems, and ecosystems – a diversity that is at the heart of the wonder that makes mountains special places. This diversity and unifying character of mountains can be used to empower mountain futures and make mountains more central to future global sustainable agendas.

8.2 Connecting the 'mountain research dots'

A massive amount of effort and funding has allowed valuable research on mountains to happen. This knowledge archive has been used to inform the book, although understanding how these individual insights fit together is crucial. These sometimes disparate and disconnected glimpses into the functioning of mountain systems are connected through organisations such as the Mountain Research Initiative and the Global Mountain Partnership that form vital umbrellas for the world's mountain research community. The Mountain Partnership and the Mountain Research Initiative play a leading role in improving and sustaining healthy environments and the livelihoods of mountain people so they are included in the implementation of the United Nations agendas, including the 2030 Agenda for Sustainable Development, the UN Decade on Ecosystem Restoration, and the 5 Years of Action for

Mountain Diversity: Dynamics of High-Altitude Environments.
DOI: https://doi.org/10.1016/B978-0-443-22251-1.00009-5

the Development of Mountain Regions that runs to 2027. As documented through Chapter 6, there are many challenges for mountains, but longer-term sustainable futures are possible, and such linking organisations have a vital role to play in making the most of the existing knowledge and creating innovation happening to support future mountain systems. The Mountain Research Initiative provides a vital forum and opens up creative spaces, where the mountain research community can come together. A future where science connects with society can be used to navigate the multiple future challenges mountains are facing, particularly as these are looking more acute as planetary boundaries continue to be crossed (Williams et al., 2025).

Considerable efforts have been made over the recent decades to establish mountain research networks, arising from the recognition that mountains are providers of ecosystem services essential to human well-being. Examples of such initiatives include the Global Mountain Biodiversity Assessment, the Network for European Mountain Research, the Canadian Mountain Assessment, the Albertine Rift Conservation Society, and Mountain Invasion Research Network. Notwithstanding recent efforts, long-term ecological research studies are still scarce in tropical mountains, despite their disproportionately high biodiversity, degree of endemism, and societal importance. Future research collaborations across disciplines should be encouraged to move research on mountains forward with larger steps and higher impact than seen so far.

8.3 Supporting mountains as intertwined social–ecological systems

Like any part of the world, there is no single corner of a mountain that does not carry the fingerprints of human societies; this impact can be completely transformative, such as extensively terraced mountain slopes, although it is often quite diffuse and indirect, such as enhanced glacial melting due to global climate change. Sometimes this is so deep and embedded that it is hard to see the transformation as the natural baselines have long gone. Viewing mountains as being culturally shaped, rather than natural, allows us to craft appropriate management interventions that support the social–ecological system, or, at very least, do not work against it! This view will also showcase the importance to humanity and planetary functioning, an importance not just for those living in mountains, but also for those living far downstream in the catchments. Indeed, mountain processes, mountain changes, and how these are connected across the wider landscape remain relatively poorly understood.

This dearth of understanding is not just for those mountain systems in the Global South; the paucity of data extends to many mountain regions (Chamberlain et al., 2013), and currently precludes joined-up management approaches that could maximise the benefits and mitigate the challenges. Detailed monitoring should be a future priority if we are to make improved predictions on the impacts of environmental change on mountain societies and ecosystems. However, these really need core support from Governments and international funding agencies to help establish thresholds and support the development of effective conservation and management policies that benefit mountain societies. These societies continue to be incredibly dynamic, and while some mountain populations are growing quickly, others are declining: in some areas, agricultural activity is expanding, and in other areas, agriculturalists are being replaced by new immigrants 'working from home'. It is essential to empower this dynamic and the shifting nature of mountain communities to ensure participation in the decision-making processes and development initiatives reflect their changing needs and priorities, and hence are likely to succeed.

Appreciating the spatial and temporal complexity of mountain socio–ecological systems and applying this understanding at scales that are appropriate is a positive way forward. A fundamental start for this appreciation is that mountain socio–ecological systems are nested and connected across time and space. By viewing mountains as social–ecological systems, we can use our understanding of the spatial connections that support sustainable mountain development. Support for these development and conservation efforts would also benefit from collaboration amongst geoscientists, biodiversity scientists, and social scientists by rethinking conservation planning and considering the historical, current, and future land use patterns of both human and other species (Määttänen et al., 2024). With increasing attention on ecological resilience, epitomised by high biodiversity levels and the importance of functioning ecosystems, mountains need to be prioritised in conservation efforts. More 'joined-up' approaches that allow us to mitigate biodiversity loss while decreasing the impacts of climate and land-use change are central to calls coming out from many conservation organisations and intergovernmental panels such as IPCC, IPBES, WCRP, and Future Earth, where the linkages between climate change and biodiversity and ecosystem services are increasingly connected (Mäkinen-Rostedt et al., 2025).

Given the pervasive and deep-rooted transformations that have happened across the world's mountains, there is an urgent need for ecological restoration that is embedded within the social–ecological system. This restoration needs to work within our understanding of geographic variation in heterogeneous socio–ecological systems. Such innovation needs to be individually tailored, but also work at scale across the highland to lowland axis and be collectively developed across this socio–ecological space at culturally appropriate scales. For example, soil loss from a mountain socio–ecological system will result in fertility gain for those downstream. A future where upland land owners or catchment managers are paid for soil and hydrological services, produced and delivered to the lowland populations, would strengthen mountain socio–ecological systems, benefitting mountains and those living within the wider catchment.

Sustainable management of mountain natural resources, including water, forests, and biodiversity, is essential for the long-term resilience of mountain communities. Traditional resource management systems, such as community forests and watershed management committees, offer valuable insights into effective conservation practices and how these have evolved through the ages. During the 19th century, most of the Mediterranean mountains recorded their maximum population densities, and all possible areas were cultivated, including very steep slopes. This resulted in severe erosion problems and the development of torrential, braided rivers with downstream flooding. During the 20th century, widespread farmland abandonment in the mountain areas caused the expansion of shrublands and forests at the expense of the old cultivated fields. The consequences of such a woody encroachment are relatively well known: landscape homogenisation, abrupt declines in grassland productivity, decreased streamflow, reduced erosion rates, and the loss of cultural landscapes. The loss of traditional landscapes is a major cause of concern because such landscapes have consequences for biodiversity, soil erosion, landscape quality, and ecosystem services (García-Ruiz et al., 2020).

Ecological restoration strategies often aim to (1) construct a more complex and heterogeneous landscape, (2) recover part of the cultural landscape lost, (3) increase biodiversity, (4) reduce the probability of large wildfires, and (5) enhance the survival of extensive stockbreeding systems that also contribute to biodiversity (García-Ruiz et al., 2020). It is increasingly recognised that managing soils is important not only for food production and cultural services but also for environmental regulation and Earth system functioning (Robinson et al., 2012; Trivedi et al., 2018). Mountain agricultural systems encompass a wide range of farming practices, from terraced rice paddies in the Himalayas to agroforestry systems in the Andes, where farmers cultivate a diverse range of crops, including traditional varieties of grains, tubers, and vegetables, adapted to varying altitudes and microclimates. These systems are characterised by their unique socio-economic, environmental, and cultural dynamics, which influence agricultural productivity, livelihood strategies, and food security in mountain regions, which contribute to food security and resilience in the face of environmental uncertainties. However, the prospective value chains of the mountain agroforestry products face many challenges, including (1) socio-economic constraints of the farmers mainly because of high initial adoption costs, limited information on benefit-cost of agroforestry practices, limited knowledge on full benefits of agroforestry, and limited markets and marketing information; (2) institutional constraints because of unclear policy to support agroforestry, the lack of extension services and undefined administrative boundaries; and (3) inadequate scientific knowledge, expertise and technologies to address management complexity of agroforestry systems. Good farming land is often limited in mountains, and in some areas, there have been declining yields; farming innovation can be helped by new techniques such as deep tillage, push-pull technology, forecasting, and market information via mobile phone coverage, and embedded exchange networks. Changes can be positive, but these are often still very localised and incremental, and really need to be upscaled to make the innovations more transformational across mountain systems. This process must focus on people who live in the mountains to find solutions that work for them, their culture, and their situation.

8.4 Novel mountain environments—Glacial retreat ecosystem expansion

As shown throughout this book, mountain systems are changing, and this change is happening disproportionately faster than other systems around the world (Cuni-Sanchez et al., 2025; Pepin et al., 2022). This change is epitomised by decreasing glaciers and emerging post-glacial ecosystems that have a largely unrecognised and largely unquantified role. Glaciers are important elements of the hydrological cycle, presently storing 135,000 km^3 of water and releasing fresh water during melt seasons that often coincide with dry periods. Global ice melt will impact water supply in many regions, as well as down catchment impacts, all the way to increased sea level rise. Again, understanding these wide-scale connections, from mountains to oceans, allows them to be managed as common goods that are affected by global threats such as climate change (Bosson et al., 2023).

Glacier retreat and hydrological impacts have been well documented. However, research has overlooked the socio-cultural dimensions of their retreat, particularly in the Chilean Andes. This is a crucial oversight as emerging post-glacial ecosystems are mostly located on lands owned and governed by local or public communities (Bosson et al., 2023). The Cerro El Plomo glaciers, near Santiago, Chile, are arguably the most visible and socio-culturally significant in the Santiago region. The retreat of the Cerro El Plomo glaciers presents not only a water security risk, but also important socio-cultural implications. Clearly, culture is not static, but dynamic and evolving, and communities in the Santiago region are coping with and adapting to a changing environment. However, this adaptation must be supported by collaborative work between glaciologists, social scientists, and local communities to better understand the localised implications of glacier loss and adaptation options (Carey & Moulton, 2023).

The glaciers located on Cotopaxi, an active volcano in the Ecuadorian Andes, have lost a total ice area of 53% in the last 48 years; a trend that is well correlated to observed climate changes in the region. However, the area losses have been enhanced by the activity during 2015-2016 and more recently by the last volcanic pulse, 2022–2023. The main effect of this volcanic activity is the glacier surface albedo change due to ash deposition. During the last eruption of Cotopaxi between August 2015 and January 2016, almost 50% of the total glacier area was covered by ash. Apart from the ash deposition effects, several new hot spots have been detected near the top of Cotopaxi, which are also affecting ice melt (Barbosa et al., 2022). Under continuous glacier retreat, wetlands close to glaciers will lose their buffered seasonality and might switch to a precipitation-driven regime. As glacier impact reduces and precipitation impact increases, future droughts will be more impactful. As the glacier's impact quickly dilutes further downstream, the impact of glacier disappearance will be experienced in downstream wetlands. Communities which rely on glacier melt for wetland irrigation may benefit in the short term from increased melt, but they will become more vulnerable as meltwaters start to decline.

It is not just glaciers that are at the forefront of rapid change. The process of woody encroachment is a controversial topic among ecologists, engineers, geographers, and other scientists. Rising atmospheric CO_2 concentration can also increase future forest encroachment, particularly in water-stressed environments (McDowell et al., 2020). In Switzerland, in common with many temperate mountain regions, forests are expanding. When forests expand, they also colonise various other habitats that can be important for biodiversity, such as alpine grasslands and glacier moraines (Klaus et al., 2023; Pretzsch et al., 2023). First and foremost, we have to understand the long-term history of these open areas. For example, in the currently forested Blue Mountains of Australia, occupied by the Dharug people, the local and traditional Aboriginal people of the Blue Mountains extend back some 20,000 years, when the area was a largely grassland environment with patches of open forest. In the Hessen mountains, Germany, palaeoecological insights document the changing composition of *Tilia, Quercus, Corylus,* and *Carpinus* that can feed into the contemporary management of deciduous forests (Stobbe & Gumnior, 2021). Indeed, understanding mountain systems as being dynamic and a product of long human-environment interactions is being embraced and embedded in contemporary management decisions (Gillson et al., 2025).

Increased tree density and canopy closure following fire exclusion are impacting open habitat species that are facing a potentially severe loss of habitat as alpine grasslands are colonised by forest and shrubs. Of those species predicted to be most sensitive to climate change in Europe, the Water Pipit is likely to represent the biggest potential conservation problem as it is very much a mountain specialist throughout most of Europe, and it also seems completely intolerant of increased forest cover (Chamberlain et al., 2013). The characteristic Chamois (Plate 8.1) is a species that lives in alpine pastures and in mountainous regions across Europe (Herrero et al., 2023). This species was exploited as a resource by Palaeolithic human groups (Rodríguez-Gómez et al., 2022) and is still extensively hunted. The future of Chamois is being impacted by these ecosystem shifts, with conservation and management implications that depend on demography and life history traits that still need to be fully clarified (Corlatti et al., 2022).

Many areas under change need an expansion of forest and wider ecosystem management in upstream areas; this need is often apparent following disastrous impacts. For example, following the Slovenian floods in 2023, headwater torrent supervision is now part of the forest service with technical support and a GIS-integrated forest management system aimed at reducing the risks of flooding. Mountain systems as living labs can be used as a capacity-building tool to showcase how different sectors can effectively work to manage the dynamic mountain shifts. For example, the Forest Living Lab Soteska brings together forest experts, owners, contractors, infrastructure, and local municipal government and communities. Capacity building

Plate 8.1 Characteristic chamois around Branjochkreuz above Innsbruck, Austria. *(Photograph Rob Marchant)*.

is essential for the future. Three major questions can support these joined-up approaches: (1) What can we do with cultural landscapes that are being rapidly recovered with shrubs and forests? (2) Is farmland and stockbreeding abandonment the best option for improving biodiversity and sustainability, and (3) is it preferable to actively manage the landscapes to enhance their natural heterogeneity with the presence of light human activity? These are critical questions that have not been definitively addressed to date, largely due to the complexity of the ecological, geographical, social, and economic factors that contribute to management decisions regarding mountain areas and their downstream catchments.

8.5 Future perspectives

At the heart of Mountain systems is their diversity! Despite this, mountain regions share common core features: high geological diversity, complex landscape structures, climatic and fire regimes – both past and present – further contributing to their ecological and geological uniqueness. Understanding these environmental foundations is key to interpreting both the distribution of mountain ecosystems and the applied aspects of mountain systems explored in previous chapters. There are many knowledge gaps and areas of uncertainty about how species and cultures will respond to current and future climate change. Accounting for the spatio-temporal connectivity of mountain biota to mountain environments will provide a better understanding of the patterns we have and a basis for the effective conservation of mountain biodiversity in response to climate, social, and economic changes. Indeed, understanding the underlying physical foundations is essential for identifying both the challenges facing mountain environments and potential solutions – we must know the mountain system to work with it rather than against it! An interdisciplinary understanding of the different drivers and directions of changes in settlement, land-cover and land-use, and the associated environmental histories and interactions with various cultures, technologies, and subsistence strategies through time and across space for the world's Mountains, allows for a nested temporal understanding that tracks the evolution of socio–ecological systems and quantify the character of the ecosystem services provided by mountain ecosystems at; household, village, catchment, national, and international levels. By synthesising and visualising existing data from the past and adding new state-of-the-art data from targeted investigations, we can assess socio–ecological evolution and understand the resilience of socio–ecological systems to past, present, and future climate change (Gillson et al., 2025). Such an interdisciplinary approach will fundamentally advance understanding of past human-environment interactions and highlight the relevance of these data for present-day conservation and management issues.

Mountains play a vital role in ensuring food security, preserving cultural heritage, and maintaining ecosystem services in high-altitude regions. Mountain community diversity is a source of strength, resilience, and innovation, rooted in centuries-old traditions and adaptive strategies. Among landscapes, mountain ranges can be considered the undervalued ecological backbone of Europe (Canedoli et al., 2020). Indeed, developing the analogy, mountains form the world's skeleton! Embracing and supporting this diversity and the challenges associated with mountains is essential for promoting inclusive development, environmental sustainability, and cultural heritage preservation in mountain regions. By recognising the socio-cultural dynamics, challenges, and opportunities inherent in mountain communities, policymakers, development practitioners, and civil society actors can foster vibrant and resilient mountain societies for generations to come. By supporting this diversity, it is possible to enhance the resilience, innovation, and sustainability of mountain farming systems, and we can contribute to the well-being of mountain communities and the conservation of mountain ecosystems for future generations. Implementing sustainable practices, such as responsible tourism and land-use planning, is crucial. Engaging communities in conservation initiatives ensures that the unique relationship between people and mountains is respected, fostering a harmonious coexistence that benefits both biodiversity and human well-being.

Mountain communities and mountain food production systems, be it agroforestry, terraced agriculture, or pastoral, face the unprecedented task of feeding a growing world population while simultaneously avoiding harmful environmental and social effects. There has to be recognition that the farmer is making a positive contribution to aspects, such as water regulations, carbon sequestration, and biodiversity conservation, that have benefits far outside of the farm boundaries, and as such should be suitably compensated for this service production and export (Platts et al., 2023). The global relevance of mountains must be brought to the fore to ensure that mountains receive the attention they deserve. Mountains have provided food, water, shelter, and secure essential ecosystem services for hundreds of millions of people in both lowlands and highlands (Schneiderbauer et al., 2021). There is currently a societal disconnect between urban spaces (Plate 8.2) and their surrounding rural or 'natural' areas (Xue et al., 2023). These surroundings not only provide cities with ecosystem services but can also pose a risk to mountain city dwellers (Schneiderbauer et al., 2021), particularly those large cities close to urban centres (Plate 8.2).

Mountains are the most climate-sensitive ecosystems worldwide (Seddon et al., 2016), and many mountain regions face unique problems that require bespoke solutions. Many of the challenges detailed in this book combine to form a complex layer that can impact in an additive, multipurpose, or subtractive manner for mountain areas and their communities. Understanding the potential additive and interactive effects of climate and land use land cover change on biodiversity and ecosystem functions is key in our attempts to predict and mitigate the consequences of global change on the world's mountains, and along other climatic gradients. Mountain people are the custodians of unique cultures, traditions, and knowledge systems that have been shaped by the challenging environments they inhabit. Mountain people are stewards of biodiversity and have unique and valuable local knowledge, traditions, and cultural practices that must shape effective land management strategies and sustainable food systems. For this potential to be realised, there has to be greater investment in understanding change and investment in supporting connections across disciplines, cultures, and geographic space. Despite the resilience and adaptability of these communities, they face significant challenges from climate change, economic

Plate 8.2 There is a global growth in urban populations, with urban centres close to mountain systems having extra challenges of building land and issues of air quality, such as Santiago, Chile (*top left*), Chefchaouen (*top left*), Meknes (*top right*), and Innsbruck (*top right*). *(Photographs Rob Marchant)*.

marginalisation, environmental degradation, and cultural erosion. Mountain systems and their people are the engines of the global supply of water, food, and fuel, and need to be cherished and central to the future of world development.

References

Barbosa, A. M., Francelino, M. R., Thomazini, A., Schaefer, C. E. G. R., Anjos, L. H. C., Pereira, M. G., & Lyra, G. B. (2022). The thermal regime and mineralogical attributes of highland volcanic-ash soils from the Cotopaxi volcano, Ecuador: Absent permafrost and little pedogenesis. *Geoderma Regional, 29*, e00496. https://doi.org/10.1016/j.geodrs.2022.e00496.

Bosson, J. B., Huss, M., Cauvy-Fraunié, S., Clément, J. C., Costes, G., Fischer, M., Poulenard, J., & Arthaud, F. (2023). Future emergence of new ecosystems caused by glacial retreat. *Nature, 620*(7974), 562–569. https://doi.org/10.1038/s41586-023-06302-2.

Canedoli, C., Ferrè, C., El Khair, D. A., Comolli, R., Liga, C., Mazzucchelli, F., Proietto, A., Rota, N., Colombo, G., Bassano, B., Viterbi, R., & Padoa-Schioppa, E. (2020). Evaluation of ecosystem services in a protected mountain area: Soil organic carbon stock and biodiversity in alpine forests and grasslands. *Ecosystem Services, 44*, 101135. https://doi.org/10.1016/j.ecoser.2020.101135.

Carey, M., & Moulton, H. (2023). Inequalities of ice loss: A framework for addressing sociocryospheric change. *Annals of Glaciology, 64*(91), 67–76. https://doi.org/10.1017/aog.2023.44.

Chamberlain, D. E., Negro, M., Caprio, E., & Rolando, A. (2013). Assessing the sensitivity of alpine birds to potential future changes in habitat and climate to inform management strategies. *Biological Conservation, 167*, 127–135. https://doi.org/10.1016/j.biocon.2013.07.036.

Corlatti, L., Iacolina, L., Safner, T., Apollonio, M., Buzan, E., Ferretti, F., Hammer, S. E., Herrero, J., Rossi, L., Serrano, E., Arnal, M. C., Brivio, F., Chirichella, R., Cotza, A., Crestanello, B., Espunyes, J., Fernández de Luco, D., Friedrich, S., Gačić, D., & …, Šprem, N. (2022). Past, present and future of chamois science. *Wildlife Biology, 2022*(4). https://doi.org/10.1002/wlb3.01025.

Cortés, A. J., Garzón, L. N., Valencia, J. B., & Madriñán, S. (2018). On the causes of rapid diversification in the páramos: Isolation by ecology and genomic divergence in espeletia. *Frontiers in Plant Science, 871*. https://doi.org/10.3389/fpls.2018.01700.

Cuni-Sanchez, A., Aneseyee, A. B., Baderha, G. K. R., Batumike, R., Bitariho, R., Imani, G., Jha, N., Kaganzi, K. R., Kaplin, B. A., Klein, J. A., Leite, A., Marchant, R. A., Martin, E. H., Mcharazo, F., Mwangi, B., Ngute, A. S. K., Nkengurutse, J., Nkurunziza, A., Olaka, L., & …, Zafra-Calvo, N. (2025). Perceived climate change impacts and adaptation responses in ten African mountain regions. *Nature Climate Change, 15*(2), 153–161. https://doi.org/10.1038/s41558-024-02221-w.

García-Ruiz, J. M., Lasanta, T., Nadal-Romero, E., Lana-Renault, N., & Álvarez-Farizo, B. (2020). Rewilding and restoring cultural landscapes in Mediterranean mountains: Opportunities and challenges. *Land Use Policy, 99*, 104850. https://doi.org/10.1016/j.landusepol.2020.104850.

Gillson, L., Seddon, A., Mottl, O., Zhang, K., Kirsten, K., Gell, P., Marchant, R. A., Schwörer, C., Razanatsoa, E., Lane, P. J., Courtney-Mustaphi, C. J., & Dearing, J. (2025). Exploring the interface between planetary boundaries and palaeoecology. *Global Change Biology, 31*(1). https://doi.org/10.1111/gcb.70017.

Herrero, J., García-Serrano, A., Félez, C., Herrero, F., Machuca, A., Ponz, B., & Sancho, S. (2023). A low altitude forest dwelling Pyrenean chamois population increases the potential habitat of this subspecies. *Forest Systems, 32*(3). https://doi.org/10.5424/fs/2023323-20521.

Klaus, G., Cordillot, F., & Künzle, I. (2023). Espèces et milieux menaces en Suisse. *Ofce fédéral de l'environnement (OFEV)*. https://www.bafu.admin.ch/bafu/fr/home/themes/biodiversite/publications/publications-biodiversite/especes-menacees-suisse.html.

Kreft, H., & Jetz, W. (2007). Global patterns and determinants of vascular plant diversity. *Proceedings of the National Academy of Sciences of the United States of America, 104*(14), 5925–5930. https://doi.org/10.1073/pnas.0608361104.

Määttänen, A. M., Kemppinen, J. and Hällfors, M., 2024. Geodiversity is an inseparable but underutilized part of connectivity in nature conservation amid climate change. https://doi.org/10.22541/au.172977173.33452815/v1.

Mäkinen-Rostedt, K., Oreschnikoff, A., & Uusikylä, P. (2025). Expert knowledge and institutional interplay: Exploring the epistemic linkages of the IPCC, IPBES, WCRP, and Future Earth. *Environmental Science & Policy, 171*, 104181. https://doi.org/10.1016/j.envsci.2025.104181.

McDowell, N. G., Allen, C. D., Anderson-Teixeira, K., Aukema, B. H., Bond-Lamberty, B., Chini, L., Clark, J. S., Dietze, M., Grossiord, C., Hanbury-Brown, A., Hurtt, G. C., Jackson, R. B., Johnson, D. J., Kueppers, L., Lichstein, J. W., Ogle, K., Poulter, B., Pugh, T. A. M., Seidl, R., & …, Xu, C. (2020). Pervasive shifts in forest dynamics in a changing world. *Science, 368*(6494). https://doi.org/10.1126/science.aaz9463.

Pepin, N. C., Arnone, E., Gobiet, A., Haslinger, K., Kotlarski, S., Notarnicola, C., Palazzi, E., Seibert, P., Serafin, S., Schöner, W., Terzago, S., Thornton, J. M., Vuille, M., &

Adler, C. (2022). Climate changes and their elevational patterns in the mountains of the world. *Reviews of Geophysics, 60*(1). https://doi.org/10.1029/2020rg000730.

Platts, P. J., Schaafsma, M., Turner, R. K., Burgess, N. D., Fisher, B., Mbilinyi, B. P., Munishi, P. K. T., Ricketts, T. H., Swetnam, R. D., Ahrends, A., Ashagre, B. B., Bayliss, J., Gereau, R. E., Green, J. M. H., Green, R. E., Jeha, L., Lewis, S. L., Marchant, R., Marshall, A. R., & ..., Balmford, A. (2023). Inequitable gains and losses from conservation in a global biodiversity hotspot. *Environmental and Resource Economics, 86*(3), 381–405.

Pretzsch, H., del Río, M., Arcangeli, C., Bielak, K., Dudzinska, M., Forrester, D. I., Klädtke, J., Kohnle, U., Ledermann, T., Matthews, R., Nagel, J., Nagel, R., Ningre, F., Nord-Larsen, T., & Biber, P. (2023). Forest growth in Europe shows diverging large regional trends. *Scientific Reports, 13*(1). https://doi.org/10.1038/s41598-023-41077-6.

Robinson, D. A., Hockley, N., Dominati, E., Lebron, I., Scow, K. M., Reynolds, B., Emmett, B. A., Keith, A. M., de Jonge, L. W., Schjønning, P., Moldrup, P., Jones, S. B., & Tuller, M. (2012). Natural capital, ecosystem services, and soil change: Why soil science must embrace an ecosystems approach. *Vadose Zone Journal, 11*(1). https://doi.org/10.2136/vzj2011.0051.

Rodríguez-Gómez, G., Pérez-Fernández, E., Fernandez, P., Arsuaga, J. L., Díez, C., & Arceredillo, D. (2022). Palaeoecology of the Southern chamois from Valdegoba Cave (Burgos, Spain) and its exploitation by the Neanderthals. *Lethaia, 55*(4). https://doi.org/10.18261/let.55.4.3.

Schneiderbauer, S., Fontanella Pisa, P., Delves, J. L., Pedoth, L., Rufat, S., Erschbamer, M., Thaler, T., Carnelli, F., & Granados-Chahin, S. (2021). Risk perception of climate change and natural hazards in global mountain regions: A critical review. *Science of the Total Environment, 784*. https://doi.org/10.1016/j.scitotenv.2021.146957.

Seddon, A. W. R., Macias-Fauria, M., Long, P. R., Benz, D., & Willis, K. J. (2016). Sensitivity of global terrestrial ecosystems to climate variability. *Nature, 531*(7593), 229–232. https://doi.org/10.1038/nature16986.

Stobbe, A., & Gumnior, M. (2021). Palaeoecology as a tool for the future management of forest ecosystems in Hesse (Central Germany): Beech (Fagus sylvatica L.) versus Lime (Tilia cordata Mill.). *Forests, 12*(7), 924. https://doi.org/10.3390/f12070924.

Trivedi, P., Singh, B. P., & Singh, B. K. (2018). Soil carbon: Introduction, importance, status, threat, and mitigation. In *Soil carbon storage: Modulators, mechanisms and modeling* (pp. 1–28). Elsevier. https://doi.org/10.1016/B978-0-12-812766-7.00001-9.

Williams, A., Perego, P., & Whiteman, G. (2025). Boundary conditions for organizations in the Anthropocene: A review of the planetary boundaries framework 10 years on. *Journal of Management Studies, 62*(4), 1811–1846. https://doi.org/10.1111/joms.13150.

Xue, J., Zong, L., Yang, Y., Bi, X., Zhang, Y., & Zhao, M. (2023). Diurnal and interannual variations of canopy urban heat island (CUHI) effects over a mountain–valley city with a semi-arid climate. *Urban Climate, 48*, 101425. https://doi.org/10.1016/j.uclim.2023.101425.

Index

Page numbers followed by "*f*" and "*t*" indicate, figures and tables respectively.

Printed and bound by CPI Group (UK) Ltd, Croydon, CR0 4YY
14/09/2026
02216410-0002